Introduction to
Mathematical Oncology

CHAPMAN & HALL/CRC
Mathematical and Computational Biology Series

Aims and scope:
This series aims to capture new developments and summarize what is known over the entire spectrum of mathematical and computational biology and medicine. It seeks to encourage the integration of mathematical, statistical, and computational methods into biology by publishing a broad range of textbooks, reference works, and handbooks. The titles included in the series are meant to appeal to students, researchers, and professionals in the mathematical, statistical and computational sciences, fundamental biology and bioengineering, as well as interdisciplinary researchers involved in the field. The inclusion of concrete examples and applications, and programming techniques and examples, is highly encouraged.

Series Editors

N. F. Britton
Department of Mathematical Sciences
University of Bath

Xihong Lin
Department of Biostatistics
Harvard University

Nicola Mulder
University of Cape Town
South Africa

Maria Victoria Schneider
European Bioinformatics Institute

Mona Singh
Department of Computer Science
Princeton University

Anna Tramontano
Department of Physics
University of Rome La Sapienza

Proposals for the series should be submitted to one of the series editors above or directly to:
CRC Press, Taylor & Francis Group
3 Park Square, Milton Park
Abingdon, Oxfordshire OX14 4RN
UK

Published Titles

Published Titles (continued)

Chapman & Hall/CRC Mathematical and Computational Biology Series

Introduction to Mathematical Oncology

Yang Kuang

Arizona State University
Tempe, Arizona, USA

John D. Nagy

Scottsdale Community College
Scottsdale, Arizona, USA

Steffen E. Eikenberry

University of Southern California
Los Angeles, California, USA

CRC Press
Taylor & Francis Group
Boca Raton London New York

CRC Press is an imprint of the
Taylor & Francis Group, an informa business

A CHAPMAN & HALL BOOK

CRC Press
Taylor & Francis Group
6000 Broken Sound Parkway NW, Suite 300
Boca Raton, FL 33487-2742

Printed on acid-free paper
Version Date: 20160114

International Standard Book Number-13: 978-1-58488-990-8 (Hardback)

Visit the Taylor & Francis Web site at
http://www.taylorandfrancis.com

and the CRC Press Web site at
http://www.crcpress.com

Contents

Preface

On December 23, 1971, President Richard Nixon signed into law the National Cancer Act, effectively declaring a never-ending war on cancer. Despite decades of significant investments and the relentless efforts of the medical and pharmaceutical communities, progress in efficacy of treatment for most cancers remains surprisingly limited. In 2003, the director of the National Cancer Institute issued an ambitious challenge "to eliminate the suffering and death from cancer, and to do so by 2015." Although overly optimistic, this heroic goal was supported by the American Association for Cancer Research in 2005. Sadly, most types of cancer remain incurable and the death rate for cancers in the United States dropped only about 10 percent from 1975 to 2011.

Needless to say, new research directions and methods are desperately needed to win this ongoing war on cancer. Cancer is largely an evolving process that expresses different challenges in different patients in the same stage of disease and in the same patient at different stages. Therefore, consistent calls have been made for individualized treatment strategies, often referred to as personalized medicine. Central to any such individualization of medical intervention is the requirement for deep understanding of tumor dynamics at a variety of levels, from the whole organism to individual molecules.

Mathematical oncology is therefore emerging as a foundational discipline for modern treatment innovations. It promises to provide tools for both quantification of key parameters from patient-specific clinical data and for customizing cancer treatment using carefully formulated mathematical models. Over the past decade, mathematical oncology has grown into an exciting field evolving at a breathtaking pace. This textbook was conceived when the two of us (YK and JDN) decided to offer a graduate level course on mathematical medicine at Arizona State University (ASU) in 2007. We have since taught it many times at ASU, including a variety of mathematical models of cancer and viral infections. The book was originally set to be delivered in 2008 with the title, *Dynamical Models in Medicine*. However, teaching mathematical medicine the first time in earnest rapidly taught us that the book project was too broad and ambitious. After delving deep into the wonderland of mathematical medicine via research publications, we decided to limit our scope to mathematical oncology. As the years went by we continuously iterated and evolved the contents, eventually deciding to focus on biologically well motivated and mathematically tractable models that can inspire a deeper understanding of cancer biology and help design better cancer treatments. Due to these limitations, many timely and important topics of mathematical oncology are left untouched. We sincerely apologize to any colleagues who may feel slighted by such omissions, and sincerely hope that this text provides a gateway through which readers can find their way to the many excellent studies not found in these pages.

Indeed, the purpose for writing this book was primarily pedagogical. Although we hope that the book will be useful to professionals, it is intended to be used as a textbook for both graduate and upper-division undergraduate courses in mathematical oncology. It therefore contains many exercises and research projects of varying levels of difficulty.

Chapter 1 provides a brief introduction to the general theory of medicine and how mathematics can be essential in its understanding. Chapter 2 introduces the readers to some well-known, practical, and insightful mathematical models of avascular tumor growth and some mathematically tractable treatment models based on ordinary differential equations. Chapter 3 continues the topic of avascular tumor growth in the context of partial differential equation (PDE) models by incorporating the spatial structure, and Chapter 4 expands the topic of avascular tumor growth in a PDE context by incorporating physiological structure such as cell size. Chapter 5 focuses on the recent active multi-scale modeling efforts on prostate cancer growth and treatment dynamics. Chapter 6 exposes readers to more mechanistically formulated models, including cell quota-based population growth models with applications to real tumors and validation using clinical data. Chapters 7 through 12 present abundant additional historical, biological, and medical background materials for advanced and specific treatment modeling efforts.

The text may be used in a variety of ways, allowing instructors to emphasize specific topics relevant to clinical cancer biology and treatment. A sample single semester undergraduate level course may cover the first three chapters plus Chapter 5. A more ambitious graduate level course may cover the first six chapters plus selected readings from latter material and cited literature. A full year sequence on mathematical oncology may cover most of the chapters contained in this book.

A book such as this could not exist without creative input from many sources besides its authors. First and foremost, we would like to thank our students and colleagues, who in the last eight years have provided us with many helpful inputs and whose interest in mathematical medicine inspired us. We would also like to thank the School of Mathematical and Statistical Sciences at ASU for allowing us to repeatedly teach the course on mathematical medicine and for providing a first rate environment for our research efforts in mathematical medicine. We are grateful to CRC editor Sunil Nair and his team, for their unlimited patience and support during this long period of textbook development.

Last but not the least, we would like to thank our families for their constant support for this seemingly never-ending book project. In particular, YK would like to thank his wife, Aijun Zhang, for frequently reminding him that it is important to finish this book soon. YK would also like to thank his daughters

Youny and Belany and son Foris for being great kids so that he can spend more time on writing this book. A special thank is due to Youny for her creative cover design for this book. JDN begs the forgiveness of his wife and daughter, Bethel and Grace, who had to share his attention with this book and endure many absences, including a months-long trip to Finland. A significant fraction of this text is a result of their patience and understanding. SEE would like to thank his co-authors for their long mentorship and perseverance in this project. He would also like to thank his family, in particular his mother and father for their unconditional support in all endeavors, his brother Keenan and sister Greta for the many friendly arguments that lead to insight, and his fiancée Lindsey, for sharing her time with this book with gentle good humor.

Yang Kuang, John D. Nagy, and Steffen E. Eikenberry

July, 2015[1]

[1]MATLAB® is a registered trademark of The MathWorks, Inc. For product information, please contact:
The MathWorks, Inc.
3 Apple Hill Drive
Natick, MA 01760-2098 USA
Tel: 508-647-7000
Fax: 508-647-7001
E-mail: info@mathworks.com
Web: www.mathworks.com

Chapter 1

Introduction to Theory in Medicine

1.1 Introduction

On September 5, 1976, a man named Mabalo Lokela was admitted to the Yambuku Mission Hospital in what is now the Democratic Republic of the Congo [8, Ch. 5]. Gravely ill, Lokela was suffering an intense fever, headache, chest pain and nausea. He vomited blood and had bloody diarrhea. Medical personnel recognized the signs of a hemorrhagic fever, but they were still largely in the dark. A variety of pathogens cause hemorrhagic fever, and there was no time to determine which one was causing this particular case. Lokela was clearly in serious trouble. Unfortunately, his health care workers failed to recognize that they were also in serious trouble.

Roughly speaking, Lokela's tissues, including his skin, were melting away, causing massive internal bleeding. After a few days he had exhausted the clotting factors in his blood, and he began "bleeding out." By then, the hospital staff could do little more than watch him die.

Ominously, just after Lokela's burial, a number of his friends and family began experiencing similar symptoms. Eventually, 20 of them contracted the same disease. Two survived. While these 20 people suffered, the hospital in Yambuku started admitting patient after patient with the same sickness. Eventually, hospital staff, too, began to fall ill. The epidemic spread like wildfire. Within weeks of the outbreak, astonished, and frankly terrified, scientists and medical professionals around the world scrambled to understand what was happening in Yambuku. Initially they focused on two questions: what was the pathogen, and how did it pass from person to person? Surprisingly quickly they discovered that the pathogen was unknown to science (at that time) and that it jumped between hosts via body fluids. Unaware of this latter point early in the outbreak, many Yambuku medical workers contracted the disease because they failed to protect themselves from their patients' blood. The hospital became an amplifier of the epidemic as the pathogen spread from patient to patient, carried at times by the medical workers themselves. Quickly, however, hospitals throughout Sub-Saharan Africa were made aware of the disease and taught how to handle it. The epidemic died out nearly as rapidly as it flared.

In this way the world learned of Ebola hemorrhagic fever, also called Ebola virus disease (EVD). Since then the world has seen a number of other Ebola outbreaks. In 2014 an Ebola epidemic spread more extensively to West Africa, specifically to Sierra Leone, Liberia and Guinea. This extremely long lasting outbreak was probably started by a 2-year-old boy died in December 2013 in the village of Meliandou, Guinea. It caught the world off guard because, unlike previous outbreaks, this one did not flare and rapidly die out on its own, largely due to the poor health infrastructure and the lack of standard practices to prevent the outbreak in the affected countries. As of November 1, 2015, over 28,607 cases have been reported, of which over 11,314 patients have died, making it the deadliest Ebola epidemic thus far according to WHO Ebola situation reports [19]. Despite this tragic human suffering and loss, the dynamics of this epidemic can be reasonably predicted by simple mathematical models correctly modeling the human behavior change dynamics after some initial period of time [2].

Against the background of the tragedies in Yambuku and West Africa we see the functioning of modern medicine. Here we use the word *medicine* in its most general sense—it means "the art of preventing or curing disease" and "the science concerned with disease in all its relations" [15].[1] In Yambuku, the two "arms" of medicine, curative and public health care, complemented each other beautifully. As its name implies, curative medicine focuses on cures or treatments for diseases—the Yambuku Mission Hospital staff trying to keep Mabalo Lokela alive, for example. In contrast, the goal of public health is to prevent disease. In Yambuku, public health professionals (perhaps) slowed the epidemic by identifying the pathogen and recommending techniques to prevent infection among hospitals in the region. When these two arms of medicine work together, the effectiveness of medical intervention is maximized. Although popular culture tends to focus on curative medicine— nearly all health-related movies and shows are set in hospitals with physicians and surgeons treating individual patients, and our experience tends to suggest that most young students entering so-called "pre-med" undergraduate studies are unaware of the existence of public health as an arm of medicine—a strong argument can be made that public health is primary. As the developing world shows us daily, inadequate public health care makes curative medicine superfluous. Without clean water, for example, the number of cases of infectious disease simply swamps curative efforts (see Section 1.3 below).

The definition above also suggests that medicine is both art and science. The art of medicine typically refers to clinical practice.[2] In the clinic, medical professionals work with individual patients out of necessity—each patient presents a unique case. In contrast, medical *science* seeks broad patterns and

---

[1]In addition to the meaning used here, "medicine" can also mean "a drug," or health care not associated with surgery.

[2]The word "clinic" here refers generally to areas where medical doctors work—hospitals, private offices, and actual clinics, among others.

causative relationships within the chaos of individual cases. These patterns exist both within and among patients.

As the tragedies of Yambuku and West Africa show, medical *science* informs, or should inform, the practice of the medical *art*. Health professionals in the clinic rely on discoveries made by their scientific colleagues. At least, they should. When they do, we refer to the practice as *evidence-based medicine*. Our goal in this book is to explore how mathematics, dynamical models in particular, have in the past and can in the future advance the practice of evidence-based medicine, specifically as it applies to *oncology*, the science and art of studying and treating tumors.

1.2 Disease

Central to all of medicine, and its founding scientific discipline of physiology (see below), is the concept of homeostasis, a concept that includes both equilibrium and disequilibrium. For example, we say that mammals homeostatically regulate body temperature because they maintain a constant body temperature in disequilibrium with the environment. A dead mammal in a thermally invariant environment will maintain a constant body temperature, but not homeostasis.

Antithetical to homeostasis is the concept of disease. By the standard definition [15], disease is "an interruption, cessation, or disorder of body function, system or organ." A more modern outlook would take this down to the level of cells and even molecules. Since almost all organs, systems, cells and molecules work to maintain homeostasis, it might be tempting to define disease as a threat to homeostasis. However, this definition would not apply to diseases of the reproductive system, which functions not to maintain homeostasis but to perpetuate the genes. Nevertheless, homeostatic mechanisms exist, ultimately, in support of the reproductive system in metazoans (multicellular animals).

Like the word medicine, "disease" can be used with subtly different meanings. The word also applies to a sickness with "at least two of these criteria: recognized [causative] agent(s), identifiable group of signs and symptoms, or consistent anatomic alterations" [15]. A symptom is something a patient feels that indicates disease, whereas a sign is an outward, objective manifestation of disease. According to these definitions, sore throat is a common *symptom* of a cold, whereas fever is a common *sign* of bacterial infection. A collection of signs and symptoms characteristic of disease is called a syndrome. For example, HIV infection is a disease characterized by acquired immunodeficiency syndrome (AIDS), signs of which include loss of certain types of immune cells and the presence of various opportunistic infections and cancers, like *Pneu-*

mocystis carinii pneumonia and lymphoma, among others.

1.3   A brief survey of trends in health and disease

We adopt the view that any mathematical model of disease dynamics must connect in some way to the clinic. Otherwise the exercise is either pure mathematics, in which case its origin as a model of disease is hardly relevant, or it reduces to a triviality. This viewpoint justifies our decision to start with a survey of biomedical science and not mathematical techniques. Our goal is to help both mathematics and science students interested in theory to develop the skills and understanding necessary to contribute in a meaningful way to pathology—literally, the study of suffering—with the ultimate goal of alleviating some of that suffering. It is a daunting task that no one should enter without a clear understanding of what they are up against. Therefore, we include here a survey of the patterns of disease around the world.

About half the people on the planet will die from infection, cancer, coronary artery disease or cerebrovascular disease (primarily strokes).[3] However, simple lists like this are misleading because patterns of disease are strongly influenced by socioeconomics at all levels, from individuals to nations (Fig. 1.1). Speaking generally, poor countries contend primarily with infectious diseases, particularly pulmonary (including tuberculosis), diarrheal, HIV, malaria and neonatal (among infants) infections. Infectious diseases are less significant in wealthier countries, where chronic diseases mostly associated with senescence—cancer, CAD, cerebrovascular disease, chronic obstructive pulmonary disease (COPD) and dementia—are the major killers. This pattern is reflected in life expectancy; in countries the WHO classifies as "high income," $> 2/3$ of the inhabitants reach the age of 70, compared to "low income" countries, where $< 1/4$ do, largely due to high death rates among children.

That infectious disease is the leading killer tells a somewhat misleading story because the concept "infectious disease" includes many more disease entities than a simpler concept like "CAD" does. To microorganisms our bodies represent a vast reservoir of natural resources. Not surprisingly, an enormous diversity of microbes in all but two of the standard kingdoms have evolved to exploit those resources.[4] Humans are particularly susceptible to infections in the lungs. Tuberculosis, a disease caused by the bacterial pathogen *My-*

---

[3]This statement is based on data from the World Health Organization's statistics service. For example, in 2004, infectious disease killed approximately 11 million people worldwide (about 19% of total deaths that year), cancer 7.4 million (12.5%), CAD 7.2 million (12.2%) and cerebrovascular disease 5.7 million (about 10%).

[4]We know of no plants or archaeans that act as human pathogens.

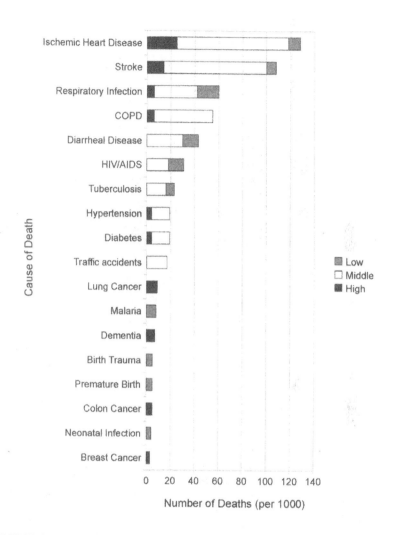

FIGURE 1.1: Main causes of death in the world in 2004, broken down by general income class (high income countries = black, middle income countries = white, low income countries = grey). Data from the World Health Organization (WHO) *Top 10 Causes of Death Fact Sheet*. NOTE: Data do not reflect all causes of death; it is a sum of the top 10 causes of death in each economic class.

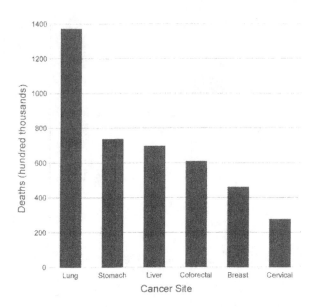

FIGURE 1.2: Worldwide mortality from the five leading causes of death from neoplasia. Height of each bar represents total number of deaths (in 100,000s) in the world in 2004. Data from WHO cancer fact sheet, February 2009.

cobacterium tuberculosis, commonly, but not exclusively, resides in the lungs, as does an astonishing variety of bacterial, fungal and viral agents. For this reason, respiratory infection is a major killer in all nations, regardless of income status. Next most vulnerable to infection is the gut, especially the lower intestine. Here, however, diarrheal disease is a major killer only among the poorer nations. Poor and middle-income countries also suffer disproportionately from HIV, *Mycobacterium*, neonatal, and *Plasmodium* infections. This last agent is a class of eukaryotic (nucleated) parasites that cause malaria, which is a particularly devastating disease. Even though it ranks sixth in the world list of infectious killers, its impact is perhaps greater than all the others combined. The disease itself is very common—WHO estimates approximately 173 million cases occurred in 2008, compared to 2.7 million tuberculosis and the same number of new HIV cases.[5] Although largely survivable, malaria causes debilitating, sometimes recurring fevers that limit an individual's ability to work and provide for the family. There is a strong negative correlation between national wealth and malaria incidence, and some have suggested that

---

[5]In 2008, WHO estimates that 33.4 million people were living with HIV, which unlike most malaria infections is chronic. Therefore, there are about 5 times more *new* malaria cases in a single year than there are *existing* HIV cases.

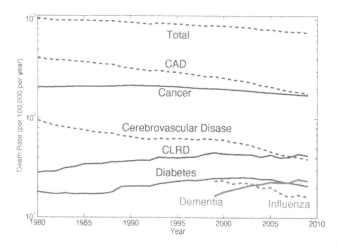

FIGURE 1.3:　Trends in age-adjusted mortality rate in the United States, 1980-2007, all races and both sexes combined. Data from National Center for Health Statistics 2009 report [13].

the relationship is causal.

Neoplasia, literally "new growth," includes all tumors, both benign and malignant. This latter class is synonymous with cancer. As the names imply, most deaths from neoplasia occur from malignant tumors, but not all. "Benign" tumors are not always benign; pheochromocytomas, for example, are rarely malignant but can still kill by secreting hormones that generate hypertension (high blood pressure). Like infectious disease, neoplasia comprises a vast array of distinct disease entities, and patterns of both incidence and mortality vary with economic status. In most high-income countries like the United States, deadly cancers arise primarily in the lung, breast, colon/rectum and prostate. However, middle and low income countries bear a much greater burden of stomach and liver cancer. So the global cancer mortality pattern as shown in Figure 1.2 represents only the total cancer death rates for some major cancers in the world, which may not resemble cancer mortality pattern of any actual countries.

In the last century, wealthy countries like the United States have made tremendous strides in combating infectious disease. In the last few decades, effective prevention strategies and drugs combating CAD and cerebrovascular disease have significantly decreased the number of deaths from these causes as well, although they remain leading killers. On the other hand, deaths from diabetes mellitus and chronic lower respiratory diseases (CLRD) are moving in the opposite direction. Cancer still kills essentially the same number as always over the same time period (Fig. 1.3).

This observation and the fact that a higher proportion of people die from neoplasia than ever before suggests that we are losing the "war on cancer."

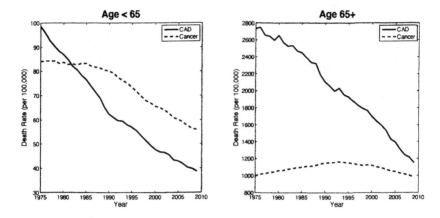

FIGURE 1.4:   Trends in age-adjusted mortality rate from heart disease and neoplasia (all sites) in the United States, 1975-2007 for deaths among individuals < 65 vs. ≥ 65 years of age, all races and both sexes combined. Figure reprinted from the Surveillance, Epidemiology and End Results (SEER) Cancer Statistics Review, 1975-2007 with the kind permission of the National Cancer Institute.

In fact, a subtle relationship among leading causes of death obscures the fact that we are clearly winning. By 2007, cancer death "rates" (essentially the proportion of individuals killed by cancer in a fixed time interval) have fallen to 70% of their 1975 value among Americans younger than 65 at time of death (Fig. 1.4). While the number dying from cancer in older age groups is essentially unchanged over the same time period, deaths from CAD, the leading killer, have dropped tremendously (Fig. 1.4). In fact, the drop in CAD deaths among all age groups should result in an *increase* in cancer deaths as those who before would have died from heart attacks (and strokes) survive and become increasingly susceptible to cancer as they age. But, in fact, cancer death rates have decreased slightly, which itself is significant.

Trends in 5-year survival statistics also support the notion that, generally speaking, we are treating cancer patients increasingly effectively. As shown in Figure 1.5, 5-year survival is improving for all four leading killers: lung, colorectal, breast and prostate cancers, particularly the latter two. Colorectal cancer, while still very deadly with generally poor prognosis, is now considerably more survivable than in 1975; so is lung cancer, but its prognosis remains extremely poor.

Lung cancer warns us that, despite recent successes, we still have far to go. Pancreatic and esophageal cancers, along with lung, are in the list of extremely difficult cancers to treat successfully, despite recent modest improvement. In some cases the story is more complex; the odd pattern of survivability in Kaposi sarcoma, for example (Fig. 1.5), is explained largely by its relationship

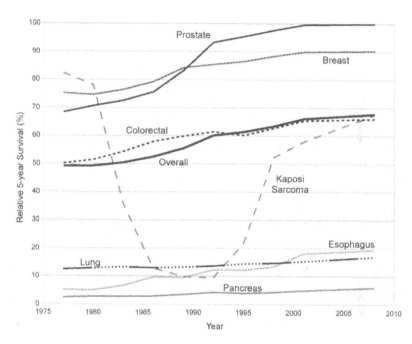

FIGURE 1.5: Relative 5-year cancer survival in the United States, all cancer types, all races, both sexes combined. Data represent average survival for patients diagnosed within the period between data points, with the data plotted on the right-hand endpoint of the period. Data from the National Cancer Institute's SEER Cancer Statistics Review, 1975-2007.

to HIV/AIDS. Kaposi sarcoma is extremely rare, arising only in individuals who are already immune-compromised. Therefore, it is a common opportunistic disease characterizing AIDS. When the HIV epidemic flaired in the early 1980s, Kaposi sarcoma rode along. Then in the late 1990s, fairly effective treatments for HIV became widely available, which consequently diminished mortality from Kaposi sarcoma because the disease is rarely fatal in immune-competent individuals.

1.4   The scientific basis of medicine

Modern curative medicine is founded on the scientific disciplines of anatomy and physiology, the studies of form and function, respectively, of living systems.[6] However, emphases are shifting. As a direct result of the molecular biology revolution, the scale of focus continues to become finer and finer. Traditionally, disease processes were understood at the organ system level. With the advance of microscopy, including improved instruments and staining techniques, tissue- and cell-level processes were added to our descriptions and explanations of disease. Now, new medical insights arise mostly from molecular biology and so-called "-omics" technologies. Modern medical practice requires competence at all levels of biological organization from organ systems to DNA.

For example, consider coronary artery disease (CAD), a leading cause of death in the world (see Section 1.3). At the organ system level we think of the disease as being confined largely to the cardiovascular system—the heart, blood and blood vessels. We describe its most common cause as atherosclerosis, a narrowing of coronary (heart) arteries caused by plaque development within the vessels. Eventually, blood flow to the heart becomes blocked, most often by coronary thrombosis (a blood clot blocking an artery of the heart), causing infarction (tissue death caused by lack of blood flow) of a portion of the myocardium (muscular wall of the heart)—hence, myocardial infarction, or MI. At the tissue level, the arterial plaques, deposits of lipid and calcium carbonate mixed with smooth muscle cells, originate from inflammation of the artery wall, roughly speaking. These basic research discoveries led directly to the application of antiinflammatory drugs as preventative treatments for CAD. At the molecular level, it appears that low density lipoproteins (LDLs) somehow cause the inflammation. Not only does that observation explain why high blood LDL levels correlate positively with CAD, it led to lifestyle

---

[6]Despite the argument in the introduction section, in this course we largely ignore public health, founded on the science of epidemiology. Although fascinating, this topic is well treated in numerous other texts. See [1] and [4] for introductions to this field.

recommendations and new drugs, primarily statins, that have helped people in developed nations avoid premature deaths. How the body handles LDLs is in part determined genetically, so we recognize DNA as another potential target for therapy.

This CAD example, although greatly abbreviated, already exhibits many of the basic biology disciplines related to curative medicine—anatomy, physiology, histology (study of tissues), cell biology,[7] molecular biology and genomics (roughly, the study of all genes and DNA sequences in an organism and their variations among organisms). In addition, a competent biomedical scientist must have proficiency in basic chemistry, biochemistry, biophysics and classical physics. (Hemodynamics, the study of blood flow dynamics, requires plenty of fluid physics, for example). In mathematical medicine, we add the powerful tools of dynamical systems and stochastic processes to the mix, among other mathematical tools.

1.5 Aspects of the medical art

On its clinical side, medicine's primary concerns include diagnosis, prognosis and treatment. Diagnosis refers to the identification of disease processes and their causes. For example, in Mabalo Lokela's case, the disease process was hemorrhagic fever and the cause was Ebola Ebola virus infection. Prognosis, one of the most difficult aspects of medicine, involves estimation of the likely course of a disease and its threat to health or survival. A person infected with the same strain of Ebola virus that killed Mabalo Lokela faces a grim prognosis, for example—the probability of survival is less than 10%, and death usually occurs within two weeks of the onset of symptoms.

Treatment, of course, means an attempt to alleviate symptoms. One way to do that is to effect a cure, or permanent reversal of the disease process, usually by eliminating its cause. However, the goal of treatment is not always a cure. In many cases, a disease cannot be cured. In such situations, treatment may simply manage the disease so that it no longer progresses or presents a threat. The Highly Active Antiretroviral Treatment (HAART) applied to HIV patients works in this way. Although it cannot cure HIV disease, in most cases HAART allows HIV patients to live, if not a normal life, one close to it. In extreme cases, like terminal cancer, for example, it is not even possible to manage the disease. In these cases, standard practice calls for

---

[7]Here we avoid the outdated term "cytology," which seems to focus primarily on cell anatomy. The Medline Plus online medical dictionary defines cytology as "a branch of biology dealing with the structure, function, multiplication, pathology and life history of cells," but most modern biologists studying cell function call themselves cell or molecular biologists.

palliative treatment—an attempt to alleviate symptoms to make the patient as comfortable as possible.

1.6 Key scientific concepts in mathematical medicine

In 1973, Theodosius Dobzhansky, a very highly respected Russian émigré geneticist from Rockefeller University and the University of California, Davis published an article in *The American Biology Teacher* with the deliberately provocative title, "Nothing in Biology Makes Sense Except in the Light of Evolution" [5]. Although Dobzhansky's argument in that paper may not support the claim's universality as it is now interpreted, the subsequent three decades of research in biology and medicine largely have. Certainly, evolutionary concepts, especially the theory of natural selection, have become a founding pillar of biomedical research and the medical practice, even if its practitioners are not always aware of the fact. Antibiotic and cytotoxic treatment regimens for infectious agents and cancer, respectively, are now routinely framed around evolutionary principles, either implicitly or explicitly. If natural selection is ignored, such treatments fail at an unacceptably high rate due to the evolution of resistance. At a more mechanistic level, molecular data generated by the "genomics revolution" show clearly that the genome is not a collection of carefully crafted genes constructing a perfectly adapted organism. Rather, genes are typically copies of existing genes jury-rigged to generate novel, often imperfectly functioning proteins. This commonly results in unnecessarily extravagant genetic regulatory systems with potential failure modes that a human engineer designing a genetic system from the ground up could eliminate. Evolutionary theory helps make sense of this otherwise confusing design philosophy, and more excitingly is beginning to be used to explain diseases, predict their course and rationally search for treatments.

If evolution is one supporting pillar of biomedical science, genetics is an equally important second. Here we use the term "genetics" in its most general sense—the study of heredity and variation in organisms. Under this definition, essentially all diseases have a genetic component because a person's response to disease is unique and largely explained by heredity. Also, as we will see, if "nothing in biology makes sense except in the light of evolution," nothing in evolution makes sense except in the light of genetics. That is not to imply that genetics is the queen of biology because, as we have seen, genetics makes no sense without evolution. Instead, the two are so deeply entwined that appeal to one without the other either makes no sense or is unnecessarily trivialized.

1.6.1   Genetics

In 1865, an obscure paper entitled *Versuche über Pflanzen-Hybriden* [*Experiments in Plant Hybridization*] was read at two meetings of the equally obscure Brünn Natural Science Society in the city of Brünn, Moravia (now Brno, Czech Republic). Its author was a monk and teacher from the Augustinian Order of St. Thomas, a local monastery dedicated to teaching science. He was an amateur without an advanced science degree, and the results disagreed with what was already known about how traits were passed from parent to offspring. So after the paper was published in the society's proceedings the following year, it was roundly ignored. One of the leading authorities on heredity at the time, Carl von Nägeli, dismissed the results out of hand, despite a series of explanatory letters sent to him by the paper's author. So the paper died of neglect, and its author never published again on heredity.

In 1900, three biologists—Hugo de Vries, Carl Correns, and Erick von Tschermak—discovered the basic rules of heredity independently of each other. Everlasting fame would certainly have been theirs, except that these same rules had already been worked out completely some 35 years earlier by an obscure monk who published his results in the equally obscure Proceedings of the Brünn Natural Science Society. The author was by then nearly 20 years in his grave, but it took that long for the scientific culture to mature sufficiently to allow his work's significance to be recognized. Today no one but specialists remember who Carl von Nägeli, Hugo de Vries, Carl Correns or Erick von Tschermak were, although they also made significant contributions. But literally millions around the world know the name of the Augustinian monk from Brünn—Johann Mendel, given the name Gregor when he took his monastic vows.

Mendel discovered a special class of traits determined by a very simple genetic system. Although traits of this type are generally uncommon across all organisms, Mendel found seven examples in his experimental species, the garden pea (*Pisum sativum*). He discovered that each of these seven traits were determined in a given individual by two immiscible "factors," one inherited from each parent. These factors come in two possible forms, which today we call alleles. In modern terminology, we say that each of these simple traits is determined by a single gene that has two allelic variants. A list of the precise alleles an individual carries is called its genotype, and the expression of the trait is called the phenotype. Individuals carrying two copies of the same allele are homozygous for the trait, whereas those with two different alleles are referred to as heterozygous. The transmission properties of these seven traits are further simplified because all exhibit dominance and recessiveness. A dominant allele, by definition, is expressed in the heterozygote, while a recessive allele is not. We call the phenotype associated with each allele the dominant and recessive phenotypes, respectively. All seven traits on which Mendel focused show one dominant and one recessive phenotype.

Mendel's stature and the importance of his discoveries can easily lead one to the false conclusion that Mendel is the architect of our modern theory of genetics. He is not. He correctly interpreted the laws of transmission of only the simplest class of inherited traits, which are rare. Mendel never claimed that these patterns hold for *all* traits. In fact, toward the end of his paper he identifies a number of exceptions. For example, Mendel found that flower and seed color in beans (*Phaseolus*) does not follow the same pattern as he saw in peas. Since Mendel, we have found a number of patterns that nearly conform to his "laws," but not quite. For example, snapdragons (*Antirrhinum magus*) with red flowers crossed with white-flowered plants have pink-flowered offspring. This type of blending in heterozygotes is now called incomplete dominance. Mendel also pointed out that flower color in a variety of species is wildly variable, and then came remarkably close to the modern notion of polygenic inheritance—that is, traits determined by multiple genes—to explain it. Another common pattern not explained entirely by Mendel's hypotheses are codominant traits, in which two alleles are dominant at the same time. The classic example is ABO blood type in humans. Individuals with blood type AB are heterozygous and are blood type A as well as blood type B; therefore, by definition both phenotypes are dominant, a pattern Mendel never addressed. In fact, so few traits exhibit Mendel's patterns that they must be considered the exception rather than the rule, a fact most biologists at the time, including Mendel himself, recognized.

So, if Mendel's patterns are so rare, why do we revere him—and we do revere him—as the "father of modern genetics?" Mendel's hypotheses were initially considered flawed in part because all competent biologists at the time knew obvious counterexamples to Mendel's observations, but also because they completely disagreed with widely accepted contemporary theories based on the incorrect idea that the genetic material was a sort of fluid. In contrast, Mendel suggested that the material of heredity acted like particles passed from each parent to the child. This—the particulate nature of inheritance— was Mendel's great contribution, and it opened the door to the modern genetic theory. The rules he discovered for the seven simple traits in peas represent the threshold separating ancient and modern ideas of heredity. We therefore codify these rules today as the following two principles:

Principle 1 (Segregation) *For the seven traits Mendel studied, each parent passes exactly one of its two copies of the gene for that trait to its offspring. The copy chosen is a random draw, with equal probability of either copy being chosen. In other words, the gene copies segregate randomly among gametes.*

Principle 2 (Independent Assortment) *The Principle of Segregation acts independently on the seven traits Mendel studied.*

Even though here we limit these statements to the traits Mendel studied in peas, the principles extend to some other traits in many other species. Traits and their controlling genes that obey the Principles of Segregation and

Independent Assortment are called *Mendelian*. Let A_i and a_i be distinct alleles of a gene indexed by i. Suppose two individuals heterozygous for n traits breed. If these traits are all Mendelian, then this breeding can be represented symbolically as

$$A_1a_1\,A_2a_2\,\ldots\,A_na_n \times A_1a_1\,A_2a_2\,\ldots\,A_na_n.$$

(Here the symbol $\times$ represents a breeding cross.) All possible combinations of genotypes in the offspring and their respective expected proportions are given by

$$\left(\frac{1}{2}A_1 + \frac{1}{2}a_1\right)^2 \left(\frac{1}{2}A_2 + \frac{1}{2}a_2\right)^2 \ldots \left(\frac{1}{2}A_n + \frac{1}{2}a_n\right)^2. \tag{1.1}$$

By convention we would write A_iA_i or a_ia_i for homozygotes instead of A_i^2 or a_i^2. (Here the letters represent alleles and act like units; they are not quantities.) Since allele order does not matter in the genotype, the probability mass function for a single offspring to be homozygous for k (and heterozygous for $n - k$) traits is binomial:

$$\Pr\{\#\text{ homozygous traits} = k | \#\text{ traits} = n\} = \binom{n}{k}\left(\frac{1}{2}\right)^n.$$

As we have already seen, not all traits are Mendelian. Among the exceptions that Mendel noted are polygenic traits which are determined by the interaction of multiple distinct genes. In some cases, polygenic traits are expressed as continuous variables—human height, for example—in which case we call them quantitative. In others, the genes modify expression of a qualitative trait, in which the trait is given a non-numerical value—like fur color in mice. Even among relatively simple traits determined largely by a single gene, the Mendelian pattern is rare. For example, distinct traits determined by genes on the same chromosome, called linked traits, often fail to assort independently. Sex-linked traits are a special case; the genes are carried on only one of the sex chromosomes. Another example is meiotic drive, in which gametes carrying a particular allele for a particular gene are favored over other gametes carrying another allele for that gene. Meiotic drive represents only one mechanism disrupting Mendel's Principle of Segregation. In general, any mechanism interfering with either of Mendel's principles generates transmission distortion.

Two years after De Vries, Correns and von Tshermak rediscovered Mendel, Walter Sutton of Columbia University discovered "the physical basis of the Mendelian law of heredity" [16, pg. 39]. By carefully examining cells from the plains lubber grasshopper, *Brachystola magna*, Sutton discovered that every chromosome (in this species, and it turns out in many other organisms including humans) has a twin with the same size and location of the centromere, the structure holding the two halves of the chromosome together. Twin chromosomes are called homologous, which is a general term for structures or genes that descend from a single common ancestor. Homologous chromosomes therefore carry copies of the same genes. Any cell with two complete sets of chromosomes is called diploid, as are organisms in which the majority of cells are diploid. Cells with only one set are referred to as haploid. Cells in which the number of copies varies, like many cancer cells, are called aneuploid.

Sutton also discovered that during meiosis—the process by which animals produce gametes (sex cells: egg and sperm)—homologous chromosomes segregate independently, one into each gamete. As a result, gametes from a diploid organism are haploid. Sutton therefore referred to meiosis as reduction division because it reduced the number of chromosomes by half. When the gametes fuse to form a zygote, the offspring must get one copy of each homologous chromosome from each parent, thereby reconstituting diploidy of the offpsring. So, chromosomes come in pairs, one inherited from each parent, and they segregate randomly and independently into gametes. It was obvious to Sutton that chromosomes are physical objects that obey Mendel's two principles; therefore, equation (1.1) describes their behavior during meiosis and subsequent fertilization. Here was a physical manifestation of Mendel's abstract "factors."

Fifty-one years after Sutton's paper appeared, two researchers from the University of Cambridge's Cavendish Laboratory—an American named James Watson and a Brit named Francis Crick—published a tiny paper in the journal *Nature*[8] that finally provided the physical foundation for Mendel's mechanism of hereditary [17]. Although Sutton's chromosomes behaved like Mendel's factors, the number of chromosomes in any cell was always small, on the order of tens, compared to the number of traits apparently requiring genes, numbering at least thousands. Viewed naïvely, Watson and Crick's paper does nothing obvious to resolve this problem; it simply provides a structure for deoxyribonucleic acid (DNA). However, DNA was known at that time to be the material of heredity, and every competent biologist in the world immediately recognized that Watson and Crick found the mechanical connection between physics and chemistry on one hand and phenotype on the other.

A gene, as understood today, is a "packet" of information coded in the DNA. The most well-known genes are structural genes, so called because they determine the structure of a specific biological molecule, usually a protein or RNA molecule. But they also contain control elements that govern when and how many copies of the molecule are to be constructed at any given time. In general, genes are abstract concepts, not objects. In eukaryotes—cells with a

---

[8]A companion paper by Maurice Wilkins, Alec Stokes and Herbert Wilson [20] provided support for Watson and Crick's hypothesis using what turned out to be controversial x-ray crystalography. In essence, the molecule's structure is worked out in part by studying how it defracts x-rays. An outstanding biochemist, Rosaline Franklin, provided much of the spark, including a key x-ray crystalograph, that led Wilkins' team to their discoveries. Wilkins received the Nobel Prize with Watson and Crick. Franklin died of ovarian cancer before the prize was conferred; however, she was never nominated before she died, although Watson and Crick were.

true nucleus and membrane-bound organelles—an entire gene cannot be accurately described as existing at a specific locus (location on a chromosome), as was once thought and still often taught. For example, a single protein coding gene comprises both localized structural and regulatory sequences along with dispersed regulatory regions. The primary structure (amino acid sequence) of the gene's protein product is determined by a coding region adjacent to a set of regulatory sequences collectively called the promoter that in part controls the gene's activity. Other regulatory sequences controlling the gene's activity, called enhancers, are dispersed on the same chromosome hundreds or thousands of base pairs away from the coding region and sometimes found even on other chromosomes. Together, all these elements determine both structure and intracellular concentrations of their protein products.

Speaking tersely, a protein coding eukaryotic gene "turns on" when a molecule called a transcription factor binds to the promoter and assembles a protein complex that constructs a ribonucleic acid (RNA) copy of the DNA. The key element of this complex is a molecule called RNA polymerase. The polymerase copies the coding region along with "leader" and "trailer" sequences called untranslated regions (UTRs). This copy process is called transcription and yields a physical copy of the gene called messenger RNA (mRNA). The mRNA copy contains the information carried in the coding region along with regulatory sequences in the UTRs that control how many proteins are made.

After the copy process is complete, the mRNA is modified and ushered out of the nucleus into the cytosol. There, an enormous molecular complex made of protein and other RNA molecules called a ribosome assembles itself around the mRNA. The ribosome then builds a protein to the precise specifications encoded in the mRNA coding region. The ribosome works its way along the mRNA molecule, reading the information stored there while it catalyzes the construction of the protein in a process called translation. Afterwards, the protein typically undergoes further processing and folding into its functional form. Inasmuch as proteins are the major molecular machines performing and regulating cell functions, as well as forming a nontrivial component of cell structure, a newly minted protein assumes a role in determining its cell's anatomy and physiology—in short, its phenotype. The cell's interaction with other cells and noncellular elements in its immediate environment is determined by its phenotype and that of the cells with which it interacts. These interactions then determine the phenotypes of tissues, which determine the phenotypes of organs, organ systems and finally the organism itself. Our understanding of this entire process, and much else besides, traces its intellectual ancestry back to Watson and Crick's paper of 1953.

1.6.2 Evolution

Consider a collection of self-replicating entities—concretely, think of organisms or cells. If these entities (i) vary in their characteristics, (ii) that variation is heritable, and (iii) some variants have, by virtue of their heritable characteristics, consistently higher reproductive success than do others, then evolution by natural selection follows as a necessary consequence. This argument traces its historical roots back to Charles Darwin, although it was articulated in this clean way by Richard Lewontin [11] and John Endler [7] (see [9] for a historical treatment). Although the argument, as stated this way, is elegant, it has some technical ambiguities that need clarification. In particular, "consistently higher reproductive success," also called fitness, seems clearer than it actually is. In reality, reproductive success is at least partially random. Therefore, by chance some "less fit" individuals could out-reproduce more fit individuals even though "on average" that won't happen. So Lewontin's and Endler's conditions apply only in a "mean field" sense. Nevertheless, if reproducing entities vary, that variation is heritable and consistently associated with differential fitness, then natural selection can be expected to produce evolutionary change. Randomness and other evolutionary forces may swamp or negate its effects, but natural selection is still the rule.

Why does evolution and natural selection matter to someone interested in medicine? First of all, replicating entities are ubiquitous in disease systems. Cancer cells, pathogens and parasites all reproduce. So do the human beings who suffer from disease. These replicating entities vary, and that variation is heritable. Cancer cells within a single tumor are often genomically unstable and therefore quite genetically diverse. Replicating cancer cells cannot help but pass these genetic alterations on to their daughters. Pathogens and parasites also vary genetically, both among and within hosts. Again, replicating microbes pass their genetic characteristics to their "offspring." Certainly humans vary, and just as certainly that variation is largely heritable. So, reproduction with heritable variation is without doubt a ubiquitous property of entities of interest to medicine.

If any of this heritable variation correlates with a fitness advantage *vis-á-vis* disease, then natural selection almost certainly plays a role in the disease process. Consider tumor progression, for example. Tumor development is largely explained by a mapping between genetic alterations and differential reproductive advantage among tumor cells. In addition, some tumor cell variants resist cytotoxic (cell killing) chemo- and radiotherapy, and therefore have a selective advantage over other variants under treatment. Precisely analogous statements can be made for every known pathogen—certain variants can outcompete others within and between hosts, and we always find some strains that resist treatment. As for humans, much of our variation is associated with our immune systems, which has obvious fitness implications in the face of infectious disease and cancer. Heritable diseases like Duchenne muscular dystrophy or ataxia telangiectasia, which typically kill before the age of 25, have equally obvious fitness consequences. Some inherited disorders, like sickle-cell anemia and thalassemia, have both fitness costs and benefits. Both are examples of inherited anemias exhibiting major and minor syndromes. The major syndromes kill the very young from severe anemia and other complications. The minor form also exacts a fitness cost, either from

direct morbidity (a milder form of anemia) or dangerous complications arising from comorbidity (another disease that arises in someone who already has
one of these anemias). But there is also a fitness benefit—individuals with the
minor form resist complications from malaria. These are just a few examples
of how heritable variation in fitness affects disease, and we find it extremely
challenging to find a single counter-example, a disease in which heritable variation has no effect on fitness in any organism (or virus) in the disease system
in any way. Therefore, we conclude that evolution by natural selection is a
common, probably ubiquitous, aspect of human disease.

Here we adopt the somewhat narrow view that organic evolution, or just
evolution, is *change in **allele frequencies** in a population over time*. The
frequency of a given allele, say A, is simply the fraction of all alleles in a
given population that are A. Suppose a single gene has exactly m different
alleles, $A_1, A_2, \ldots A_m$, and the organisms carrying these genes are diploid.
Denote the proportion of individuals in the population with genotype $A_i A_j$ as
$P_{ij} \in [0, 1]$, where the alleles' order is considered. Biologists would therefore
say the genotype frequency of $A_i A_j$ is P_{ij}. (Actually, geneticists typically
disregard allele order, so the conventional genotype frequencies would be P_{ii}
for homozygotes and $2P_{ij}, i \neq j$ for heterozygotes with the understanding
that $P_{ij} = P_{ji}$.) Then the allele frequency of the ith allele is

$$p_i := \sum_j P_{ij}. \tag{1.2}$$

If we view p_i as a function of time, either discrete or continuous, then by
definition no evolution occurs in a time interval $[t_0, t_1]$ if and only if

$$p_i(t) = p_i(t_0) \, \forall t \in (t_0, t_1], i \in \{1, 2, \ldots, m\}. \tag{1.3}$$

Condition 1.3 is expected to fail if different genotypes confer different (mean)
fitnesses. As an illustration that introduces the main mathematical ideas, consider a closed population (no immigration or emigration) with non-overlapping
generations. We view time, therefore, as discrete; that is, $t \in \{0, 1, 2, \ldots\}$.
Suppose that every individual with genotype $A_i A_j$ contributes w_{ij} offspring
to the next generation's breeding population. We assume that w_{ij} is an invariant property of the genotype, which is typically justified by arguing that
the population is infinitely large so that statistical fluctuations in the number
of offspring born to a single individual are "averaged out." If the effects of all
other genes are evolutionarily neutral, then we can equate w_{ij} with fitness.
This definition makes no assumptions about the source of fitness variation—
the gene could affect juvenile survival, reproductive success or both. Finally
we assume that in every reproductive bout, an offspring's genotype is determined by two independent random draws from the population of alleles,
with a single draw choosing allele A_i with probability p_i. Biologically, this
assumption means that mating is random with respect to gene A, and again
the population is large enough that statistical fluctuations are expected to be

averaged out. Under these assumptions the number of individuals that carry genotype $A_i A_j$ in the breeding population at time $t+1$ will be

$$p_i(t)p_j(t)w_{ij}. \tag{1.4}$$

We define the *mean fitness* of this population as

$$\bar{w}(t) = \sum_i \sum_j p_i(t)p_j(t)w_{ij}, \tag{1.5}$$

which is also the total number of breeding adults contributed to the next generation. Furthermore,

$$p_i(t+1) = \sum_j \frac{p_i(t)p_j(t)w_{ij}}{\bar{w}(t)}. \tag{1.6}$$

Now suppose there is no differential fitness; that is, set $w_{ij} = \hat{w}$ for all $i, j \in \{1, 2, \ldots, m\}$, where $\hat{w}$ is a constant. Then

$$p_i(t+1) = \frac{\sum_j p_i(t)p_j(t)}{\sum_i \sum_j p_i(t)p_j(t)} = p_i(t), \tag{1.7}$$

because $\sum_i p_i = 1$. By condition 1.3, no evolution has occurred. An inductive argument based on these observations leads to the following celebrated result:

THEOREM 1.1 (Hardy-Weinberg-Pearson)
Consider a reproductive population with non-overlapping generations. Suppose (i) w_{ij} is constant for all $i, j \in \{1, 2, \ldots, m\}$ (no differential fitness; i.e., no natural selection); (ii) the proportion of genotype $A_i A_j$ among offspring after any breeding bout is $p_i p_j$, where allele order is maintained, and p_i is the frequency of allele A_i in the breeding population (mating is random); (iii) there is no immigration or emigration; (iv) no mutations occur in the gene A; and (v) fitness is invariant over time (the population is infinitely large). Then in a population with non-overlapping generations, $p_i(t) = p_i(t_0) \forall t \in \{t_0 + 1, t_0 + 2, \ldots\}$ (no evolution will occur).

Condition (ii) is equivalent to assuming that the genotype frequencies (with allele order considered) are given by each term in the expansion of

$$(p_1 + p_2 + \ldots + p_m)^2.$$

If we are biologists and disregard allele order in genotypes, then $A_i A_j$ is the same genotype as $A_j A_i$, $i \neq j$. With this convention, the results of Theorem 1.1 and the corollary above imply that the equilibrium genotype frequencies in a population as described in the theorem are

p_{ii} for genotypes $A_i A_i, i \in \{1, 2, \ldots m\}$;

$2p_i p_j$ for genotypes $A_i A_j, i \in \{1, 2, \ldots, m-1\}, j \in \{i+1, \ldots m\}$.

These results (and more) were first obtained by Karl Pearson in 1904 [14], and less general versions were formulated independently by Godfrey Hardy [10] and Wilhelm Weinberg [18] in 1908. For some reason tradition ignores Pearson—Theorem 1.1 is traditionally called the "Hardy-Weinberg" theorem, and the equilibrium frequencies are called the Hardy-Weinberg equilibrium. In fact, Hardy only focused on the special case in which $m = 2$, which reduces the expected equilibrium frequencies of genotypes AA, Aa and aa to p_1^2, $2p_1(1-p_1)$ and $(1-p_1)^2$, respectively, where here we use the convention that $A_1 = A$ and $A_2 = a$. This simplification is how the topic is usually introduced in general biology textbooks.

To a biologist, the usefulness of the Hardy-Weinberg theorem comes from its hypotheses (assumptions). Relaxation of any introduces the possibility that $p_i(t) \neq p_i(t_0)$ at some time $t > t_0$, implying evolution would be possible. The random mating assumption (hypothesis ii), for example, can be violated in a number of different ways, a notorious one occurring when organisms choose mates differentially based on phenotypes conferred by the gene in question. This situation is called sexual selection, and most biologists view it as a special case of natural selection. Inbreeding is another violation, although alone inbreeding is not a mechanism of evolution. Evolution by gene flow occurs when hypothesis iii fails, mutational drive arises by relaxation of hypothesis iv, and finally genetic drift results from failure of hypothesis v. In the diseases we study in this book, natural selection, mutational drive and genetic drift are the dominant evolutionary mechanisms.

As a practical matter, biologists often use the Hardy-Weinberg theorem to determine if a particular gene is evolving in a natural population. If the population is not evolving (the Hardy-Weinberg theorem applies), then $P_{ij}(t) = p_i(t)p_j(t)$ since allele frequencies would be constant. If, on the other hand, $P_{ij}(t) \neq p_i(t)p_j(t)$, then at least one of the mechanisms of evolution must be operating.

1.7 Pathology—where science and art meet

The goal of biomedical science is to develop a correct theory of disease. It can be approached, and is by many practitioners, as pure science. Ultimately, however, all biomedical theory must apply to health and disease in some way. The point of transition from medical science to medical art is typically the discipline of pathology, which according to a very influential text [3] is "a bridging discipline involving both basic science and clinical practice and is devoted to the study of the structural and functional changes in cells, tissues, and organs that underlie disease." Note the scales in the biological hierarchy on which pathology focuses: from cells to organs. Taking a reductionist view, as we did earlier, organ function and malfunction can often be understood at the tissue level, which itself derives from interacting phenotypes among cells and so on down to the genes. So, by necessity pathology involves genetics. As genetics and evolution are deeply intertwined, evolutionary principles can also inform our understanding of pathology.

At its most general level, pathology attempts to define and explain the association between disease processes and their anatomical and physiological manifestations. Diseases by definition are deviations from normal function of the body at some level. That deviation is caused by some set of agents—for example, mutant genes, infectious agents, toxins, allergens and other environmental irritants, congenital malformations, and autoimmune responses. How these agents cause disease is the pathology subdiscipline of etiology, and the process by which the etiological agent and host physiology interact to produce disease is referred to as pathogenesis. The derangement of cells and tissues characteristic of disease generate changes in morphology, and the study of these changes, their cause and effects are the subdisciplines of histopathology and cytopathology, respectively. The clinical impact describes a disease's presentations, prognosis and (sometimes) susceptibility to treatment. The physiological derangements generating the signs and symptoms of a disease is referred to as pathophysiology. As a practical matter, the mechanisms by which a disease threatens health are so tightly intertwined with the process generating the disease that pathophysiology often subsumes much of pathogenesis. These disease expressions, addressed from a scientific standpoint, are used by clinicians to both diagnose and plan treatment. Here, in the realm of pathology, theory built on mathematical modeling can and is having a great impact on our attempts to understand and treat disease.

References

[1] Brauer F, Castillo-Chávez C: *Mathematical models in population biology and epidemiology.* New York: Springer, 2001.

[2] Chowell G, Simonsen L, Viboud C, and Kuang Y: Is west africa approaching a catastrophic phase or is the 2014 ebola epidemic slowing down? different models yield different answers for Liberia. *PLoS Currents* 2014, 6.

[3] Cotran RS, Kumar V, Collins T: *Pathologic Basis of Disease.* St. Louis: Sanders, 1999.

[4] Diekmann O, Heesterbeek JAP: *Mathematical epidemiology of infectious diseases: Model building, analysis and interpretation.* New York: Wiley, 2000.

[5] Dobzhansky T: Nothing in biology makes sense except in the light of evolution. *Amer Biol Teacher* 1973, 35:125-129.

[6] Dunn LC: Mendel, his work and his place in history. *Proc Amer Phil Soc* 1965, 109:189-198.

[7] Endler J: *Natural Selection in the Wild*. Princeton: Princeton University Press, 1986.

[8] Garrett L: *The Coming Plague: Newly Emerging Diseases in a World out of Balance*. Chapter 5. New York: Penguin, 1994.

[9] Godfrey-Smith P: Conditions for evolution by natural selection. *J Phil* 2007, 104:489-516.

[10] Hardy GH: Mendelian proportions in a mixed population. *Science* 1908, 28:49-50.

[11] Lewontin R: The units of selection. *Ann Rev Ecol Syst* 1970, 1:1-18.

[12] Mendel G: Versuche über Pflanzen-Hybriden. *J Brünn Nat Hist Soc* 1866, 4:3-47.

[13] National Center for Health Statistics: *Health, United States, 2009: With Special Feature on Medical Technology*. 2010. Hyattsville, MD.

[14] Pearson K: Mathematical contributions to the theory of evolution. XII. On a generalized theory of alternative inheritance, with special reference to Mendel's laws. *Phil Trans Royal Soc London A* 1904, 203:53-86.

[15] Pugh MB: *Steadman's Medical Dictionary*. [27th ed.]. Philadelphia: Lippincott Williams and Wilkins, 2000.

[16] Sutton WS: On the morphology of the chromosome group in *Brachystola magna*. *Biol Bull* 1902, 4:24-39.

[17] Watson JD, Crick FHC: Molecular structure of nucleic acids: A structure for deoxyribose nucleic acid. *Nature* 1953, 171:737-738.

[18] Weinberg W: Über den Nachweis der Vererbung beim Menschen. *Jahresh Wuertt Ver varerl Natkd* 1908, 64:369-382.

[19] World Health Organization: *Ebola Situation Reports*. http://apps.who.int/ebola/ebola-situation-reports, accessed on November 5, 2015.

[20] Wilkins MHF, Stokes AR, Wilson HR: Molecular structure of nucleic acids: Molecular structure of deoxypentose nucleic acids. *Nature* 1953, 171: 738-740.

Chapter 2

Introduction to Cancer Modeling

2.1 Introduction to cancer dynamics

Describing cancer as "uncontrolled cell growth," as is frequently done, is a bit like calling a rocket "a stick that moves." Indeed, rockets move and cancers grow without a lot of regulation, but something is obviously amiss. First of all, "cell growth," i.e., cells becoming larger, is not the problem. Cell replication (proliferation) is. But even saying "cancer is uncontrolled cell proliferation" still fails to uniquely characterize cancer because many other diseases—elephantiasis, Huntington's disease, Alzheimer's disease, diverticulitis and atherosclerosis to name a few—also are characterized by uncontrolled proliferation.

A more insightful description views cancer as a loss of tissue homeostasis. By definition, homeostasis in metazoa refers to "the maintenance of relatively stable internal physiological conditions ... under fluctuating environmental conditions" [24]. A dynamicist immediately recognizes this as a type of stability. From the pathologist's perspective, in the transition to cancer something about the tissue changes, causing cell proliferation that appears "uncontrolled." This phenomenon leads to neoplasia but not necessarily to cancer. Malignancy of the tumor is conferred by two additional properties: the ability to invade surrounding tissue, and the ability to spawn new tumors elsewhere in the body, a process called metastasis. These spawn new tumors are called secondary tumors or metastases (mets). So, besides modeling oncogenesis, or the process by which tumors arise, mathematical oncologists are also interested in modeling how neoplasia becomes malignant (carcinogenesis), how tumors grow, invade, metastasize and generally cause disease (pathophysiology), and how they can be managed clinically. In this chapter we focus on simple models of growth of already extant tumors.

2.2 Historical roots

Tumor growth models have their historical roots in the work of Ludwig von Bertalanffy and Benjamin Gompertz. Although neither indicated in print any particular interest in tumors, both studied general growth equations that were later successfully applied to actual tumor data, along with many ecological applications. Gompertz is by far the older of the two, publishing in 1825 what has become a foundational treatise establishing the concept variously called "Gompertz's Law" or "Gompertzian growth" [14]. Von Bertalanffy's contribution came some 130 years later, which he reviewed nicely in another classic paper published in 1957 [6]. Although Gompertz's work came earlier, we start with von Bertalanffy so that we can immediately make contact with what is probably a familiar object to both mathematicians and biologists—the logistic equation.

2.2.1 The von Bertalanffy growth model

The fundamental questions, "why does an organism grow at all, and why, after a certain time, does its growth come to a stop?" [6, pg. 217] mark the starting point of von Bertalanffy's work. Replacing "organism" with "tumor" shows why his work so obviously applies to oncology. Von Bertalanffy launches his attack from the well-established observations that many metabolic processes, like pulse rate and basal metabolic rate in mammals, scale not with body mass or volume but rather with something more akin to surface area. The reason why is not always clear, but often metabolic measures can be fit very well with a model of the form

$$M = kW^{\frac{2}{3}},$$

with M the metabolic process, W with organism's mass and k a constant. The cube root takes a volume measure, proportional to mass, into a linear unit, which when squared produces something akin to surface area; with a proper choice of k, it is precisely the surface area. However, von Bertalanffy noted that not all processes scale as a 2/3 power of the mass. In some cases—oxygen consumption rate in certain insects, for example—the metabolic measure is proportional to mass, and in still others—metabolic rates in certain snails and flatworms—it somehow scales with something between mass and surface area, perhaps the surface area of an organism changing shape as it grows. The general relation covering all these cases would be

$$M = kW^{\lambda},$$

$2/3 \leq \lambda \leq 1$.

To model growth von Bertalanffy starts with the conservation equation: growth equals "births" minus "deaths." For our purposes here, "births" equate

to cell proliferation and "deaths" means necrosis or apoptosis. (Von Bertalanffy equated them to anabolism and catabolism, respectively.) If we assume
that these process are essentially independent metabolic mechanisms, then we
can immediately write down the general model for growth in mass $(W(t))$:

$$\frac{dW}{dt} = \alpha W^\lambda - \beta W^\mu, \tag{2.1}$$

where the first term represents proliferation and the second cell death. Equation (2.1) can be called the "generalized von Bertalanffy model," although
Marusic et al. [23] refer to it as the "general two-parameter" model.

One can immediately generate a number of well-known population growth
models as special cases of this general model. For example, setting $\lambda = \mu = 1$
yields the exponential model,

$$\frac{dW}{dt} = (\alpha - \beta)W,$$

where we can interpret α and β as per-capita birth and death rates, respectively. The model predicts either exponential growth or decay depending on
the sign of $\alpha - \beta$. If we let $\lambda = 1$ and $\mu = 2$, we recover the logistic model
typically associated with Verhulst:

$$\frac{dW}{dt} = \alpha W - \beta W^2 = \alpha W \left(1 - \frac{W}{K}\right), \tag{2.2}$$

where α is the ecologist's "intrinsic rate of natural increase," which represents
the growth rate when competition is minimized, βW^2 is the density-dependent
death rate, and $K = \alpha/\beta$ can be interpreted as the "carrying capacity." As
is well known, if $\alpha > 0$, the only biologically interesting case, then this model
has two fixed points: the origin and K. The first is always a source and the
second is globally asymptotically stable (see Section 2.7 below). Generalizing
a bit, we can let μ vary and fix λ at unity, which produces an autonomous
version of Bernoulli's equation,

$$\frac{dW}{dt} = \alpha W - \beta W^\mu, \tag{2.3}$$

which has also been applied to population modeling.

Von Bertalanffy's immediate concerns did not include these special cases
per se, although he did note the exponential case. Instead, von Bertalanffy
hypothesized that in most physiologically relevant situations death processes
are proportional to mass, in which case his model (2.1) becomes

$$\frac{dW}{dt} = \alpha W^\lambda - \beta W, \tag{2.4}$$

which with initial condition $W(0) = W_0$ has the solution,

$$W(t) = (\alpha/\beta - [\alpha/\beta - W_0^{(1-\lambda)}]e^{-(1-\lambda)\beta t})^{\frac{1}{1-\lambda}}, \quad \lambda \neq 1. \tag{2.5}$$

Assuming that proliferation scales with body surface area, one recovers the "classical" von Bertalanffy model,

$$dW/dt = \alpha W^{\frac{2}{3}} - \beta W. \tag{2.6}$$

Applied to cancer, equation (2.6) may represent an avascular tumor in which proliferation is nutrient limited, and only cells at some relatively small distance from the surface have sufficient nutrient to divide. Cell mortality, on the other hand, is constant regardless of depth, and therefore nutrient content. We explore this geometry in more detail below and in the next chapter.

Following Thieme [37], if we let $S(t)$ represent length in a single dimension and assume that the tissue or body does not change shape as it grows, then $W(t) = \gamma S(t)^3$, where γ is a constant. Making this substitution into model (2.6) yields the following linear ordinary differential equation (ODE):

$$\frac{dS}{dt} = \frac{1}{3}(\hat{\alpha} - \beta S), \tag{2.7}$$

which has solution

$$S(t) = \frac{\hat{\alpha}}{\beta} + \left(S_0 - \frac{\hat{\alpha}}{\beta}\right) e^{-\frac{\beta}{3}t}, \tag{2.8}$$

where $S(0) = S_0 > 0$ and $\hat{\alpha} = \alpha(\gamma^{-1/3})$. From this it is easily seen that $\hat{S} = \hat{\alpha}/\beta$ is a globally asymptotically stable fixed point and that solutions $W(t)$ have a sigmoid shape (Fig. 2.2C).

Relaxing the restrictions on λ and μ complicates things considerably, but it is still easy to show that in the most general model (2.1), tumor size tends to the general carrying capacity $(\alpha/\beta)^{1/(\mu-\lambda)}$ as long as $\lambda < \mu$ (see exercises).

2.2.2 Gompertzian growth

As with von Bertalanffy, Gompertz [14] was not explicitly interested in tumor growth, or any kind of growth for that matter. Rather, he was interested in death. What is sometimes called his "law of population growth" derived from an attempt to improve and simplify calculations of life annuities. In particular, he sought a correct description of the number of people alive as a function of their age x, which he denoted $L(x)$ (translated into modern notation). Gompertz argued that human life tables show two distinct behaviors. In some cases (countries or range of ages), $L(x)$ decays exponentially, or if x is discrete as in real life tables, decreases as a geometric series. Death rates in these situations must therefore be constant across age classes since

$$L(x) = L_0 e^{-\alpha x} \Rightarrow \frac{dL}{dx} = -\alpha L,$$

where α (the death rate) is a positive constant. On the other hand, some life tables (or portions of them) show a distinctly different pattern, in which $L(x)$

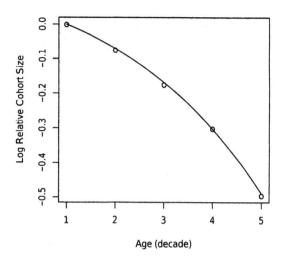

FIGURE 2.1: Example of Gompertzian cohort dynamics in a region in which human death rates appear to be increasing exponentially with time. Circles: Data from Swedish life tables cited by Gompertz [14, pg. 524] for ages from 10 to 50. Curve: Equivalent Gompertz curve (solution to equation (2.10) with parameters: $a = 0.06$, $q = 1.38$, and $L_0 = 6013$ and time scaled in decades). "Log relative cohort size" $= \ln(L(t)/L(10))$.

declines faster than exponentially, implying that death rate must be increasing with age in these situations. He expressed this notion mathematically as follows, rendered into modern notation:

$$\frac{dL}{dx} = -aq^x L, \tag{2.9}$$

where a and $q > 1$ are constants. Solving equation (2.9) with initial data $L(0) = L_0$ yields

$$L(x) = L_0 \exp\left(\frac{a}{\ln q}(1 - q^x)\right). \tag{2.10}$$

This decay is faster than exponential in the sense that $\ln(L(x)/L_0) = a(1 - q^x)/\ln q$ is not a linear function of age x, but rather is concave down (Figure 2.1).

In Gompertz's mind, "[t]his equation between the number of the living, and the age [equation (2.10)], becomes deserving of attention, not in consequence of its hypothetical deduction, ... [but] because I derive the same equation from various published tables of mortality ... and in fact the hypothesis itself was derived from an analysis of ... experience" [14, pg. 519]. So, Gompertz sought an equation that fits data, not a mechanistic understanding of the forces of mortality. Nor was he after a model of population growth. By the mid-20th century, however, biologists began to apply his insight to their models of population growth that included mechanistic detail [40, 41].

Population models employing Gompertz's insight in the most transparent way take the following form:

$$\begin{cases} \dfrac{dN}{dt} = G(t)N(t), \\[2mm] \dfrac{dG}{dt} = -\alpha G(t), \end{cases} \tag{2.11}$$

with initial data $N(0) = N_0 > 0, G(0) = G_0 > 0$. Here, $N(t)$ is the number of individuals in the population and $G(t)$ represents per capita growth rate (birth rate minus death rate at time t) and is loosely related to Gompertz's notion of the "power to avoid death." Really, however, this model assumes that for some unspecified reason, the population growth rate decays exponentially over time at rate α, either from increasing death rates, as Gompertz assumed, or decaying reproduction rates or both. We assume, therefore, that $\alpha > 0$.

From the second equation in system (2.11), $G(t) = G_0 e^{-\alpha t}$, where G_0 is some measure of the initial population's net fecundity. So we can write the model as a nonautonomous ODE:

$$\frac{dN}{dt} = G_0 e^{-\alpha t} N(t), \tag{2.12}$$

which has solution

$$N(t) = N_0 \exp\left(\frac{G_0}{\alpha}(1 - e^{-\alpha t})\right) = N_0 e^{\alpha^{-1}(G_0 - G(t))}. \tag{2.13}$$

Most authors refer to expressions of the form of equation (2.13) as the Gompertz equation, although some apply the name to the differential equation generating it. Figure 2.2 compares the Gompertz curve to a solution of the classic von Bertalanffy model.

Note that the nontrivial equilibrium in model (2.13) depends on the initial condition; that is,

$$\lim_{t \to \infty} N(t) = N_0 \exp(G_0/\alpha). \tag{2.14}$$

If we interpret this limit as a carrying capacity, as we did in the von Bertalanffy model, then we are faced with the biologically implausible conclusion that the carrying capacity depends on the initial population size.

One can mask this problem, and at the same time transform model (2.11) into a single, autonomous equation, with the following argument [25]. First, let $\lim_{t \to \infty} N(t) \equiv K < \infty$, assuming such a limit exists, and note that by (2.14), $G_0 = \alpha \ln\left(\frac{K}{N_0}\right)$. This and equation (2.13) imply that $G(t) = \alpha \ln\left(\frac{K}{N(t)}\right)$, which gives us the following 1-D, autonomous ODE:

$$\frac{dN}{dt} = \alpha N(t) \ln\left(\frac{K}{N(t)}\right). \tag{2.15}$$

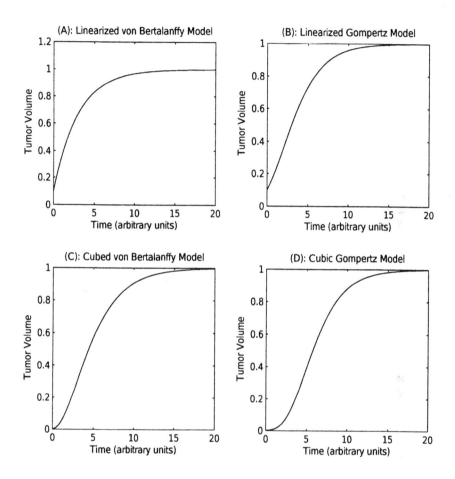

FIGURE 2.2: Comparison of von Bertalanffy and Gompertz growth curves with similar kinetic parameters. Panels (A) and (C) plot the solution to the linearized von Bertalanffy model, equation (2.8), and its cube, respectively, for $\hat{\alpha} = \beta = 1$. Panels (D) and (B) plot the solution to the Gompertz model, equation (2.13), and its cube root, respectively, for $\alpha = 0.4$ and $G_0 = \alpha \ln N_0^{-1}$. In both plots, the initial linear measure was 0.1 units. Note that "tumor volume" and "diameter" are not scaled consistently; they have been rescaled to force their attractors to unity.

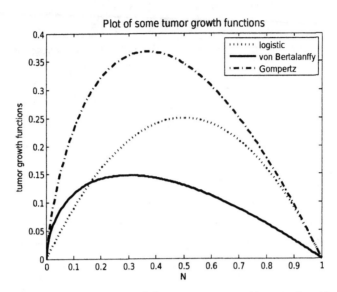

FIGURE 2.3: Comparison of three tumor growth rate functions: logistic $N(1 - N)$, von Bertalanffy $N^{2/3} - N$, Gompertz $-N \ln(N)$.

Again, following Thieme [37] the substitution $x(t) = \ln(K/N(t))$ into (2.15) yields

$$\frac{dx}{dt} = -\alpha x, \tag{2.16}$$

from which one can see that

$$\ln | \ln(K/N(t))| = \ln | \ln(K/N_0)| - \alpha t. \tag{2.17}$$

Equation (2.17) suggests, then, that a combination of measurable quantities—namely the tumor size $N(t)$ and its asymptotic limiting size K—yields a linear function of time *if* tumor growth is Gompertzian. This observation plays an important role in fitting the Gompertz model to actual data (see Section 2.3). It is also clear from equations (2.15) and (2.17) that $\hat{N} = K$ is a globally asymptotically stable fixed point, and that like the classic von Bertalanffy model, the Gompertz model exhibits sigmoidal growth (Fig. 2.2). Notice that maximum per capita growth rate varies among the three models (von Bertalanffy, Gompertz and classical logistic) (Fig. 2.3).

One may also notice that

$$\frac{d(\ln(N))}{dt} = \frac{1}{N}\frac{dN}{dt} = G(t) = \frac{-1}{\alpha}\frac{dG}{dt}. \tag{2.18}$$

From this we obtain that for some constant β,

$$\ln(N) = (-G(t) + \beta)/\alpha, \tag{2.19}$$

which is equivalent to saying that $G(t) = \beta - \alpha \ln(N)$. This gives us a third alternative form of the Gompertz model:

$$\frac{dN}{dt} = \beta N - \alpha \ln(N)N = N(\beta - \alpha \ln(N)). \qquad (2.20)$$

Note that this function is not defined for $N = 0$, so we must assume that the tumor has a certain size before applying this model. A key advantage of this formulation is that the carrying capacity, $K = e^{\beta/\alpha}$, need not be assumed but rather becomes a function of the kinetic parameters α and β.

2.3 Applications of Gompertz and von Bertalanffy models

In a perfect world, one in which we understood all tumor biology, we could write a model that would predict the growth curve of a tumor into the future given its current state. Of course, we cannot do that because our knowledge is woefully incomplete. Contributing to our ignorance is the fact that repeated measures of growing tumors—required for any detailed study of dynamics— are very difficult to obtain. Animal models with implanted tumors offer the best hope of advance because the researcher knows precisely when the tumor "originated," and relatively invasive techniques are available. But even in the best case, obtaining accurate size measures of irregular, three-dimensional masses without influencing the physiology of either tumor or host remains problematic.

Largely because of this difficulty, early theorists sought "growth laws" that attempt to represent general tumor kinetics without reference to any particular tumor. The first hypothesis that proposed a coherent growth law suggested that tumors grow exponentially, an idea following naturally from the view that cancer is "uncontrolled proliferation." The idea is further supported by the analogy with bacteria. In lab cultures, bacterial colonies grow exponentially and then abruptly stop when they exhaust some limiting resource. As early as the 1930s, however, it was becoming clear that the exponential hypothesis was wrong. If tumor volume increases exponentially, then so must a one-dimensional measure of its size (in essence akin to tumor "diameter," but remember that tumors are irregular solids at best). In other words, log plots of both volume and "diameter" over time are linear if and only if the tumor is growing exponentially. However, careful studies of tumor growth in animal models show that tumor "diameter," not its logarithm, tends to be linear, implying a decreasing volumetric (mass) growth rate over time (Fig. 2.4).

An early hypothesis (1930s) suggested that such a declining growth rate, complete with a "linear" phase of growth, can be explained by a growing

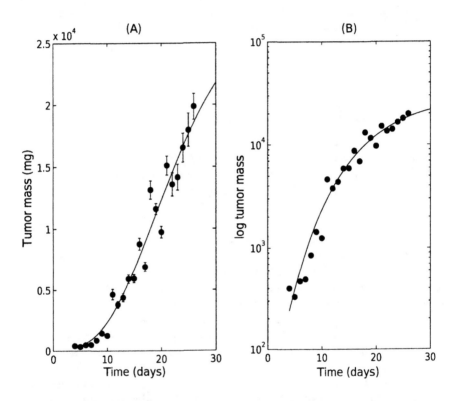

FIGURE 2.4: Growth kinetics of Fortner Plasmacytoma 1 tumors. Points represent mean mass of subcutaneous tumor implants in mice, and curve is the least-squares, best-fit Gompertz model (equation (2.13); $N_0 = 18.4$, $G_0 = 0.789$, $\alpha = 0.107$). Error bars represent ± 1 standard error of the mean at each point. Data from Simpson-Herren and Lloyd [32].

mass obtaining nutrient only on its surface (see Chapter 3 for details). Indeed this hypothesis was verified later in cell culture. Although this hypothesis cannot strictly apply to most clinically relevant tumors, which absorb nutrients from their vascular infrastructure, the notion has sparked a great deal of progress. In the 1960s, for example, Anna K. Laird [20, 21], in a series of papers that still rank among the finest contributions on the subject, proposed a hypothesis that could apply to all tumors, angiogenic or not. In particular, she suggested that doubling time increases (growth rate decreases) for some reason as the tumors age. It could be due to a lengthening cell cycle, an increasing death rate, or both. Whatever the mechanism, this notion led Laird to recommend the Gompertz model as a description of tumor growth, since one can interpret that model as a population in which per capita growth rate is decreasing exponentially over time. To test this idea, Laird attempted to fit the Gompertz curve to growth data from around 20 different tumors in mice, rats and rabbits, including solid and diffuse tumors, both implanted and autochthonous. In each case she used a successive approximation technique to obtain least-squares best-fit estimates for α, G_0 and N_0 in equation (2.13). In general, this procedure generated far better descriptions of the data than any exponential curve could, although not always. Later, Simpson-Herren and Lloyd [32], using techniques similar to Laird's, demonstrated that the Gompertz model provides an excellent empirical description of viral-induced mammary tumors in a particular tumorigenic strain of mice (Fig. 2.4).

More recently, Rygaard et al. [30] (see also [7]) successfully used the Gompertz model to describe the growth of human lung cancers xenografted into immune-deficient mice. After implantation, tumors were measured with calipers and the linear measures were transformed into volumes using a standard relation. Assuming that volumetric growth of these tumors would be Gompertzian, Rygaard et al. transformed the data with the equation, $\ln | \ln(K/V_n)|$, where $\{V_n(t)\}$ are the volume measurements from n mice taken at time points $t \in \{0, 1, 2...m\}$. The means across mice for each of the m time points were then plotted against time, and a regression line was fit to the data. The slope parameter estimated α, and the results generate a remarkably good correlation (see figure 1C of [30]), suggesting an excellent fit between data and the Gompertz model.

Such studies, especially Laird's, are the foundations of statement after statement in the mathematical oncology literature that real cancer growth is Gompertzian. Nevertheless, one would be wise to retain some skepticism for three reasons. First, while the mathematical model can be fit to data from animal models and *aggregated* data from human tumors, it probably does not describe any clinical tumor particularly well, certainly not from inception to final clinical outcome. Second, a good fit between the Gompertz model and data does not really limit the number of other mathematical forms that might also describe the data equally well or better. Finally, a good fit does not imply a correct mechanistic description of tumor growth.

Any attempt to fit the Gompertz curve to tumor data must face the awkward fact that the asymptotic tumor size K generally cannot be measured. Tumors routinely become life threatening well short of their asymptotic size limit in both humans and animal models. Therefore, anyone wishing to apply equations (2.17) or (2.20) to real tumors must either find a way to independently estimate K (or α and β, which lack clear biological meaning) or avoid it altogether. Rygaard et al. chose the first strategy, setting K to the value that produced the "most linear-looking data." With this procedure, of course, it becomes a simple task to generate an excellent correlation coefficient since two parameters, K and α, are free to vary. But with such a flexible family of curves, the fit may be meaningless. Laird successfully avoided estimating K but at the cost of introducing a third free parameter. The extra parameter makes a good but vacuous fit to data that much easier. What's worse, one of these parameters, G_0, has a biologically nebulous interpretation that defeats independent verification, even in principle.

In autochthonous human tumors these problems are exacerbated by lack of data. Most cancer patients begin treatment immediately upon diagnosis. Those few who refuse are rarely monitored with serial follow-up imaging studies, and even then determining tumor mass from clinical imaging technology presents significant challenges. So, very few untreated tumors have ever been adequately studied. Worse, even in the few exceptions, followups are relatively sparse. For example, in one study [34], Spratt and colleagues obtained data on tumor growth for 32 women suffering untreated breast cancer. On average each patient's data set consisted of 3.4 images, with a maximum of six. Facing this dearth of time series data, Spratt et al. took the best available course—they aggregated patient data to increase statistical power, as is typical in such studies. In particular, they fit equation (2.17) to each patients' tumor measures in the same way Rygaard did in mice. But unlike Rygaard, Spratt et al. had far fewer time points to work with for any single patient. Also, timing of followup images, initial (at presentation) tumor sizes and tumor growth rates all varied. So averaging tumor growth across subjects as Rygaard et al. did was impossible. Instead, Spratt et al. measured "fit" between model and data by regressing each individual patient's linearized data against time and summing regression mean squared errors across all patients. Although Spratt et al. fixed K (essentially a guess), α was allowed to vary across patients, so many degrees of freedom were available to produce a good fit.

Despite all this, one could argue that the model, in fact, *has* been fit to data; it can produce solutions with the correct qualitative behavior. Therefore, the model, as a hypothesis, has been tested and verified. Nevertheless, both evidence and theory warn that these verifications may be more apparent than real. On the evidentiary side, the few untreated tumors followed by sufficient serial imaging tend to suggest that natural tumors alternate between rapid and slow growth at irregular intervals [29, 35]. Therefore real growth curves exhibit many more inflections than predicted by the Gompertz equation.

Why, then, does the Gompertz model successfully describe real tumors?

As we saw earlier, these successes tend to be based on fitting the curve to changes in *average* tumor size over time, Laird's work being an important exception. Even if individual tumors grow in a "jerky" fashion, on average one can expect a sigmoidal curve based on current theory.

Consider the following two commonly argued hypotheses explaining saltatory growth in tumors. ("Saltation" literally refers to hopping or dancing; here it means growing in spurts.) The first suggests that tumors experience ever-changing environmental conditions over time. They often bleed or otherwise leak fluid from poorly-constructed vascular networks. Pressure inside tumors is often much higher than the surrounding tissue, which can cause poorly constructed blood vessels to collapse, resulting in ischemic infarction of part of the tumor. As vessels are rebuilt or pressure is relieved, restored blood flow sparks another round of rapid tumor growth. Cycles of such events would be observed as saltatory growth.

Alternatively, selection on cell phenotypes within the tumor should continually replace less adapted clones with better-adapted, faster-growing mutants. Since these mutants arise at random times, growth rates of individual tumors "jump" at random times, also resulting in saltatory growth. Note that these two hypotheses are not mutually exclusive.

In either case, smooth sigmoidal growth "on average" is conceivable. In small tumors, in general the derivative of the growth rate should be increasing because better adapted clones are evolving and (or) because small tumors are largely free from the ecological catastrophes generated by complex, poorly integrated vasculature. Still speaking generally, as tumors age they become more susceptible to catastrophic events, and the host's physiology is manifestly failing. Evolution continues to select for better adapted mutant strains, but it cannot overcome the increasing ecological degradation as the host physiology fails. In this degrading environment, selection may switch from favoring more rapidly proliferating (what ecologists refer to as r-selected) clones to favoring clones that use resources efficiently (K-selected). So *on average* tumor growth rate falls as the tumor ages.

That the Gompertz model can be fit to "average" tumor growth is therefore not surprising since the model generates a sigmoid growth curve [29]. However, many other relevant population growth models also predict sigmoidal growth, including von Bertalanffy and assorted models obtained from its generalization like the Verhulst and Bernoulli equations, along with the Beaverton-Holt and Ricker models among others (see Thieme [37] Chs. 4-6 for an excellent review). So sigmoidal growth does not necessarily imply the Gompertz equation, a point made by Castro et al. [7]. Indeed, Spratt et al. in the work described above found that the autonomous Bernoulli model fit human breast cancer data better, according to their criteria, than did the Gompertz. We find ourselves, therefore, in the same position as Sewall Wright did in 1926 [41, pg. 114] when he wrote of the Verhulst, or logistic, model of population growth:

It may perhaps be questioned whether the capacity which this mathematical form undoubtedly possesses for fitting growth and population curves has any very deep significance. ...there will [likely] be increasing adverse pressure as growth goes on, leading to damping off and reversal of curvature, and ultimately, if conditions are uniform, to an asymptotic approach to an upper limit. In this we have simply described a general S-shaped curve, which may take various forms depending on the exact nature of the adverse process. Anything growing under constant or even changing conditions, provided the changes are sufficiently gradual [or, we might suggest, averaged], can hardly be expected to grow in any other way. Any flexible mathematical formula which gives this general shape can hardly fail to give an empirical means of fitting such a curve. The logistic curve [or Gompertz, as the case may be] is, perhaps, the most convenient but it is by no means the only mathematical form with this property.

Finally, many critics of the Gompertz model make the valid claim that it has a weak physiological basis (for example [38]). On one hand, as we saw in Section 2.2.2, the function G in model (2.11) can represent cellular aging mechanisms. On the other hand, there are no clear biological mechanisms involved that would produce an exponential decline in growth rate. This sort of ambiguity limits the model's applicability to anything but empirical descriptions and hypotheses. The von Bertalanffy model, which has a much cleaner derivation from biological first principles directly applicable to tumor biology, is probably a better theoretical foundation, although perhaps not the best empirically.

The von Bertalanffy and Gompertz models are only two of many simple models that have been applied to tumor growth (Table 2.1), and there is disagreement about which is the best. The disagreement probably arises due to the complexity of the disease. For example, Marušić et al. [23] compared how well models fit data for a variety of simple one- or two-dimensional models (including all those in Table 2.1, among others). Although they found that von Bertalanffy and logistic models consistently performed poorly (based on their fitting criteria), the models that gave the best fit varied by tumor type. They studied only two tumor cell lines implanted into athymic mice, and found that the classical Gompertz was the best fit for one line while the general 2-parameter model gave the best fit for the other. This result also failed to agree entirely with earlier data from multicell spheroids [22], a common culture model of tumors (see Chapter 3). Therefore, there probably is no single answer to the question, which is the best model to fit cancer data. However, there are alternatives to these simple models, including one by Gyllenberg and Webb [15] that brings both Gompertz and von Bertalanffy models into a single elegant framework.

TABLE 2.1: A partial list of scalar models that have been applied to cancer dynamics. In each case, parameters α, β, γ, λ, μ and k are positive constants. Parameter p is a constant.

Name	Model
Classical von Bertalanffy	$N' = \alpha N^{2/3} - \beta N$
General von Bertalanffy	$N' = \alpha N^{\lambda} - \beta N$
Bernoulli	$N' = \alpha N - \beta N^{\mu}$
General growth model	$N' = \alpha N^{\lambda} - \beta N^{\mu}$
Logistic (Verhulst)	$N' = \alpha N - \beta N^2$
Hyper-logistic	$N' = (\alpha/\beta)N^{1-p}(\beta - N)^{1+p}$
Gompertz	$N' = N(\beta - \alpha \ln N)$
General Gompertz	$N' = N^{\lambda}(\beta - \alpha \ln N)$
Hyper-Gompertz	$N' = N(\beta - \alpha \ln N)^{1+p}$
Holling II growth	$N' = \frac{\alpha N}{k+N} - \beta N$
Piantadosi	$N' = \frac{\alpha N}{(k+N^{\gamma})^{1/\gamma}} - \beta N$

2.4 A more general approach

In science, competing theoretical models often are united when someone demonstrates that they are nothing more than special cases of a more global theoretical approach. As it turns out, this applies to the Gompertz and von Bertalanffy models, as shown by Gyllenberg and Webb [15], who bypass insignificant mathematical details to clarify how a simple *biological* hypothesis generates a family of solutions all with Gompertz- or von Bertalanffy-like S-shaped growth.

In particular, Gyllenberg and Webb explore the consequences of two simple observations from real tumors: (1) actively proliferating cells can enter a quiescent state where they stop dividing; and (2) quiescence tends to be more common in large compared to small tumors. Quiescence appears to be a cellular response to stress, like hypoxia, nutrient limitation, or increased hydrostatic pressure. Such stresses typically increase, often nonlinearly, with tumor size. It can be reversible or irreversible, as reviewed by Skipper [33].

Gyllenberg and Webb propose a simple, but rather general, model of the transition into and out of quiescence.[1] Let $P(t)$ and $Q(t)$ be the number of proliferative and quiescent cells, respectively. Define $N(t) = P(t) + Q(t)$.

---

[1]Their model is very similar to, but more general than, a model proposed by Panetta and Fister that included chemotherapy. See [11, 26].

Then the Gyllenberg-Webb model takes the following form:

$$\begin{cases} \dfrac{dP}{dt} = (\beta - \mu_p - r_0(N))P + r_i(N)Q, \\[3mm] \dfrac{dQ}{dt} = r_0(N)P - (r_i(N) + \mu_q)Q, \end{cases} \tag{2.21}$$

with initial conditions,

$$P(0) = P_0 > 0, \quad Q(0) = Q_0 \geq 0. \tag{2.22}$$

Cells proliferate at per capita rate $\beta > 0$, and proliferative and quiescent cells die at rates $\mu_p \geq 0$ and $\mu_q \geq 0$, respectively. Finally, cells transition to and from the quiescent compartment at rates $r_0(N)$ and $r_i(N)$, respectively, where both functions are continuous and defined for $N \geq 0$. Since stressed cells tend to enter a quiescent state, and stresses tend to increase with tumor size, Gyllenberg and Webb assume that $r_0(N) \geq 0$ and nondecreasing for all $N > 0$; therefore, $\lim_{N \to \infty} r_0(N)(\equiv l_0)$ exists or unbounded (i.e. the limit is ∞). Hence

$$0 \leq l_0 \leq \infty. \tag{2.23}$$

They also assume that stressed cells tend to stay quiescent, so $r_i(N) \geq 0$, and nonincreasing, so $\lim_{N \to \infty} r_i(N)(\equiv l_i)$ also exists, and

$$0 \leq l_i < \infty. \tag{2.24}$$

Suppose quiescence to be a permanent condition, and for now assume that quiescent cells never die. Alternatively, equate "quiescence" with necrosis and assume the necrotic material remains as part of the tumor. Gyllenberg and Webb show that under either interpretation, an S-shaped growth curve is expected, as summarized in the following proposition [15]:

PROPOSITION 2.1
Let $\mu_q = 0$ and $r_i(N) = 0$ for all $N \geq 0$. Also suppose $l_0 > \beta - \mu_p > 0$. Then $N(t)$ is monotonically increasing, bounded and has an inflection point $\hat{N}$, where $\hat{N}$ satisfies

$$r_0(\hat{N}) = \beta - \mu_p, \tag{2.25}$$

if it exists. If $\hat{N}$ exists, then $N(t)$ is convex on $N_0 \leq N \leq \hat{N}$, and concave on $\hat{N} \leq N < \infty$. If $\hat{N}$ does not exist, then $N(t)$ is concave on $N_0 \leq N < \infty$. Finally,

$$\lim_{t \to \infty} P(t) = 0. \tag{2.26}$$

PROOF Since μ_q and $r_i(N)$ are both identically 0, then

$$\frac{dN(t)}{dt} = (\beta - \mu_p)P(t) \tag{2.27}$$

for $t \geq 0$. By assumption, $\beta > \mu_p$, so $N(t)$ is monotonically increasing. Differentiating again yields

$$\frac{d^2 N(t)}{dt^2} = [\beta - \mu_p - r_0(N(t))]\frac{dN(t)}{dt}. \qquad (2.28)$$

Since $N'(t) > 0$ and $r_0(N)$ is nondecreasing for all $t, N \geq 0$, respectively, the inflection point, if exist at $\hat{N}$, it must satisfy $r_0(\hat{N}) = \beta - \mu_p$. By assumption $\beta - \mu_p < l_0$; therefore, continuity conditions guarantee the existence of $\hat{N}$ if and only if $N(0) < \hat{N}$, and also that the solution is (1) concave for all $N(t) \geq \hat{N}$, (2) convex for all $N(t) < \hat{N}$, and 3) concave for all $N \geq N(0)$ if $\hat{N}$ does not exist. Integrating (2.28) from $t_0 \geq 0$ gives,

$$\frac{dN(t)}{dt} = \int_{N(t_0)}^{N(t)} [\beta - \mu_p - r_0(s)]ds + (\beta - \mu_p)P(t_0), \qquad (2.29)$$

for $t > t_0$. Therefore, since $N(t)$ is monotonically increasing, $l_0 > \beta - \mu_p > 0$ and equation (2.29) holds for any t_0, it must be true that $N(t)$ is bounded (otherwise, the integral term in the above equation will tend to $-\infty$) and hence $\lim_{t\to\infty} dN(t)/dt = 0$. This result and equation (2.27) imply that

$$\lim_{t\to\infty} P(t) = 0.$$

$$\square$$

Biologically, the third hypothesis of the proposition, namely that $\beta - \mu_p < l_0$, assumes a threshold tumor size above which cells transition to quiescence or necrosis faster than their net proliferation rate. If this were not true, then it is easy to see that $N(t)$ would be convex and the tumor would grow without bound.

Certainly the restrictive assumptions of proposition 2.1 limit the biological usefulness of the result. They of course can be relaxed, and doing so allows one to explore conditions leading to S-shaped growth curves, as the following proposition from [15] shows:

PROPOSITION 2.2
Suppose $\mu_q > 0$, $r_0(N) > 0$ for $N \geq 0$, and let the following conditions hold:

$$\frac{l_0}{\beta - \mu_p} > 1 + \frac{l_i}{\mu_q}, \qquad (2.30)$$

$$r_0(0) < (\beta - \mu_p)\left(1 + \frac{r_i(0)}{\mu_q}\right), \qquad (2.31)$$

$$\beta - \mu_p < r_i(N) + r_0(N), \quad N \geq 0, \qquad (2.32)$$

$$\frac{d}{dN}\left(\frac{r_i(N)}{r_i(N) + r_0(N) - \beta + \mu_p}\right) < 0, \quad N \geq 0. \qquad (2.33)$$

Then there exists a unique, globally asymptotically stable equilibrium, (P^*, Q^*), such that $P^*, Q^* > 0$. Furthermore, $N(t) \to N^*$ as $t \to \infty$, where N^* is the (unique) solution of

$$r_0(N^*) = (\beta - \mu_p)\left(1 + \frac{r_i(N^*)}{\mu_q}\right), \tag{2.34}$$

and growth fraction $G(t) = P(t)/N(t)$ tends to $G^* = \mu_q/(\beta - \mu_p + \mu_q)$.

We leave the proof of proposition 2.2 as an exercise at the end of this chapter. The proof in the paper of Gyllenberg and Webb [15] missed some technical details (such as the details to show the solutions are bounded) and the condition (2.33) which is needed to ensure the positive steady state is unique. The key to prove this proposition is to convert the model to a system of equations involving only P and N. Conditions (2.30)-(2.33) together ensure that the isoclines of the resulting $P - N$ system will intersect at the origin and another point which produces the only positive steady state. The condition (2.32) can be used to eliminate the possibility of periodic orbits.

The great advantages of this approach arise from its minimalist specification of the transition functions $r_0(N)$ and $r_i(N)$. It admits a rigorous conclusion without the encumbrance of irrelevant formal details. It also emphasizes that an argument between "Gompertz" or "von Bertalanffy" models is largely irrelevant—a simple, general model with straightforward biological interpretations produces the same behavior. This approach brings up the question—what really are the correct forms of r_0 and r_i, and how do they vary across tumor types? Certainly, more biological details and realism can be added to this model. But model (2.64) is a more stable theoretical platform on which to found theory than those that are mere curve-fitting or mechanistically ambiguous models like Gompertz. The Gyllenberg and Webb model also provides an excellent tool to study the effect of treatment, as described in the next section.

For more realistic and mathematically challenging avascular tumor models, one can incorporate additional biological processes and many such models are discussed in the classical book of Adam and Bellomo [1] and the comprehensive review article of Araujo and McElwain [4]. Observe that the model (2.64) assumes that dead cells are removed from tumor instantly by undisclosed process. Removing this assumption yields a more realistic three dimensional version of model (2.64) which is studied in detail in Alzahrani et al. [2].

2.5 Mechanistic insights from simple tumor models

Despite, or perhaps because of, their simplicity, the models we explore in this chapter have had an enormous impact on modern chemotherapy, in particular dose scheduling, multi-drug techniques and adjuvant chemotherapy (chemotherapy used to support another treatment modality). (For reviews, see [8, 9].) Although we will cover these topics in more detail later, it is instructive to set the foundations here.

Chemotherapy entered its modern, theory-based guise with Howard Skipper's *log-kill hypothesis* [33], which is simply the notion that per capita death rate of cancer cells is constant for a fixed drug concentration. If the drug is given as a bolus dose, then this hypothesis claims that a fixed fraction of the tumor is killed for a defined time period. This phenomenon is also called fractional kill, or fractional cell kill. The fraction is a function of dose, but not tumor size. These ideas are intertwined with Skipper's exponential growth models, which assume a constant per capita growth rate regardless of tumor size. The hypothesis also implicitly assumes that drug sensitivity does not vary across cells or time, and that under treatment, a cancer cell's life span is exponentially distributed. Based on this observation, repeated doses must be administered to continue to reduce the size of the tumor to zero or near zero.

Although the log-kill hypothesis performed well in animal models, in vivo human tumors typically refuse to respond as the hypothesis predicts. Seeking a better hypothesis, Richard Simon and Larry Norton [25] began following a lead that Skipper himself suggested, namely that faster-growing tumors respond better to chemotherapy than do slower-growing neoplasms. Combining this observation with the notion that the Gompertz model describes a tumor's life history better than does Skipper's exponential model, Simon and Norton hit on the idea that response to therapy would be inversely related to tumor size, since, by the Gompertz model, per capita growth rate is inversely related to tumor size. This suggestion has since become the *Norton-Simon hypothesis* [31].

The Norton-Simon hypothesis has informed chemotherapy treatment schedules for decades. Its fundamental contribution is that treatment schedules which decrease tumor size fastest will always be superior to any other schedule, even if the same drug and overall dose is given, within the limits of patient tolerance. No mathematics is required to see why this is true—if smaller tumors respond best to chemotherapy, the best chemotherapy schedule is clearly the one that delivers most of its drug when the tumor is as small as possible. Therefore, treatment should be scheduled to deliver large early doses to knock the tumor down rapidly, followed by frequent, large-dose episodes to hit the small, fast-growing tumor and micrometastases . Norton and Simon refer to this as maximizing *dose density* .

Although Norton and Simon focus almost exclusively on the Gompertz model to support their suggestion, the hypothesis does not depend on the Gompertz model *per se*. Rather, the hypothesis simply requires two things: (1) that chemotherapeutic efficacy is inversely related to per capita growth rate, and (2) that tumor growth rate is a decreasing function of tumor size. As such, Norton and Simon note that the hypothesis is really an empirical notion and therefore does not depend on the model's formalism, but rather the qualitative behavior of its solutions. In other words, any model with sigmoidal solutions, including the Gompertz, von Bertalanffy or Gyllenberg-Webb models among others, would work.

2.6   Sequencing of chemotherapeutic and surgical treatments

A natural next step for theoreticians is to build mechanistic models that may help improve cancer treatment. We present here an interesting example from Kohandel et al. [19], who study sequencing of surgical and chemotherapeutic treatments for ovarian cancer. Ovarian cancer is one of the most common cancers in women under age 65. Most ovarian cancers are detected after the disease has spread throughout the abdomen since symptoms are often vague and easily misdiagnosed. The main treatment for ovarian cancer comprises a combination of surgery (laparotomy) and chemotherapy, with the former performed first so the disease may be staged as well as treated. Since surgery often fails to remove all cancer cells, adjuvant chemotherapy is called for. Alternatively, oncologists can opt to apply chemotherapy before surgery (this is called neoadjuvant chemotherapy), thereby shrinking the tumor and perhaps increasing the efficacy of surgery. Which is the better course remains an open question.

One recent study compared success of the treatment sequences in women with advanced ovarian cancer [18]. In this study, 336 women were given chemotherapy first while 334 underwent surgery first. The study only enrolled women with stage IIIc/IV ovarian cancer—that is, metastatic tumors greater than 2 cm in size had spread into the abdomen—and a large proportion of women in the study had very bulky tumors. The researchers found no difference between the two treatment sequences either in time to progression of the disease or survival time following treatment. In other words, an average advanced ovarian cancer patient may experience the same expected outcome whether they have chemotherapy or surgery first. But this is the *expected* outcome of an *average* patient. It is not necessarily true that a *given* patient should be ambivalent about the different treatment sequences. This is the essence of **personalized medicine**—what is the best course of action for *this particular patient*? One thing has become clear: the answer is not found in a textbook. However, the means for finding the answer may be, and mathematical modeling is a way forward.

The hope here is that treatment outcomes of adjuvant (surgery first) and neoadjuvant (surgery second) chemotherapy depend on quantifiable patient-specific factors. These factors enter into our models as growth- and treatment-related parameters, allowing us to express the problem mathematically, as follows. Assume a malignant ovarian tumor of size N_0 is diagnosed at time $t = 0$. There are two possible treatment sequences:

1. At a time $t_0 > 0$, surgery is carried out to remove a fixed fraction of tumor cells, and immediately afterward an adjuvant chemotherapy with a predetermined killing rate is administered. At time $t_f > t_0$, the tumor has final size N_{SC}. Let the fraction removed be $1 - e^{-k_s}$, where k_s is a non-negative constant.

2. At a time $t_0 > 0$, a neoadjuvant chemotherapy with predetermined killing rate is administered. At a later time (t_f), surgery is performed to remove a fixed fraction of tumor cells (again, $1 - e^{-k_s}$), resulting in a tumor with final size N_{CS}.

The relevant question is, in what scenarios is $N_{CS} < N_{SC}$?

Following Kohandel et al. [19], we consider a simple model that incorporates cancer cell growth and the effects of chemotherapeutic and surgical treatments under various cell-kill hypotheses. Let $P(t, N)$ describe the pharmacokinetic and pharmacodynamic effects of the drug on the cancer and $f(N)$ be the tumor growth model. For the standard treatment—surgery followed immediately by adjuvant chemotherapy—the number of cells at time $t > 0$ may be approximated by following scalar differential equation:

$$\begin{cases} \dfrac{dn_1}{dt} = f(n_1), \ n_1(0) = N_0, \ t \in [0, t_0] \\[2mm] \dfrac{dN_1}{dt} = f(N_1) - P(t, N_1), \ N_1(t_0) = e^{-k_s} n_1(t_0), \ t \in [t_0, t_f], \end{cases} \qquad (2.35)$$

where $n_1(t)$ and $N_1(t)$ represent tumor size before and after surgery, respectively. Then $N_{SC} = N_1(t_f)$. In contrast, for the neoadjuvant chemotherapy treatment we assume that the number of cells at time $t > 0$ is governed by following scalar differential equation:

$$\begin{cases} \dfrac{dn_2}{dt} = f(n_2), \ n_2(0) = N_0, \ t \in [0, t_0] \\[2mm] \dfrac{dN_2}{dt} = f(N_2) - P(t, N_2), \ N_2(t_0) = n_2(t_0), \ t \in [t_0, t_f], \end{cases} \qquad (2.36)$$

yielding $N_{CS} = e^{-k_s} N_2(t_f)$, where $n_2(t)$ and $N_2(t)$ represent tumor sizes before and after the onset of chemotherapy, respectively.

Below we compare two cell-kill hypotheses: the log kill, in which $P(t, N) = c(t)N$; and the Norton-Simon hypothesis, with $P(t, N) = c(t)f(N)$. We assume that the function $c(t)$ is proportional to drug concentration at time t, and the function $f(N)$ is the Gompertz model (2.15). Kohandel et al. [19] show that *under the first (log-kill) hypothesis, sequencing chemotherapy before surgery results in a smaller tumor size compared to surgery followed by chemotherapy.*

PROPOSITION 2.3
Assume the tumor grows according to Gompertz equation, with

$$f(N) = -\beta N \ln(N/K),$$

and $P(t, N) = c(t)N$. *Let* $c_1(t) \equiv -\int_{t_0}^t c(s)e^{\beta s}ds$. *Then*

$$N_{CS} = K \exp\left[e^{-\beta(t_f - t_0)}\left(c_1(t_f) + \ln \frac{N_0}{K}\right) - k_s\right],$$

and

$$N_{SC} = K \exp\left[e^{-\beta(t_f - t_0)}\left(c_1(t_f) + \ln \frac{N_0}{K} - k_s\right)\right].$$

Therefore

$$\frac{N_{CS}}{N_{SC}} = \exp\{-k_s(1 - e^{-\beta(t_f - t_0)})\} < 1.$$

PROOF First, we compute N_{CS}. For convenience, we extend the definition of $c(t)$ to $[0, t_f]$ by setting $c(t) = 0$ for $t \in [0, t_0]$. Let $x(t) = \ln N_2(t)$. Then

$$\frac{dN_2}{dt} = f(N_2) - P(t, N_2) = -\beta N_2 \ln \frac{N_2}{K} - c(t)N_2$$

is equivalent to

$$\frac{dx}{dt} = -\beta x + \beta \ln K - c(t).$$

Hence

$$\frac{de^{\beta t}x(t)}{dt} = -e^{\beta t}(c(t) - \beta \ln K).$$

Integrating the both sides of the above equation from t_i, $0 \le t_i \le t_0$, to t yields

$$e^{\beta t}x(t) - e^{\beta t_i}x(t_i) = -\int_{t_i}^t c(s)e^{\beta s}ds + \beta \ln K \int_{t_i}^t e^{\beta s}ds,$$

which results in

$$e^{\beta t} \ln N_2(t) - e^{\beta t_i} \ln N_2(t_i) = -\int_{t_i}^t c(s)e^{\beta s}ds + (e^{\beta t} - e^{\beta t_i}) \ln K.$$

We have

$$N_2(t) = N_2(t_i)^{e^{-\beta(t - t_i)}} e^{(1 - e^{\beta(t_i - t)}) \ln K} e^{e^{-\beta t}(c_1(t) - c_1(t_i))}. \tag{2.37}$$

Let $t_i = 0$. For $t \ge t_0$, the above expression is equivalent to

$$N_2(t) = K \exp\left[\left(c_1(t) + \ln \frac{N_2(0)}{K}\right) \exp(-\beta t)\right].$$

Assume that surgery performed at time t_f kills fraction $1 - e^{-k_s}$ of the tumor cells. Then with $N_2(0) = N(0)$ we obtain

$$N_{CS} = K \exp\left[-k_s + \left(c_1(t_f) + \ln \frac{N(0)}{K} \right) \exp(-\beta t_f) \right]. \qquad (2.38)$$

Observe that

$$N_2(t_0) = K \exp\left[e^{-\beta t_0} \ln \frac{N_2(0)}{K} \right]. \qquad (2.39)$$

Next, we compute N_{SC}. In this case, the surgery is performed first at time t_0. From Eq. (2.39), we have

$$N(t_0) = K \exp\left[-k_s + e^{-\beta t_0} \ln \frac{N(0)}{K} \right]. \qquad (2.40)$$

Chemotherapy is applied on the time interval $[t_0, t_f]$. Let $t_i = t_0$ in Eq. (2.37), and change N_2 to N. We have that for $t \in [t_0, t_f]$,

$$N(t) = N(t_0)^{e^{-\beta(t-t_0)}} e^{(1 - e^{\beta(t_0 - t)}) \ln K} e^{-e^{-\beta t} c_1(t)}. \qquad (2.41)$$

This result with Eq. (2.40) yields

$$N(t) = \left(K \exp\left[-k_s + e^{-\beta t_0} \ln \frac{N(0)}{K} \right] \right)^{e^{-\beta(t-t_0)}} e^{(1 - e^{\beta(t_0 - t)}) \ln K} e^{-e^{-\beta t} c_1(t)}. \qquad (2.42)$$

Hence

$$N_{SC} = N(t_f) = K \exp\left[e^{-\beta t_f} \left(c_1(t_f) - k_s e^{-\beta t_0} + \ln \frac{N(0)}{K} \right) \right]. \qquad (2.43)$$

Dividing Eq. (2.38) by Eq. (2.43) gives us

$$\frac{N_{CS}}{N_{SC}} = \exp\{ -k_s(1 - e^{-\beta(t_f - t_0)}) \} < 1.$$

Therefore, $N_{CS} < N_{SC}$ under our biological assumptions, so chemotherapy before surgery results in a smaller tumor at treatment's end compared to surgery followed by chemotherapy. ∎

In contrast to the log-kill assumption made above, the next result shows that if the Norton-Simpson hypothesis is correct, sequencing chemotherapy before surgery *may or may not* result in a smaller tumor. We leave the proof as an exercise.

PROPOSITION 2.4

Assume the tumor grows according to Gompertz equation, with

$$f(N) = -\beta N \ln(N/K),$$

and $P(t, N) = c(t)f(N)$. Let $c_2(t) \equiv \beta \int_{t_0}^t c(s)ds$. Then

$$N_{CS} = K \exp\left[e^{-\beta t_f + c_2(t_f)} \ln \frac{N_0}{K} - k_s\right],$$

and

$$N_{SC} = K \exp\left[e^{-\beta(t_f - t_0) + c_2(t_f)}\left(e^{-\beta t_0} \ln \frac{N_0}{K} - k_s\right)\right].$$

Therefore

$$\frac{N_{CS}}{N_{SC}} = \exp\{-k_s(1 - e^{-\beta(t_f - t_0) + c_2(t_f)})\}.$$

Hence, if $\int_{t_0}^{t_f} c(s)ds < t_f - t_0$, *then* $N_{CS} < N_{SC}$.

2.7 Stability of steady states for ODEs

Here we provide a summary of mathematical results useful in analyzing one dimensional (i.e., scalar) ordinary differential equation models. A *steady state* x^* (also called an equilibrium, critical point or fixed point, among other terms) of a scalar ordinary differential equation $x' = f(x)$, where $f(x)$ is continuously differentiable on an interval I, is a solution of $f(x) = 0$ in I. If $x(0) = x^*$, then $x(t) \equiv x^*$ is a solution of $x' = f(x)$.

A steady state x^* of $x' = f(x)$ is *stable* if for any $\varepsilon > 0$, there is a $\delta > 0$ such that if $|x(0) - x^*| < \delta$, then $|x(t) - x^*| < \varepsilon$ for $t > 0$. If in addition, $\lim_{t\to\infty} x(t) = x^*$, then we say $x = x^*$ of $x' = f(x)$ is *asymptotically stable*. A steady state $x = x^*$ of $x' = f(x)$ is *unstable* if it is not stable. The following local stability theorem is easy to prove and use.

THEOREM 2.1

Assume $f(x)$ is continuously differentiable on an interval I, and $x^* \in I$ is a steady state of the one dimensional ordinary differential equation $x' = f(x)$. If $f'(x^*) < 0$, then $x = x^*$ is asymptotically stable. If $f'(x^*) > 0$, then x^* is unstable.

Applying this theorem to the logistic growth model, $x' = \alpha x(1 - x/K) \equiv f(x)$ and assuming as before that $\alpha > 0$, gives us $f'(0) = \alpha > 0$ and $f'(K) = -\alpha < 0$, implying that that origin as a steady state is unstable while the steady state $x(t) = K$ is asymptotically stable.

The following global stability theorem includes the above local stability result as special cases. It is applicable to most single species population growth models.

THEOREM 2.2

Assume $f(x)$ is continuously differentiable on an interval I and has n distinct roots $x_i^*, i = 1, 2, ...n$ such that $x_i^* < x_j^*$ if $i < j$. The following statements are true.

1. *If $f(x) > 0$ for $x \in (x_{i-1}^*, x_i^*)$, $i > 1$, then for $x_0 \in (x_{i-1}^*, x^*)$, the solution of $x' = f(x)$ with $x(0) = x_0$ tends to x_i^* as t tends to ∞.*

2. *If $f(x) < 0$ for $x \in (x_i^*, x_{i+1}^*)$, $i \geq 1$, then for $x_0 \in (x_i^*, x_{i+1}^*)$, the solution of $x' = f(x)$ with $x(0) = x_0$ tends to x_i^* as t tends to ∞.*

PROOF   We sketch the proof of case 2. The proof for case 1 is similar. Since $f(x) < 0$ for $x \in (x_i^*, x_{i+1}^*)$ and $x_0 \in (x_i^*, x_{i+1}^*)$, we see that the solution passing through x_0 is a strictly decreasing function bounded below by the steady state x_i^*. Hence it must approach a constant $c \in [x_i^*, x_{i+1}^*)$. Using the fact that the derivative of a bounded, equicontinuous, differentiable, monotone function $g(t)$ must tend to zero as $t \to \infty$, we have

$$0 = \lim_{t \to \infty} x'(t) = \lim_{t \to \infty} f(x(t)) = f(c).$$

In other words, c is a steady state. Since there is no other steady state in (x_i^*, x_{i+1}^*), we must have $c = x_i^*$.   ☐

Next we turn to systems of two ordinary differential equations of the form

$$dx/dt = F(x, y), \quad dy/dt = G(x, y). \tag{2.44}$$

Model (2.64) is such a system. We assume that the functions $F(x, y)$ and $G(x, y)$ are continuous and have continuous partial derivatives in some domain Ω of the xy-plane. This condition ensures that if (x_0, y_0) is a point in Ω, then there exists a unique solution of the system (2.44) satisfying the initial conditions $x(t_0) = x_0$, $y(t_0) = y_0$ [39]. Notice that the functions F and G of system (2.44) do not depend on the independent variable t explicitly. Such systems are referred to as autonomous. If F or G explicitly depend on the independent variable t, then the system is called nonautonomous. Differential equation models in physical and life sciences often take the form of autonomous systems since the processes modeled by such systems often have solutions whose quantitative and qualitative properties depend on initial states but not on the start time of the processes.

System (2.44) is said to have steady state $E^* = (x^*, y^*)$ if and only if $F(x^*, y^*) = 0$ and $G(x^*, y^*) = 0$. The steady state E^* of system (2.44) is *stable* if for any $\varepsilon > 0$, there is a $\delta > 0$ such that if

$$\sqrt{(x(0) - x^*)^2 + (y(0) - y^*)^2} < \delta,$$

then

$$\sqrt{(x(t) - x^*)^2 + (y(t) - y^*)^2} < \varepsilon$$

for $t > 0$. If in addition, $\lim_{t \to \infty}(x(t), y(t)) = E^*$, then we say E^* is *asymptotically stable*. A steady state E^* of system (2.44) is *unstable* if it is not stable. The following theorem provides conditions for the asymptotic stability and instability of the steady state E^*.

THEOREM 2.3
Assume that $E^* = (x^*, y^*)$ is a steady state of system (2.44). The Jacobi matrix $J(E^*)$ of the system at E^* is a 2×2 matrix defined by

$$J(E^*) = \begin{pmatrix} \frac{\partial F(x^*, y^*)}{\partial x} & \frac{\partial F(x^*, y^*)}{\partial y} \\ \frac{\partial G(x^*, y^*)}{\partial x} & \frac{\partial G(x^*, y^*)}{\partial y} \end{pmatrix}.$$

If $tr(J(E^*)) \equiv \frac{\partial F(x^*, y^*)}{\partial x} + \frac{\partial G(x^*, y^*)}{\partial y} < 0$ *and*

$$det(J(E^*)) \equiv \frac{\partial F(x^*, y^*)}{\partial x}\frac{\partial G(x^*, y^*)}{\partial y} - \frac{\partial F(x^*, y^*)}{\partial y}\frac{\partial G(x^*, y^*)}{\partial x} > 0,$$

then E^* is asymptotically stable. If $tr(J(E^*)) > 0$ or $det(J(E^*)) < 0$, then E^* is unstable.

Readers are referred to the introductory ODE text by Waltman [39] for a concise, more comprehensive reference on the stability steady states of the autonomous system (2.44).

More realistic models in the medical sciences, such as models of cancer growth with explicit nutrient dynamics or interactions with normal and immune cells, or models of cancer treatment, often involve more than two ordinary differential equations. In such scenarios, we are concerned with systems of n ordinary differential equations of the form

$$dx/dt = \mathbf{F}(\mathbf{x}), \tag{2.45}$$

where $\mathbf{x} = (x_1, x_2, ..., x_n)^T$ and $\mathbf{F}(\mathbf{x}) = (F_1(\mathbf{x}), F_2(\mathbf{x}), ..., F_n(\mathbf{x}))^T$. System (2.45) is said to have steady state $E^* = (\mathbf{x}^*)$ if and only if $\mathbf{F}(\mathbf{x}^*) = \mathbf{0}$. Assume $E^* = (\mathbf{x}^*)$ is a steady state of system (2.45). Then its Jacobian at E^* is

$$J = J(E^*) = \begin{pmatrix} \frac{\partial F_1(\mathbf{x}^*)}{\partial x_1} & \frac{\partial F_1(\mathbf{x}^*)}{\partial x_2} & \cdots & \frac{\partial F_1(\mathbf{x}^*)}{\partial x_n} \\ \cdot & \cdot & \cdots & \cdot \\ \cdot & \cdot & \cdots & \cdot \\ \cdot & \cdot & \cdots & \cdot \\ \frac{\partial F_n(\mathbf{x}^*)}{\partial x_1} & \frac{\partial F_n(\mathbf{x}^*)}{\partial x_2} & \cdots & \frac{\partial F_n(\mathbf{x}^*)}{\partial x_n} \end{pmatrix}.$$

Eigenvalues λ of J are the roots of the characteristic equation $P(\lambda) = 0$,

$$P(\lambda) \equiv det(\lambda \mathbf{I} - J) = 0, \tag{2.46}$$

where $\mathbf{I}$ is the $n \times n$ identity matrix. The steady state E^* of system (2.45) is *stable* if for any $\varepsilon > 0$, there is a $\delta > 0$ such that if

$$\sqrt{\sum_{i=1}^{i=n}(x_i(0) - X_i^*)^2} < \delta,$$

then

$$\sqrt{\sum_{i=1}^{i=n}(x_i(t) - X_i^*)^2} < \varepsilon$$

for $t > 0$. If in addition, $\lim_{t\to\infty} \mathbf{x}(t) = E^*$, then we say E^* is *asymptotically stable*. A steady state E^* of system (2.45) is *unstable* if it is not stable. The following theorem provides conditions for the asymptotic stability and instability of the steady state E^*.

THEOREM 2.4
If all roots of the the characteristic equation $P(\lambda) = 0$ are negative or have negative real parts, then the steady state E^* is asymptotically stable. If at least one of the roots of the characteristic equation $P(\lambda) = 0$ is positive or has positive real part, then the steady state E^* is unstable.

Theorem 2.5 provides the conditions for what is typically referred to as *local asymptotic stability* (as opposed to global) because it gives little information about which solutions have E^* as their (forward-time) limit. All it guarantees is the existence of a neighborhood of E^* such that all solutions with $\mathbf{x}_0$ in this neighborhood have E^* as their forward-time limit. Further work is required to determine the nature of the exhaustive set of such initial conditions, often called the *basin of attraction* of E^*.

The following result provides a reasonably practical tool to assess the properties of the eigenvalues of J. Since the characteristic equation (2.46) is a polynomial in λ, i.e.,

$$P(\lambda) = \lambda^n + a_1\lambda^{n-1} + a_2\lambda^{n-2} + \dots + a_n, \tag{2.47}$$

the **Routh-Hurwitz criteria** [10, 37] provide sufficient conditions for all roots of $P(\lambda) = 0$ to be negative or have negative real parts. Below we apply the Routh-Hurwitz criteria to the special cases of $n = 2, 3, 4$.

THEOREM 2.5
All roots of the

$$P(\lambda) \equiv \lambda^n + a_1\lambda^{n-1} + a_2\lambda^{n-2} + \dots + a_n = 0 \tag{2.48}$$

are negative or have negative real parts if any one of the following three sets of conditions are true.

1. $n = 2$: $a_1 > 0$, $a_2 > 0$.

2. $n = 3$: $a_1 > 0$, $a_3 > 0$, $a_1 a_2 > a_3$.

3. $n = 2$: $a_1 > 0$, $a_3 > 0$, $a_4 > 0$, $a_1 a_2 a_3 > a_3^2 + a_1^2 a_4$.

2.8   Exercises

Exercise 2.1:   *Escherichia coli* is a type of bacterium commonly found in the lower intestine of warm-blooded organisms. Most strains of *E. coli* are harmless, but some can cause serious disease in humans. *E. coli* are classified as bacilli, meaning they are rod- or cylinder-shaped prokaryotic organisms. On average, a single *E. coli* is $0.75\,\mu m$ in diameter and $2\,\mu m$ long. Under ideal growth conditions, *E. coli* populations can double in just over 20 minutes.

1. If ideal conditions can be maintained and there is no mortality, how long would it take for an exponentially growing population of *E. coli* to fill a room of $60 m^3$?

2. If ideal conditions can be maintained and there is no mortality, how long would it take for an exponentially growing population of *E. coli* to fill our planet earth?

Exercise 2.2:   Consider the general von Bertalanffy model, equation (2.1). Show that the origin is a source and $(\alpha/\beta)^{1/(\mu-\lambda)}$ is globally attracting if and only if $\lambda < \mu$. Analyze completely the case in which $\lambda \geq \mu$.

Exercise 2.3:   The solution to the logistic equation can be written in the form

$$y(t) = \frac{k}{1 + e^{a-bt}}. \tag{2.49}$$

Similarly, the Gompertz equation can be expressed in the following general form:

$$y(t) = ke^{-e^{a-bt}}. \tag{2.50}$$

Both, therefore, have an exponential decay term that is a general linear function of time. Pearl and Reed [27] suggested the following generalization of the logistic:

$$y(t) = \frac{k}{1 + e^{F(t)}}, \tag{2.51}$$

where $F(t)$ is a function satisfying $F'(t) = g(F(t))$ and $g(x)$ is a suitable function. If $F(t)$ is a general polynomial in t, for example, one recovers the "general logistic" function of Pearl and Reed [28]. Following Pearl and Reed's lead, Winsor [40] generalized the Gompertz equation similarly:

$$y(t) = ke^{-e^{F(t)}}. \tag{2.52}$$

1. Derive autonomous ordinary differential equation models leading to the generalized equations (2.51) and (2.52). An autonomous ordinary differential equation model is an ODE system which does not explicitly depend on the independent variable t. It takes the form of $x'(t) = F(x(t))$, where x and F are vector functions.

2. What would be the biological motivation for introducing the general functions $F(t)$ in equations (2.51) and (2.52)?

Exercise 2.4: This exercise provides a problematic attempt to show that the Gompertz model can be viewed as a limiting case of the von Bertalanffy model (page 238, [5]).

1. In the general von Bertalanffy model (2.1), let $\mu = 1$, $b = \alpha - \beta$ and $a = \beta(\mu - \lambda)$. Show that the von Bertalanffy model can be written as

$$\frac{dN}{dt} = bN^\lambda - aN^\lambda \frac{N^{1-\lambda} - 1}{1 - \lambda}. \tag{2.53}$$

2. In equation (2.53), take the limit as $\lambda \to 1$ from the left, and show that it becomes the Gompertz model (2.20).

3. Is the above argument valid? If not, where is the problem? (Hint: a and β are constants.)

Exercise 2.5: Let $y = W^{1-\lambda}$. If W satisfies the following equation

$$\frac{dW}{dt} = \alpha W^\lambda - \beta W, \tag{2.54}$$

then

$$\frac{dy}{dt} = (1 - \lambda)(\alpha - \beta y). \tag{2.55}$$

Solve the above equation in y and show that equation (2.54) with initial condition $W(0) = W_0$ has the solution,

$$W(t) = (\alpha/\beta - [\alpha/\beta - W_0^{(1-\lambda)}]e^{-(1-\lambda)\beta t})^{\frac{1}{1-\lambda}}, \quad \lambda \neq 1. \tag{2.56}$$

Exercise 2.6: Suppose the growth of a tumor of mass W satisfies the Gompertz growth function,

$$G(W) = aW(1 - b\ln(W)),$$

where a and b are positive constants.

1. Find the equilibrium mass of the tumor.

2. Find the maximum growth rate for this tumor. What is tumor's growth rate when the tumor size is very small? Sketch the graph of $G(W)$ with $a = 0.5$ and $b = 0.1$.

Exercise 2.7: Reproduce Fig. 2.1 and Fig. 2.2.

Exercise 2.8: Using the following MATLAB® program to reproduce Fig. 2.4.

```
----------------------------------------------------------------
function dataandcurve
n = [20 15 39 71 65 54 87 37 86 92 93 106 58 86 26 73 62 46 32 ...
     21 15 22 18];
sd = [226 173 362 381 709 1054 1164 2591 2567 3176 3278 3371 ...
      3598 3521 3737 3704 3686 5206 5326 4805 4463 6279 4455];
```

```
mass = [400 330 470 491 852 1440 1251 4638 3780 4377 5916 ...
     5940 8762 6927 13130 11600 9735 15120 13585 14170 16550 ...
     17970 19865];
time = 4:1:26; E = sd./sqrt(n);
G0 = 0.789; alpha= 0.107; N0= 18.4;
t = [4:.1:30];
N_t= N0*exp(G0/alpha*(1-exp(-alpha*t)));
figure(1)
set(gcf,'DefaultAxesFontSize',18)
hold on
plot(t,N_t,'k')
hold on
h = errorbar(time,mass,E,'ko');
title('(A)')
ylabel('Tumor mass (mg)')
xlabel('Time (days)')
ylim([0,2.5e4])
xlim([0,30])
figure(2)
set(gcf,'DefaultAxesFontSize',18)
semilogy(time,mass,'ko')
hold on
semilogy(t,N_t,'k')
title('(B)')
ylabel('log tumor mass')
xlabel('Time (days)')
end
```

Exercise 2.9: Assume the mass of a tumor has a weight of 0.5 grams on day 1, 1 gram on day 2, 3 grams on day 3, 4 grams on day 4, and 4.5 grams on day 5.

   1. Use the following MATLAB program to find a set of parameters that best fits the data to the logistic growth model using fminsearch to minimize mean squared error (MSE) with a termination tolerance of 10^{-6} and maximum number of iterations allowed of 200. Try a few sets of different initial values of the parameters.

```
% This program attempts to find values for r and K in logistic
% model that best fit a given set of data by using fminsearch
% with respect to mean squared error (MSE).
function estimation  %(x0)
clear all
close all
global T M r K
x0=[0.8; 5];
[min, fval]=fminsearch(@er,x0,optimset('TolX',1e-6,'MaxIter',200))
min(1)=r; min(2)=K;
[t,y]=ode23s(@logistic,[1 5],0.5);
plot(t,y,T,M,'o')
```

```
%function for ode solver
function z=er(x)
global T M r K
tt=0:1:60;
% data points from experiment
T=[1,2,3,4,5]'; M=[0.5,1,3,4,4.5]';
r=x(1); K=x(2);
y0 = [0.1]; y = y0;
[t1 y1] = ode23s(@logistic, tt, y0);
z=sum((y1(T)-M).^2); % minimize this.
function yp = logistic(t,y)
global r K
yp = y; yp(1) = r*y(1)*(1 - y(1)/K);
```

---

2. Find a set of parameters of the general von Bertalanffy model that best fit the data using fminsearch to minimize mean squared error (MSE) with a termination tolerance of 10^{-6} and maximum number of iterations allowed of 200.

3. Find a set of parameters of Gompertz model that best fit the data using fminsearch to minimize mean squared error (MSE) with a termination tolerance of 10^{-6} and maximum number of iterations allowed of 200.

Exercise 2.10: Let $b \equiv \beta - \mu_p$ in the Gyllenberg-Webb model (2.64).

1. Show that model (2.64) is equivalent to

$$
\begin{cases}
\dfrac{dP}{dt} = (b - r_i(N) - r_0(N))P + r_i(N)N, \\[2mm]
\dfrac{dN}{dt} = (b + \mu_q)P - \mu_q N.
\end{cases}
\tag{2.57}
$$

2. Show that for $b = 1, \mu_q = 0, P(0) = 1, Q(0) = 0, r_i(N) = 0$ and $r_0(N) = 1 + \ln N$, then the solution of model (2.64) is also a solution of a Gompertz model.

3. Show that, for $b = 2, \mu_q = 0, P(0) = 1, Q(0) = 1, r_i(N) = 0$ and $r_0(N) = N$, then the solution of model (2.64) is also a solution of a logistic model.

Exercise 2.11: In the PN-form of the Gyllenberg-Webb model (2.57), assume that

$$
r_0(N) = \frac{kN}{aN + 1},
$$

and

$$
r_i(N) = \frac{r}{N + m}.
$$

1. Use pplane8.m to study the phase plane dynamics of model (2.57) and formulate mathematical conjectures regarding its existence and stability of the positive steady state.

2. Write a MATLAB program to calculate a numerical solution of model (2.57) with several representative sets of parameter values.

3. Establish sufficient conditions for the local stability of the positive steady state of model (2.57).

4. Establish sufficient conditions for the global stability of the positive steady state of model (2.57).

Exercise 2.12: Show that the solution of the Gompertz model

$$\frac{dN}{dt} = -\beta N \ln \frac{N}{K}, \quad N(0) = N_0$$

takes the form of

$$N(t) = K \left(\frac{N_0}{K} \right)^{\exp(-\beta t)}.$$

Exercise 2.13: Consider the treatment model (2.35).

1. Assume the tumor grows according to the exponential growth model with $f(N) = aN, a > 0$ and $P(t, N) = c(t)N$. Show that $N_{CS} = N_{SC}$.

2. Assume the tumor grows according to the logistic growth model with $f(N) = aN(1 - N/K)$ and $G(t, N) = cN$, where a, K, c are positive constants. Show that $N_{CS} < N_{SC}$.

Exercise 2.14: Prove Proposition 2.4.

Exercise 2.15: Write a critical review of at least 400 words on the paper by Kohandel et al. on mathematical modeling of ovarian cancer treatments [19].

Exercise 2.16: In this exercise, we consider a natural extension of the Gyllenberg-Webb model (2.64). Specifically, we would like to include the dead cell population, say $D(t)$, and allow the transition function to depend on sizes of both the total population and the proliferative population. For example, if the total nutrient entering the tumor is proportional to the surface area and they are shared by only proliferative cells, then we may assume that $r_0(N, P) = f(N^{2/3}/P)$ and $r_i(N, P) = g(N^{2/3}/P)$. This may result in the following model.

$$\begin{cases} \dfrac{dP}{dt} = (\beta - \mu_p - r_0(N, P))P + r_i(N, P)Q, \\[2mm] \dfrac{dQ}{dt} = r_0(N, P)P - (r_i(N, P) + \mu_q)Q, \\[2mm] \dfrac{dD}{dt} = \mu_q Q. \end{cases} \quad (2.58)$$

Assume that

$$f(N) = \frac{kN}{aN + 1},$$

and

$$g(N) = \frac{r}{N + m}.$$

1. Write a MATLAB program to calculate the numerical solution of model (2.58) with several representative sets of parameter values.

2. Establish sufficient conditions for the local stability of the positive steady state of model (2.58).

3. Study the global stability of the positive steady state of model (2.58).

Exercise 2.17: Prove Proposition 2.2.

Exercise 2.18: In this exercise, we assume that dead cells $D(t)$ are removed from an avascular tumor at a constant rate d and extend the Gyllenberg-Webb model (2.64). We will denote $\beta - \mu_p$ by b and μ_q by μ. All parameters below are nonnegative constants. This may result in the following three-compartment model.

$$
\begin{cases}
\dfrac{dP}{dt} = (b - r_0(N))P + r_i(N)Q, \\[2mm]
\dfrac{dQ}{dt} = r_0(N)P - (r_i(N) + \mu)Q, \\[2mm]
\dfrac{dD}{dt} = \mu Q - dD,
\end{cases}
\tag{2.59}
$$

where

$$
N(t) = P(t) + Q(t) + D(t)
\tag{2.60}
$$

and initial conditions are,

$$
P(0) = P_0 > 0, \quad Q(0) = Q_0 \geq 0, \quad D(0) = Q_0 \geq 0.
\tag{2.61}
$$

We assume the following conditions for $r_0(N)$ and $r_i(N)$.
(A1): $r_0(N) \geq 0$ and $r_0'(N) \geq 0$ for all $N > 0$ and,

$$
0 < \lim_{N \to \infty} r_0(N) \equiv l_0 \leq \infty.
\tag{2.62}
$$

(A2): $r_i(N) \geq 0$, but $r_i'(N) \leq 0$, and

$$
0 \leq \lim_{N \to \infty} r_i(N) \equiv l_i < \infty.
\tag{2.63}
$$

1. Show that the solutions of model (2.71) with positive initial values will stay positive.

2. Assume that $b < l_0 < \infty$ and there is a constant $I > 0$ such that $r_i(N)N < I$ for all $N > 0$. Then, solutions of model (2.71) are bounded. Moreover, there is a constant $L > 0$ such that

$$
\limsup_{t \to \infty} N(t) \leq L.
$$

3. Assume that $r_0(N) = kN$, $k > 0$, and there is a constant $I > 0$ such that $r_i(N)N(1+N) < I$ for all $N > 0$. Then, solutions of model (2.71) are bounded.

(Hint: Details can be found in the reference [2]).

2.9   Projects and open questions

We propose some alternative approaches to the modeling efforts in the recent work of Alzahrani and Kuang [3].

Gyllenberg and Webb [15] proposed a two-compartment model of the tumor cells consisting of only proliferating and quiescent cells. They assumed that the dead cells are somehow removed from the tumor instantly which is not necessary true in reality. Their key contribution is their hypothesis that in a typical avascular multicellular tumor spheroid, proliferating cells may transit into and out of quiescence. Let $P(t)$ and $Q(t)$ be the densities of proliferating and quiescent cells, respectively. Since dead cells are pushed away instantly, $N(t) = P(t) + Q(t)$ represents the total amount of cells in the tumor. Recall that the Gyllenberg-Webb model takes the following form:

$$\begin{cases} \dfrac{dP}{dt} = (\beta - \mu_p - r_0(N))P + r_i(N)Q, \\[2mm] \dfrac{dQ}{dt} = r_0(N)P - (r_i(N) + \mu_q)Q, \end{cases} \qquad (2.64)$$

with initial conditions,

$$P(0) = P_0 > 0, \quad Q(0) = Q_0 \geq 0, \qquad (2.65)$$

proliferating cells are assumed to proliferate at a constant per capita rate $\beta > 0$, and proliferating and quiescent cells are assumed to die at constant rates $\mu_p \geq 0$ and $\mu_q \geq 0$, respectively. In reality, both rates are likely to be dependent on some limiting resource levels. Their key assumption is that the tumor cells transition to and from the quiescent compartment at rates $r_0(N)$ and $r_i(N)$, respectively, where both functions are continuous and defined for $N \geq 0$. Gyllenberg and Webb [15] made the following generic assumptions on $r_0(N)$ and $r_i(N)$.

(**A1**): $r_0(N) \geq 0$ and $r_0'(N) \geq 0$ for all $N > 0$ and,

$$0 \leq \lim_{N \to \infty} r_0(N) = l_0 \leq \infty. \qquad (2.66)$$

(**A2**): $r_i(N) \geq 0$, but $r_i'(N) \leq 0$, and

$$0 \leq \lim_{N \to \infty} r_i(N) = l_i < \infty. \qquad (2.67)$$

Alzahrani and Kuang [3] formulated an alternative model that includes features related to resource limitation effects. In their tumor model below, $N(t) = P(t) + Q(t) + D(t)$, where P and Q are as in the Gyllenberg and Webb model and D represents the dead cells inside the tumor. The resource, such as oxygen, is assumed to enter the tumor proportional to its surface area that assumed to take the form of $S(t) = aN(t)^\theta$ where θ is a scaling parameter with values in $[2/3, 1]$. Let $R(t)$ be the limiting resource and $u(R(t))$ be the proliferating cells' limiting nutrient uptake rate function, then we have

$$\frac{dR(t)}{dt} = aN(t)^\theta - u(R(t))P(t). \tag{2.68}$$

Since nutrient such as oxygen uptake is probably a process that takes place much faster than proliferation, state switching and death, we may apply the usual quasi-steady-state argument (for example, see [13]) on $R(t)$ which yields

$$u(R(t)) = \frac{aN(t)^\theta}{P(t)}. \tag{2.69}$$

If a tumor takes the form of a sphere, then $\theta = 2/3$. The model below makes an additional natural modification to the Gyllenberg-Webb model (2.64): it is assumed that proliferating cells proliferate at a rate of $f(u(R(t)))$, proliferating cells transition to quiescent cells at the rate of $g(u(R(t)))$, and quiescent cells transition to proliferating cells at the rate of $h(u(R(t)))$. For simplicity, below we assume $a = 1$ and

$$r(t) \equiv u(R(t)) = \frac{N(t)^\theta}{P(t)}. \tag{2.70}$$

For simplicity, one may assume that the proliferation rate of such newly reemerged cells from quiescent state is proportional to that of the proliferation rate of the current proliferating cells,

$$h(r(t)) = \alpha f(r(t)), \ \alpha \in [0, 1].$$

Since proliferating cells are most likely to enter quiescent state before death, one may simply assume proliferating cells do not die (i.e., $\mu_p = 0$ in model (2.64)). All parameters below are nonnegative constants. This may result in the following three-compartment model.

$$\begin{cases} \dfrac{dP}{dt} = (f(r(t)) - g(r(t)))P + \alpha f(r(t))Q, \\ \dfrac{dQ}{dt} = g(r(t))P - (\alpha f(r(t)) + \mu)Q, \\ \dfrac{dD}{dt} = \mu Q - dD, \end{cases} \tag{2.71}$$

where initial conditions are,

$$P(0) = P_0 > 0, \quad Q(0) = Q_0 \geq 0, \quad D(0) = Q_0 \geq 0. \tag{2.72}$$

In addition, the following assumptions are made

(F): $f(x)$ is continuously differentiable. $f(0) = 0$, $f'(x) > 0$ for all $x > 0$ and,

$$\lim_{x \to \infty} f(x) = f_M < \infty. \tag{2.73}$$

(G): $g(x)$ is continuously differentiable. $g(0) = g_0 > 0$, $g'(x) < 0$ for all $x > 0$ and,

$$\lim_{x \to \infty} g(x) = 0. \tag{2.74}$$

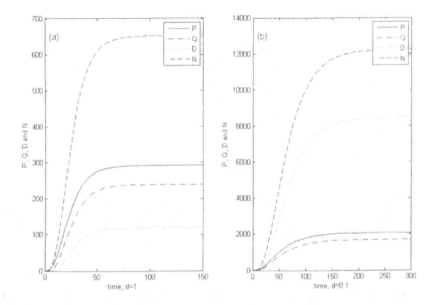

FIGURE 2.5: Two sets of solutions of model (2.71) with $f(r) = \frac{kr}{ar+1}$ and $g(r) = \frac{c}{r+m}$. Except for the dead cell removal rate d, the parameters for both panels are the same. They are $\alpha = 0.1, k = 2, a = 1, m = 2, \mu = 0.5, c = 1, \theta = 2/3$. For panel (a), $d = 1$ and for panel (b), $d = 0.1$. The initial conditions are $P(0) = 0.1, Q(0) = D(0) = 0$ for both panels. Observe that all the components as well as the tumor size grow like a sigmoid curve and tending to constants, which is the hallmark of Gompertz tumor growth.

2.9.1 Mathematical open questions

Open question 1: Assume $\alpha = 0$, prove or disapprove that the positive steady state E^* of the model (2.71) is globally asymptotically stable.

Open question 2: Assume that a tumor growth can be modeled by (2.71), prove or disapprove that a slow dying quiescent population (with smaller value of μ) will increase the size of the tumor at the steady state level (see Fig. 2.5).

2.9.2 Tumor growth with a time delay

One may consider the fact that cell proliferation takes time τ and hence incorporate it in the proliferation term in model (2.71) resulting in the following tumor growth model with time delay.

$$
\begin{cases}
\dfrac{dP}{dt} = f(r(t-\tau))P(t-\tau) - g(r(t))P + \alpha f(r(t))Q, \\[2mm]
\dfrac{dQ}{dt} = g(r(t))P - (\alpha f(r(t)) + \mu)Q, \\[2mm]
\dfrac{dD}{dt} = \mu Q - dD,
\end{cases} \tag{2.75}
$$

Similar assumptions and set up can be assumed for this model and many interesting mathematical questions can be pursued. For examples, assume that the dead cells are removed instantaneously, how the time delay length affects the sizes of both proliferating and quiescent populations? What are the conditions to ensure the tumor size is finite?

2.9.3 Tumor growth with cell diffusion

Once may also consider adding some form of cell movement into the model (2.71) resulting in a partial differential equation (PDE) based tumor growth model with or without time delay. As in the previous project statement, many interesting mathematical questions can be pursued. For examples, assume that the dead cells are removed instantaneously, how the dispersion rate affects the sizes of both proliferating and quiescent populations? What are the conditions to ensure the tumor size is finite? Interested readers are suggested to read the next chapter before pursuing this PDE based project.

References

[1] Adam JA, Bellomo B (ed): *A Survey of Models on Tumour Immune Systems Dynamics.* Berlin: Birkhauser, 1997.

[2] Alzahrani EO, Asiri A, El-Dessoky MM, Kuang Y: Quiescence as an explanation of Gompertzian tumor growth revisited. *Math Biosc* 2014, 254:76-82.

[3] Alzahrani EO, Kuang Y: Nutrient limitations as an explanation of Gompertzian tumor growth. *Discrete Cont Dyn-B* 2016, 21:357–372.

[4] Araujo RP, McElwain DLS: A history of the study of solid tumour growth: The contribution of mathematical modelling. *Bull Math Biol* 2004, 66:1039–1091.

[5] Britton NF: *Essential Mathematical Biology.* London: Springer, 2003.

[6] Bertalanffy L von: Quantitative laws in metabolism and growth. *Quart Rev Biol* 1957, 32:217–231.

[7] Castro MAA, Klamt F, Grieneisen VA, Grivicich I, Moreira JCF: Gompertzian growth pattern correlated with phenotypic organization of colon carcinoma, malignant glioma and non-small cell lung carcinoma cell lines. *Cell Prolif* 2003, 36:65–73.

[8] Dy GK, Adjei AA: Principles of chemotherapy. In: Chang AE, Hayes DF, Pass HI, Stone RM, Ganz PA, Kinsella TJ, Schiller JH, Strecher VJ, editors. *Oncology.* New York: Springer, 2006. p. 14–40.

[9] Dy GK, Adjei AA: Systemic cancer therapy: Evolution over the last 60 years. *Cancer* 2008, 113:1857–1887.

[10] Edelstein-Keshet L: *Mathematical Models in Biology.* SIAM, Philadelphia: SIAM, 2005.

[11] Fister KR, Panetta JC: Optimal control applied to cell-cycle-specific cancer chemotherapy. *SIAM J Appl Math* 2000, 60:1059–1072.

[12] Frenzen CL, Murray JD: A cell kinetics justification for Gompertz equation, *SIAM J Appl Math* 1986, 46:614–629.

[13] Geritz S, Gyllenberg M: A mechanistic derivation of the DeAngelis-Beddington functional response. *J of Theor Biol* 2012, 314:106–108.

[14] Gompertz B: On the nature of the function expressive of the law of human mortality, and on a new mode of determining the value of life contingencies. *Phil Trans Royal Soc London* 1825, 115:513–583.

[15] Gyllenberg M, Webb GF: Quiescence as an explanation of Gompertzian tumor growth. *Growth Dev Aging* 1989, 53:25–33.

[16] Gyllenberg M, Webb GF: A nonlinear structured population model of tumor growth with quiescence. *J Math Biol* 1990, 28:671–694.

[17] Kozusko F, Bajzer Z: Combining Gompertzian growth and cell population dynamics. *Math Biosc* 2003, 185:153–167.

[18] Morrison J, Haldar K, Kehoe S, Lawrie TA: Chemotherapy versus surgery for initial treatment in advanced ovarian epithelial cancer. *Cochrane Database of Systematic Reviews* 2012, Issue 8. Art. No.: CD005343. DOI: 10.1002/14651858.CD005343.pub3.

[19] Kohandel, M, Sivaloganathan, S, Oza, A: Mathematical modeling of ovarian cancer treatments: Sequencing of surgery and chemotherapy. *J Theor Biol* 2006, 242:62–68.

[20] Laird AK: Dynamics of tumor growth. *Brit J Cancer* 1964, 18:490–502.

[21] Laird AK: Dynamics of tumour growth: Comparison of growth rates and extrapolation of growth curve to one cell. *Brit J Cancer* 1965, 19:278–291.

[22] Marušić M, Bajzer Ž, Freyer JP, Vuk-Pavlović S: Analysis of growth of multicell tumor spheroids by mathematical models. *Cell Prolif* 1994, 27:73–94.

[23] Marušić M, Bajzer Ž, Vuk-Pavlović S, Freyer JP: Tumor growth in vivo and as multicellular spheroids compared by mathematical models. *Bull Math Biol* 1994, 56:617–631.

[24] Merriam-Webster Medical Dictionary [Internet]: Homeostasis. Springfield: Merriam Webster. 2013. Available from: http://www.merriam-webster.com/medlineplus/homeostasis.

[25] Norton L, Simon R, Brereton HD, Bogden AE: Predicting the course of Gompertzian growth. *Nature* 1976, 264:542–545.

[26] Panetta, JC: A mathematical model of breast and ovarian cancer treated with Paclitaxel. *Math Biosci* 1997, 146:89–113.

[27] Pearl R, Reed LJ: On the rate of growth of the population of the United Sates and its mathematical representation. *Proc Natl Acad Sci USA* 1920, 6:275–288.

[28] Pearl R, Reed LJ: On the mathematical theory of population growth. *Metron* 1923, 3:6–19.

[29] Retsky MW, Swartzendruber DE, Wardwell RH, Bame PD: Is Gompertzian or exponential kinetics a valid description of individual human cancer growth? *Med Hypoth* 1990, 33:95–106.

[30] Rygaard N, Sprang-Thomsen M: Quantitation and Gompertzian analysis of tumor growth. *Breast Cancer Res Treat* 1997, 46:303–312.

[31] Simon R, Norton L: The Norton-Simon hypothesis: Designing more effective and less toxic chemotherapeutic regimens. *Nat Clin Pract Oncol* 2006, 3:406–407.

[32] Simpson-Herren L, Lloyd HH: Kinetic parameters and growth curves for experimental tumor systems. *Cancer Chemother Rep* 1970, 54:143–174.

[33] Skipper HE: Kinetics of mammary tumor cell growth and implications for therapy. *Cancer* 1971, 28:1479–1499.

[34] Spratt JA, von Fournier D, Spratt JS, Weber EE: Decelerating growth and human breast cancer. *Cancer* 1993, 71:2013–19.

[35] Swartzendruber DE, Retsky MW, Wardwell RH, Bame PD: An alternative approach for treatment of breast cancer. *Breast Cancer Res Treat* 1994, 32:319–325.

[36] Thalhauser CJ, Sankar T, Preul MC, Kuang Y: Explicit separation of growth and motility in a new tumor cord model. *Bull Math Biol* 2009, 71:585–601.

[37] Thieme H: *Mathematics in Population Biology*. Princeton: Princeton University Press, 2003.

[38] Vinay GV, Alexandro FJ: Evaluation of some mathematical models for tumor growth. *Int J Biomed Comput* 1982, 13:19–35.

[39] Waltman, P: A Second Course in Elementary Differential Equations. Orlando, FL: Academic Press, 1986.

[40] Winsor, CP: The Gompertz curve as a growth curve. *Proc Natl Acad Sci USA* 1932, 18:1–8.

[41] Wright, S: Book review. *J Am Stat Assoc* 1926, 21:493–497.

Chapter 3

Spatially Structured Tumor Growth

3.1 Introduction

In Chapter 2, we introduced Laird's notion of Gompertzian tumor growth [9, 10] as a phenomenological idea. In contrast, Gyllenberg and Webb's model [7] (Section 2.4) is *mechanistic* in that it suggests an *explanation*—quiescence— of Gompertz-like dynamics. In 1966, Alan Burton proposed an alternative hypothesis [2]. Based on William Mayneord's [13] studies of a type of sarcoma, Burton suggested that necrosis causes Gompertz-like growth. Specifically, as tumors grow, necrotic regions form and expand faster than the tumor does.

To explore this idea, Burton modeled tumors as ideal spheres with nutrient diffusing into the interior from the surface (Fig. 3.1). He further suggested that necrosis occurs when nutrient concentration drops below a minimum threshold needed to sustain life. Since all living cells constantly consume nutrient, a decreasing nutrient gradient would exist from surface to core. Therefore, in his model (see Section 3.2), the larger the tumor becomes, the more nutrient-starved the interior cells become until eventually they begin to die. The necrotic region grows faster than the proliferative region, causing an ever-decreasing overall growth rate until equilibrium is reached. At the same time as this model was developing, Thomlinson and Gray [18] developed a similar model with a different geometry. Together, these two studies sparked one of the most important threads in mathematical oncology.

Shortly after Burton published his model, experimentalists developed a tissue model of cancer called the **multicell spheroid** (MCS), or alternatively, **tumor spheroid** or **avascular spheroid**. A "tissue model" is nothing mathematical. Rather, it is a type of cell culture that resembles a small tumor. In the early 1970s, the MCS was relatively new and exciting, quickly becoming *the* tissue model of nascent (young) tumors. One can produce them relatively easily in liquid or gel culture using certain strains of transformed cells. Typically, although not always, the progeny cells cling together to form a small, roughly spherical clump of tissue. As the spheroid grows, the interior eventually develops a necrotic region in its center (its "necrotic core"). The necrotic core increases disproportionately until it becomes large enough to halt overall growth of the MCS, typically when the spheroid reaches about 1 to 2 mm in diameter. Clearly, then, the MCS is an excellent experimental system to study Burton's model.

In an interesting historic twist, it turned out that Burton's model generally does not generate a Gompertz curve, but it does fit actual spheroid data *better* than the Gompertz model does. As early as 1938 [7], and confirmed later in numerous MCS experiments, experimentalists recognized that tumor spheroid growth does not follow Gompertz kinetics. Instead, spheroid growth has three phases: an initial exponential phase, followed by a "linear" phase (the growth curve is linear; growth *rate* is constant), followed finally by a continually declining growth rate with asymptotic approach to an upper limit. Burton (and Thomlinson and Gray) first discovered this behavior in their models, and later Harvey Greenspan [5, 6] extended Burton's model to explain this 3-phase growth behavior in detail.

3.2 The simplest spatially structured tumor model

Somewhat like the physiologically structured tumor models of Chapter 4, the models we study here are *spatially structured*. We start with the simplest possible model of this type. Imagine a perfectly spherical multicell spheroid (MCS; see introduction) of radius $R_0(t)$ at time t (Fig. 3.1). The spheroid grows in culture media in which some critical nutrient is maintained at a constant level σ_∞. Nutrient may only enter the spheroid via diffusion, so the nutrient concentration, $\sigma(\mathbf{x}, t)$, depends on both spatial position $\mathbf{x} \in \mathbb{R}^3$ and time. Assume σ is continuous with continuous first derivatives in space and time, as physics demands. Finally, we assume that cells consume nutrient at a constant rate, A, independent of the nutrient concentration, σ, and spatial position, $\mathbf{x}$.

We ask a few questions about this MCS:

1. How large is the necrotic core (at any time)? and

2. How large will the spheroid be and what proportion of it will be necrotic when it reaches equilibrium size?

3. How fast will be the radial expansion of the MCS in its early growth phase?

3.2.1 Model formulation

Following Burton [2], we assume for now that R_0 is fixed. Also, for the moment, we assume that cells survive and consume nutrient no matter how little nutrient exists. Although these simplifications are obviously inaccurate,

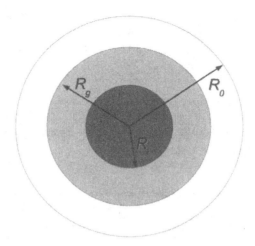

FIGURE 3.1: Idealized cross-sectional geometry of a multicell spheroid. White region represents the annulus of proliferative cells, light gray is the region of living but quiescent cells and dark gray region is the necrotic core. Variables R_0, R_g and R_i represent the outer diameter of each region, respectively.

they allow us to see important foundational structures that are obscured by more realistic, and complex, models that we will study presently.

With these assumptions, we arrive at the conservation equation (see Section 3.8 for general formulation details),

$$\frac{\partial \sigma}{\partial t} = -\nabla \cdot \mathbf{J}(\mathbf{x}, t) - m(\mathbf{x}, \sigma, t), \tag{3.1}$$

where $\mathbf{J}(\mathbf{x}, t)$ is the flux of nutrient at spatial point $\mathbf{x}$ and time t, and $m(\mathbf{x}, \sigma, t)$ is the rate at which nutrient is consumed by cells at point $\mathbf{x}$ and time t. We have already established that $m = A$. Assuming Fickian diffusion of nutrient (see section 9.2 in Edelstein-Keshet [3]) or section 1.2 in Logan [11]),

$$\mathbf{J}(\mathbf{x}, t) = -D(\mathbf{x}, \sigma)\nabla \sigma,$$

where D, which is everywhere positive, measures the diffusivity of nutrient at point $\mathbf{x}$. In this case the model (3.1) becomes

$$\frac{\partial \sigma}{\partial t} = \nabla \cdot [D(\mathbf{x}, \sigma)\nabla \sigma(\mathbf{x}, t)] - A. \tag{3.2}$$

If the spheroid is homogeneous in the sense that D is constant, then model (3.1) becomes

$$\frac{\partial \sigma}{\partial t} = D\nabla^2 \sigma - A, \tag{3.3}$$

where the functions' arguments have been removed for clarity. Since we assume ideal spheroidal geometry, we save ourselves a lot of work by adopting spherical coordinates. To that end, let r be the radial distance from the spheroid center, and θ and ϕ be the rotational coordinates. That is,

$$r = \sqrt{x^2 + y^2 + z^2},$$

and

$$x = r \sin\phi \cos\theta, \quad y = r \sin\phi \sin\theta \quad \text{and} \quad z = r \cos\phi.$$

The general Laplacian of a scalar function $f(\mathbf{x})$, $\mathbf{x} \in \mathbb{R}^3$ in spherical coordinates is

$$\nabla^2 f = \frac{1}{r^2} \frac{\partial}{\partial r}\left(r^2 \frac{\partial f}{\partial r}\right) + \frac{1}{r^2 \sin^2\phi} \frac{\partial^2 f}{\partial \theta^2} + \frac{1}{r^2 \sin\phi} \frac{\partial}{\partial \phi}\left(\sin\phi \frac{\partial f}{\partial \phi}\right).$$

In addition to perfect spheroidal geometry, we assume that the nutrient field σ is also spherically symmetrical. Therefore, the derivatives of σ with respect to angles θ and ϕ in the Laplacian disappear, and we can therefore write model (3.3) as follows:

$$\frac{\partial \sigma}{\partial t} = \frac{D}{r^2} \frac{\partial}{\partial r}\left(r^2 \frac{\partial \sigma}{\partial r}\right) - A, \tag{3.4}$$

which is the form we will study.

At the boundaries we set

$$\sigma(R_0, t) = \sigma_\infty \text{ for all } t \geq t_0, \tag{3.5}$$

and

$$\frac{\partial}{\partial r}\sigma(0, t) = 0 \text{ for all } t \geq t_0. \tag{3.6}$$

Condition (3.5) represents constant reconditioning of the media. The no flux boundary condition (3.6) at the spheroid's center arises for two reasons. First, it keeps the diffusion term from blowing up. (Expand the outer derivative in the diffusion term and take the limit as $r \to 0$.) Second, if this were not true, then the assumption of spherical symmetry of the nutrient concentration field would be violated. To see why, imagine the reflection of the radius R_0 through the opposite side of the spheroid; that is, the two radii R_0 and $-R_0$ are colinear and represent a diameter line segment through the spheroid for any combination of θ and ϕ. Designate points on one radius $r \geq 0$ and points on the other $-r$. That is, one radius is "positive," the other is "negative" and the center is the origin. Spherical symmetry is equivalent to the statement: $\sigma(r, \cdot) = \sigma(-r, \cdot)$ for every $r \in (-R_0, R_0)$ and $\theta, \phi \in (0, 2\pi)$. Now, suppose $\partial\sigma(0, t)/\partial r \neq 0$ at some time t. Recall that we assumed σ to be continuously differentiable in space and time throughout the entire spheroid. Therefore, there must be an ϵ such that $\sigma(\epsilon, t) \neq \sigma(-\epsilon, t)$, which contradicts the assumption of spherical symmetry. Therefore, condition (3.6) must hold. This no flux boundary condition is rather technical to ensure the smoothness requirement at the center. For our practical purpose, it can be replaced by a simple condition on the boundedness of solution.

3.2.2 Equilibrium nutrient profile with no necrosis

Let $\hat{\sigma}(r)$ be a steady state solution of model (3.4) with boundary conditions (3.5) and (3.6). This equilibrium nutrient distribution, $\hat{\sigma}(r)$, satisfies

$$0 = \frac{D}{r^2} \frac{\partial}{\partial r} \left(r^2 \frac{\partial \hat{\sigma}}{\partial r} \right) - A. \tag{3.7}$$

Integrating formula (3.7) twice along r, we obtain the following expression for the equilibrium nutrient concentration profile:

$$\hat{\sigma}(r) = \frac{A}{6D} r^2 - \frac{C_1}{r} + C_2, \tag{3.8}$$

where C_1 and C_2 are integration constants. The no flux boundary condition (or the condition that solutions must be bounded) requires $C_1 = 0$, and the outer boundary condition makes

$$C_2 = \sigma_\infty - \frac{A}{6D} R_0^2.$$

Therefore,

$$\hat{\sigma}(r) = \sigma_\infty - \frac{A}{6D}(R_0^2 - r^2), \tag{3.9}$$

with the stipulation that

$$\sigma_\infty \geq AR_0^2/6D. \tag{3.10}$$

We will deal with the causes and consequences of this stipulation below.

First, we note that this steady-state solution has some nice biophysical properties. The nutrient concentration decreases as one moves from the edge toward the center of the spheroid (Figs. 3.2 and 3.3), as expected since all cells consume nutrient but the only source is diffusion from the MCS's edge. As diffusivity, D, increases, total amount of nutrient within the MCS increases (Fig. 3.2), since nutrient supply increases but demand stays constant, whereas the opposite is true for nutrient consumption rate, A (Fig. 3.3).

Despite these intuitively pleasing properties, something is seriously wrong with the model. The problem was forewarned by stipulation (3.10), which is required to avoid a possibly negative nutrient content in the spheroid's core. This bizarre "negative nutrient concentration" is a consequence of a biophysically implausible assumption in our formulation of model (3.4).[1] Our mistake was to assume that cells consume nutrient at a constant rate no matter how much nutrient is there, *even if no nutrient exists*. Suppose, for

---

[1] This assumption can be found in early MCS models, including those of Burton and Greenspan; however, these authors assumed that necrosis would occur, and therefore cells would stop consuming nutrient, when the nutrient concentration reached a positive threshold. Therefore, nutrient concentrations could not become negative in their models; see the next section.

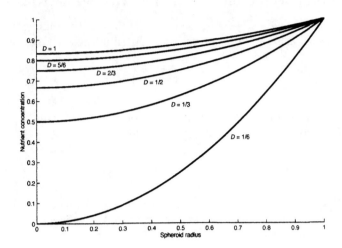

FIGURE 3.2: Solutions of model (3.4) showing steady-state nutrient profiles along the radius of an idealized MCS for various values of the diffusivity, D. In this figure, basic nutrient consumption rate A, total MCS radius R_0 and media nutrient concentration σ_∞ are all unity.

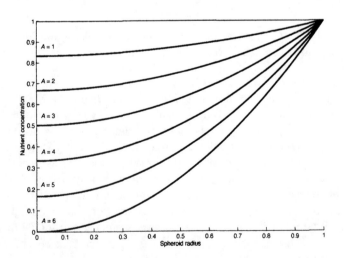

FIGURE 3.3: Solutions of model (3.4) showing steady-state nutrient profiles along the radius of an idealized MCS for various values of the nutrient consumption rate, A. In this figure, diffusivity D, total MCS radius R_0 and media nutrient concentration σ_∞ are all unity.

example, that there were no nutrient at some point in time directly in the center of the spheroid; i.e., suppose $\sigma(0,t) = 0$. Then

$$\frac{\partial \sigma}{\partial t} = -A,$$

because of the no flux boundary condition. Biologically, cells are consuming nutrient that is not there, forcing the concentration to become negative.

We can correct this problem by appealing to more plausible underlying biology. Specifically, imagine that the nutrient is absorbed into cells through a transport protein expressed on the cell surface, as many nutrients are. The simplest assumption would have nutrient and transporters interact via mass action. Mass action describes many chemical reactions very well, and nutrient transport is a chemical reaction. And although we know that transport kinetics do not obey mass action, often this assumption is a good first approximation. Since we assumed earlier that tissue density is constant, then nutrient transport would be described by

$$\alpha p(\mathbf{x},t)\sigma(\mathbf{x},t),$$

where α is the mass action constant and p is the concentration of membrane transporters. Now, assume that all cells have equal numbers of transport proteins in their membranes—an assumption one could perhaps justify with an appeal to the mean field. We expect that the effect random fluctuations have on the mean field is proportional to $1/\sqrt{N}$, where N is the number of cells. Since N in the spheroid is large—a spheroid 2 mm in diameter, a reasonable size for actual spheroids, has on the order of 1 million cells—the fluctuations should have a small overall impact. This assumption and that of homogeneous cell density make p constant. Therefore, the nutrient uptake rate can be expressed as $A\sigma$, where $A = \alpha p =$ a constant, and our model becomes

$$\frac{\partial \sigma}{\partial t} = \frac{D}{r^2}\frac{\partial}{\partial r}\left(r^2 \frac{\partial \sigma}{\partial r}\right) - A\sigma, \tag{3.11}$$

with the same boundary conditions as model (3.4).

3.2.3 Size of the necrotic core

Now that we have a "chassis" for our model, we attack the first of the scientific questions motivating the model, namely, how large is the necrotic core? To that end, we propose the following scientific hypothesis: if the nutrient concentration in a cell's environment dips below a critical level, designated σ_l, then the cell instantly dies.

Scientifically, this statement about σ_l is a hypothesis, and as such represents a reason for producing a model in the first place. The model will be useful if it generates some sort of practically testable prediction that can be used to evaluate such hypotheses. From a mathematical viewpoint, however, it is just another assumption. (It is easy to skip past this idea and miss the point of what we are doing.)

Necrosis occurs if the nutrient concentration profile, equation (3.9), goes below σ_l for some $r \in [0, R_0]$. If not, that is if $\sigma(r, \cdot) > \sigma_l$ everywhere in this interval, then solution (3.9) completely describes the nutrient profile at equilibrium. However, if $\sigma(r, \cdot) = \sigma_l$ for some $r \in (0, R_0)$, then a necrotic core exists. So, a natural definition of the radius of the necrotic core, R_i, is the following: R_i satisfies $\sigma(R_i) = \sigma_l$ if such a point exists and $0 \leq R_i \leq R_0$; otherwise, $R_i \equiv 0$ (Fig. 3.1).

By our (scientific) hypothesis, cells in a necrotic region instantly die, and therefore cease to absorb nutrient ($A = 0$ wherever necrosis occurs). So, throughout any necrotic region, $\sigma(r, \cdot) = \sigma_l$ since $\sigma(R_i, \cdot) = \sigma_l$ and $\partial \sigma / \partial r = 0$ for all $r \in [0, R_i]$. Integrating (3.7) first from 0 to R_i, then from R_i to R_0 and applying the boundary conditions (3.5) and those just described yields the following solution:

$$\sigma(r, \cdot) = \begin{cases} \sigma_l & ; 0 \leq r \leq R_i; \\ \sigma_\infty - \dfrac{A}{6D}(R_0^2 - r^2) + \dfrac{AR_i^3}{3D}\left(\dfrac{1}{r} - \dfrac{1}{R_0}\right) & ; R_i < r \leq R_0. \end{cases} \quad (3.12)$$

Notice that R_i is unique, so there are only two possible histological configurations: either no necrotic core exists, or there is a single, spherical necrotic core concentric with the spheroid. Also, if $R_i > 0$, then

$$\sigma_\infty - \sigma_l = \frac{A}{3D}\left[\frac{1}{2}(R_0^2 - R_i^2) - \frac{R_i^2}{R_0}(R_0 - R_i)\right], \quad (3.13)$$

and we can express the radius of the necrotic core R_i as a function of the nutrient concentration in the media (σ_∞), the nutrient necrosis threshold (σ_l) and the outer tumor radius, R_0, all of which are experimentally measurable. Therefore, R_i, which is itself measurable, represents a practically testable prediction generated by the model.

3.3   Spheroid dynamics and equilibrium size

We now relax the assumption that R_0 is fixed and ask, how will the spheroid grow, and how large will it become? To answer this question we require a dynamic model of spheroid growth, which can be built from another appeal to mass conservation. (Our development here is a slight variation of Greenspan's [5]). We define A as the mass of the spheroid at time t, B the spheroid's mass at some "initial" time t_0 in the past ($t_0 < t$), C the amount of mass added via cell proliferation, and D the amount of mass lost through disintegration of necrotic material between time t_0 and t. With these definitions, conservation of mass requires

$$A = B + C - D. \tag{3.14}$$

In this model we assume that when cells die, their (dead) mass does not instantly disappear. Rather, it slowly decays and washes out of the spheroid at a constant per volume rate, which we denote 3λ. (Why we multiply by 3 will become clear in a moment.) We also assume that tissue density is homogeneous throughout the spheroid; in particular, densities of living and dead tissue are equal. (In reality both are nearly the same as the density of water.) In that case, the spheroid's mass is proportional to its volume, so the conservation law (3.14) can be scaled in units of volume. That assumption, then, allows us to write

$$A = \frac{4}{3}\pi R_0^3(t),$$

$$B = \frac{4}{3}\pi R_0^3(0),$$

$$C = 4\pi \int_0^t \int_0^{R_0(t)} s(\sigma) r^2 \, \mathrm{d}r \, \mathrm{d}t',$$

$$D = \frac{4}{3}\pi \int_0^t 3\lambda R_i^3(t') \, \mathrm{d}t',$$

where $s(\sigma)$ is the per unit volume proliferation rate. (Note the assumption that proliferation depends only on nutrient availability.) Therefore, from (3.14),

$$R_0^3(t) = R_0^3(0) + 3\int_0^t \int_0^{R_0(t)} s(\sigma) r^2 \, \mathrm{d}r \, \mathrm{d}t' - \int_0^t 3\lambda R_i^3(t') \, \mathrm{d}t'. \tag{3.15}$$

Differentiating with respect to time yields the following differential equation:

$$R_0^2 \frac{\mathrm{d}R_0}{\mathrm{d}t} = \int_0^{R_0(t)} s(\sigma) r^2 \, \mathrm{d}r - \lambda R_i^3(t). \tag{3.16}$$

It should be clear now why we described the necrotic disintegration rate as 3λ.

Now we need a model for the proliferation rate, $s(\sigma)$. For illustrative purposes, and to connect with the model of Greenspan [5], we choose a very simple, if unrealistic, model as a first approximation. In particular, assume that the proliferation rate is constant throughout the region of living tissue. That is,

$$s(\sigma) = \begin{cases} s \; ; \; R_i < r \leq R_0, \\ 0 \; ; \; \text{otherwise}, \end{cases} \tag{3.17}$$

where s is a constant. Applying this assumption to equation (3.16) and adding the nutrient dynamics yields the following model of spheroid growth:

$$R_0^2 \frac{dR_0}{dt} = \frac{s}{3}(R_0^3 - R_i^3) - \lambda R_i^3, \tag{3.18}$$

$$\frac{\partial \sigma}{\partial t} = \frac{D}{r^2} \frac{\partial}{\partial r}\left(r^2 \frac{\partial \sigma}{\partial r}\right) - A(r), \tag{3.19}$$

where

$$A(r) = \begin{cases} A \; ; \; R_i < r \leq R_0, \\ 0 \; ; \; \text{otherwise.} \end{cases}$$

We can immediately recognize that diffusion and cell proliferation occur on very different time scales, and therefore somewhat rashly make a quasi-steady state argument (QSSA) and set $\partial \sigma / \partial t = 0$. But here we choose to exercise more care, placing the QSSA on a more solid logical foundation. QSSA is often referred to also as quasi-steady state assumption or quasi-steady state approximation. QSSA is also discussed in the context of chemical kinetics in Section 12.3.

To that end, rescale time using cell proliferation as the natural unit. That is, let $\tau = st$, making

$$\frac{d}{dt} = s\frac{d}{d\tau}.$$

To scale space, we recommend using the radius of a spheroid on the cusp of developing a necrotic core as the natural unit. Here's why. Imagine a tiny spheroid, so small that nutrient can easily diffuse throughout the spheroid, so well that all cells are flush with resources. Intuitively, we expect all cells to proliferate, causing the spheroid to grow. As it grows, however, nutrient becomes less and less available in the spheroid's center until eventually the nutrient concentration reaches the critical necrosis value, σ_l. At that moment the very first cell is just about to die from nutrient-deficiency necrosis, but $R_i = 0$. Therefore, from relation (3.13),

$$R_0 = \left(\frac{6D(\sigma_\infty - \sigma_l)}{A}\right)^{\frac{1}{2}} \equiv R_c. \tag{3.20}$$

Scaling space in terms of the "critical radius," R_c, we define

$$\xi(t) = \frac{R_0(t)}{R_c}, \quad \eta(t) = \frac{R_i(t)}{R_c}, \tag{3.21}$$

and the model becomes

$$\xi^2 \frac{d\xi}{d\tau} = \frac{1}{3}(\xi^3 - \eta^3) - \gamma \eta^3, \tag{3.22}$$

$$\epsilon_1 \frac{\partial \sigma}{\partial \tau} = \frac{1}{\rho^2} \frac{\partial}{\partial \rho}\left(\rho^2 \frac{\partial \sigma}{\partial \rho}\right) - \hat{A}(\rho), \tag{3.23}$$

where $\rho = r/R_c$, $\gamma = \lambda/s$, $\epsilon_1 = sR_c^2/D$ and

$$\hat{A} = \begin{cases} AR_c^2/D & ; \eta < \rho \le \xi, \\ 0 & ; \text{otherwise.} \end{cases}$$

Our intuition suggests that diffusion across a sphere of radius R_c is very efficient, and occurs on a much faster time scale than does cell proliferation; therefore, we assume that $sR_c^2 \ll D$. (Indeed, for oxygen in living tissue, $D \approx 1000 \ \mu m^2 \ s^{-1}$, and realistically $R_c \approx 150 \ \mu m$ and $s < 10^{-5} \ s^{-1}$.) On this basis we justify letting $\epsilon_1 \to 0$. Therefore, equations (3.13) and (3.20) (properly scaled) apply, allowing us to write

$$1 = \xi^2 - \eta^2 + 2\eta^3 \left(\frac{1}{\xi} - \frac{1}{\eta} \right), \quad \xi \ge 1. \tag{3.24}$$

The growth dynamics predicted by this model are naturally partitioned into two phases. The first phase occurs on the time interval $[\tau_0, \tau_1]$, where τ_1 satisfies $\xi(\tau_1) = 1$. In this phase,

$$\xi(\tau) = \xi_0 e^{\frac{1}{3}\tau}, \quad \tau_0 \le \tau \le \tau_1, \quad \xi(\tau_0) = \xi_0 < 1. \tag{3.25}$$

The moment $\xi = 1$, a necrotic core begins to form, and the second phase begins. In the second phase, $\eta(\tau) > 0$ for all $\tau > \tau_1$, and the dynamics are governed by equations (3.22) and (3.24), with initial conditions $\tau = \tau_1$ and $\xi(\tau_1) = 1$. We can integrate this model explicitly by introducing the variable

$$x = \frac{\eta}{\xi}. \tag{3.26}$$

With this change of variables in (3.22), we have

$$2\xi \frac{d\xi}{d\tau} = \frac{2}{3} \xi^2 [1 - (1 + 3\gamma x^3)]. \tag{3.27}$$

From equation (3.24), we obtain

$$\xi\eta = \xi^3\eta - \eta^3\xi + 2\eta^3(\eta - \xi), \quad \xi \ge 1. \tag{3.28}$$

Dividing both sides of the above equation by ξ^4 yields

$$\xi^{-2}x = 2x^4 - 3x3 + x. \tag{3.29}$$

Hence

$$\xi^2 = \frac{1}{1 + 2x)(1 - x)^2}. \tag{3.30}$$

Equations (3.27) and (3.30) together imply

$$\frac{9x}{(1 + 2x)(1 - x)(1 - (1 + 3\gamma)x^3)} \frac{dx}{d\tau} = 1. \tag{3.31}$$

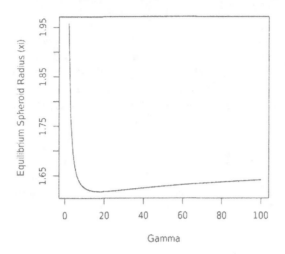

FIGURE 3.4:   Equilibrium spheroid size $\hat{\xi}$ as a function of $\Gamma = 1 + 3\gamma$ for $1 < \Gamma \le 100$.

The resulting solution is largely useless for a biologist (see exercises), but its graph shows a spheroid growing monotonically toward an asymptotic limit (for $\gamma > 0$). Biologically, this limit is of great interest, representing as it does the equilibrium spheroid size.

Note that, at equilibrium, the amount of necrosis within the spheroid depends only on γ; specifically,

$$\frac{\hat{\eta}}{\hat{\xi}} = (1 + 3\gamma)^{-\frac{1}{3}} \equiv \Gamma^{-\frac{1}{3}}. \tag{3.32}$$

From (3.24), a bit of algebra reveals that the equilibrium MCS radius

$$\hat{\xi} = \frac{\left[\Gamma^{\frac{1}{3}}\left(\Gamma^{\frac{2}{3}} - 1\right)\left(3\Gamma^{\frac{1}{3}} - 2\right)\right]^{\frac{1}{2}}}{\Gamma^{\frac{2}{3}} - 1}, \quad \Gamma > 1. \tag{3.33}$$

This function has a minimum at

$$\Gamma = \frac{(3 + \sqrt{5})^3}{8},$$

approaches ∞ as $\Gamma \to 1$, and goes to $\sqrt{3}$ as $\Gamma \to \infty$ (Fig. 3.4).

Biologically, γ represents the ratio of the rate at which necrotic debris washes out of the spheroid to the rate of proliferation, or simply the relative

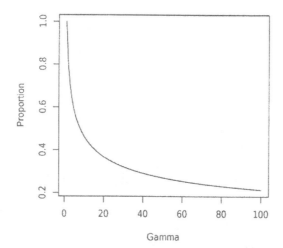

FIGURE 3.5: Necrotic radius as a proportion of the total spheroid radius from equation (3.32) for $1 < \Gamma \leq 100$.

necrotic washout rate. If this washout rate is huge ($s << \lambda$), then relation (3.32) predicts a small necrotic core (Fig. 3.5), in agreement with intuition. Also, equation (3.33) and Fig. 3.5 suggest that the equilibrium spheroid size is insensitive to small (but finite) perturbations of the washout rate when that rate is high—another intuitively pleasing observation.

At the other extreme, as the washout rate declines past its minimum, we see a transition from a spheroid that is mostly living tissue to one that is mostly necrotic (Fig. 3.5). At the same point, we start seeing large changes in spheroid radius with small changes in Γ (Fig. 3.4). As the washout rate declines toward zero ($\lambda << s$), we encounter an enormous spheroid composed almost entirely of necrotic debris at equilibrium.

3.4 Greenspan's seminal model

In 1972, H. P. Greenspan published an MCS model [5] that laid the foundation for much of modern mathematical oncology. Building on the work of Burton [2] and Thomlinson and Gray [18], Greenspan takes on another phenomenon apparent in MCS studies in addition to overall MCS growth and development of necrosis. Careful analyses of MCSs revealed that between the necrotic core and the outer proliferative rind one tends to find an annulus of quiescent cells (see Section 2.4 and Fig. 3.1). The question Greenspan addressed is, how does this layer form? He hypothesizes that tissues within the spheroid produce some sort of inhibitory chemical that causes cells to become quiescent. All other assumptions are essentially those we introduced in the previous sections.

Call the concentration of the inhibitor $\beta(\mathbf{x}, t)$. We assume for simplicity that $\beta(\mathbf{x}, 0)$ is symmetrical in the sense previously described for nutrient concentration, and all previous assumptions hold. Therefore, we can describe the inhibitor concentration field as $\beta(r, t)$. Greenspan hypothesizes that when β rises above some threshold, denoted β_l, cells become quiescent. In the simple models we will consider subsequently, the inhibitor concentration reaches this threshold at no more than one radial position. Therefore, we can define a radial position R_g analogously to R_i. That is, R_g satisfies $\beta(R_g, \cdot) = \beta_l$ if such a point exists; otherwise, $R_g = 0$. In intuitive biological terms, R_g could be called the "radius of quiescence."

From these considerations it is easy to see that the spheroid may have a number of possible histological configurations, depending on the relative sizes of R_i and R_g. In particular, we have the following possible cases:

1. $R_i = R_g = 0$. In this case, the spheroid is a solid mass of proliferative tissue.

2. $0 \leq R_g \leq R_i < R_0$. This case produces a spheroid with a necrotic core surrounded by an annulus of proliferative tissue.

3. $0 = R_i < R_g < R_0$. Here the spheroid has no necrosis, but the core is quiescent, surrounded by proliferative tissue.

4. $0 < R_i < R_g < R_0$. This produces a spheroid with a necrotic core surrounded by an annulus of living but quiescent tissue further surrounded by a rind of proliferative tissue (Fig. 3.1).

Greenspan's inhibitory chemical hypothesis was very speculative. For example, he never suggested what chemicals might be involved. He even avoided committing himself to a particular origin for the inhibitor. However, he did suggest two possibilities: the inhibitor may be produced in the necrotic core, or it could somehow be produced by living cells. We will develop the first idea and leave the second as a project.

3.4.1   The Greenspan model

If the inhibitor comes from the necrotic core, then only three histologic configurations are possible—case 3 listed above obviously cannot occur. To keep things simple in the face of uncertainty, Greenspan assumed that the inhibitor is released at a constant per unit volume rate, P, throughout the necrotic region. He also assumed a perfectly conditioned media, so as before, $\sigma(R_0, t) = \sigma_\infty$, and in addition, $\beta(R_0, t) = 0$ for all $t \geq t_0$. Biologically, this assumption implies that media conditioning is so efficient that neither spheroid nutrient consumption nor secretion of waste (=inhibitor) have any effect on the media's composition. In addition to these, all assumptions that we made in previous sections were also adopted by Greenspan. Following logic similar to that used to derive model (3.18), (3.19), Greenspan constructed the following model:

$$R_0^2 \frac{dR_0}{dt} = \frac{s}{3}(R_0^3 - \hat{R}^3) - \lambda R_i^3, \tag{3.34}$$

$$\frac{\partial \sigma}{\partial t} = \frac{D_\sigma}{r^2} \frac{\partial}{\partial r}\left(r^2 \frac{\partial \sigma}{\partial r}\right) - A(r), \tag{3.35}$$

$$\frac{\partial \beta}{\partial t} = \frac{D_\beta}{r^2} \frac{\partial}{\partial r}\left(r^2 \frac{\partial \beta}{\partial r}\right) + P(r), \tag{3.36}$$

where $\hat{R} = \max\{R_i, R_g\}$,

$$P(r) = \begin{cases} 0 \; ; R_i < r \leq R_0, \\ P \; ; \quad r \leq R_i, \end{cases}$$

and D_σ and D_β are the diffusivities of nutrient and inhibitor, respectively. Remaining notations retain their previous definitions. To boundary conditions (3.5), (3.6) and the assumption of continuous differentiability of σ we add similar conditions for β, namely that $\beta(r, t) \in C^1$ for all $r \in [0, R_0)$ and $t \geq t_0$,

$$\beta(R_0, t) = 0 \text{ for all } t \geq t_0, \tag{3.37}$$

and

$$\frac{\partial \beta(0, t)}{\partial r} = 0 \text{ for all } t \geq t_0. \tag{3.38}$$

Condition (3.37) arises because we assumed that the media is perfectly conditioned, whereas condition (3.38) is required for reasons identical to those described for σ.

Rescaling as before, we define

$$\zeta = \frac{R_g}{R_c}, \quad R_c = \frac{6D_\sigma(\sigma_\infty - \sigma_l)^{\frac{1}{2}}}{A}, \tag{3.39}$$

and retain our previous definitions of ξ and η with R_c defined as above. After applying this rescaling along with our previous definition for τ, equations (3.34) and (3.36) return to (3.22) and (3.23), respectively, and (3.36) becomes

$$\epsilon_2 \frac{\partial \beta}{\partial \tau} = \frac{1}{\rho^2} \frac{\partial}{\partial \rho} \rho^2 \frac{\partial \beta}{\partial \rho} - \hat{P}(\rho), \tag{3.40}$$

where $\epsilon_2 = sR_c^2/D_\beta$ and

$$\hat{P}(\rho) = \begin{cases} 0 & ; \eta < \rho \leq \xi, \\ \dfrac{PR_c^2}{D_\beta} & ; \rho \leq \eta. \end{cases} \tag{3.41}$$

Assuming that D_β and D_σ are of nearly the same order of magnitude, we can let $\epsilon_2 \to 0$; that is, we have a sound basis for assuming diffusive equilibrium for both inhibitor and nutrient.

In dimensionless variables we now have the following model:

$$\xi^2 \frac{d\xi}{d\tau} = \frac{1}{3} \left(\xi^3 - \max\{\eta^3, \zeta^3\} \right) - \gamma\eta^3, \tag{3.42}$$

$$0 = \frac{1}{\rho^2} \frac{\partial}{\partial\rho} \left(\rho^2 \frac{\partial\sigma}{\partial\rho} \right) - \hat{A}(\rho), \tag{3.43}$$

$$0 = \frac{1}{\rho^2} \frac{\partial}{\partial\rho} \left(\rho^2 \frac{\partial\beta}{\partial\rho} \right) - \hat{P}(\rho). \tag{3.44}$$

3.4.2 Threshold for quiescence

Before continuing, it is instructive to develop an intuitive feel for what model (3.42)–(3.44) predicts. To that end, suppose the initial spheroid is so small that all cells have access to sufficient nutrient to support proliferation ($R_i(0) = 0$). At first the spheroid radius would grow exponentially as described by solution (3.25). It would continue to obey (3.25) until it became so large that cells in the core began to starve for nutrient. At that moment, the necrotic core begins to form, and spheroid growth dynamics switch to equation (3.31). But now, in addition to retarding the spheroid's growth, the necrotic region also produces inhibitor. At first the inhibitor would have no effect on growth or histology, because its concentration must start below β_l. However, as the necrosis spreads (proportionally), the inhibitor concentration increases. At some point it may, or may not, reach β_l in the center of the core. But, if so the spheroid still would be unaffected since the core contains no living cells to be inhibited. The inhibitor cannot affect spheroid growth unless its concentration exceeds β_l somewhere in the *living* compartment. It is not at all obvious that it ever will, since intuitively it seems clear that the spheroid could reach its equilibrium (3.33) before ζ (R_g) grows larger than η (R_i). The question then is, under what conditions will the inhibitor concentration exceed the threshold β_l in the living tissue compartment?

The answer, of course, depends on the inhibitor profile. Integrating the quasi-steady state inhibitor concentration field $\beta(\rho, t)$ under the conditions that

$$\partial\beta(0, t)/\partial\rho = 0, \qquad \beta(\xi, t) = 0,$$

and assuming continuity of β and its derivative at η, yields

$$
\beta(\rho,t) = \begin{cases} \eta^3 \dfrac{\hat{P}}{3} \left(\dfrac{3}{2\eta} - \dfrac{1}{\xi} - \dfrac{\rho^2}{2\eta^3} \right) ; 0 \le \rho \le \eta, \\[3mm] \eta^3 \dfrac{\hat{P}}{3} \left(\dfrac{1}{\rho} - \dfrac{1}{\xi} \right) \quad ; \eta < \rho \le \xi. \end{cases} \tag{3.45}
$$

Therefore, a quiescent layer forms whenever

$$
\beta_l < \eta^3 \frac{\hat{P}}{3} \left(\frac{1}{\eta} - \frac{1}{\xi} \right). \tag{3.46}
$$

3.4.3 Growth dynamics of the Greenspan model

It is clear that relation (3.24) holds in the Greenspan model. In addition, from (3.46) and the definitions of R_c (3.39) and $\hat{P}$ (3.41), we can write

$$
\frac{Q^2}{2} = \eta^3 \left(\frac{1}{\zeta} - \frac{1}{\xi} \right), \quad \xi \ge \zeta \ge 1, \quad \zeta \ge \eta, \tag{3.47}
$$

where

$$
Q^2 = \frac{\beta_l D_\beta A}{P D_\sigma (\sigma_\infty - \sigma_l)}. \tag{3.48}
$$

We have already done most of the work needed to describe the growth dynamics of Greenspan's model in Section 3.3. From the heuristic description in the previous section, starting with a small MCS that is not growth limited we expect the following three phases of growth:

1. **Phase I:** *Exponential growth.* As before we assume that $\xi(t_0) = \xi_0 < 1$. MCS growth dynamics in this initial phase are therefore described by equation (3.25) until $\xi(t) = 1$, at which point the necrotic region begins to form and we pass to phase II.

2. **Phase II:** *Formation of the necrotic core.* In this phase, dynamics are governed by differential equation (3.31), which continues to apply until either the inhibitor concentration reaches the threshold described in relation (3.46) in the living layer, or the MCS reaches its equilibrium size before that happens. If β never exceeds β_l in the proliferative (living) region, then the MCS dynamics are covered completely in Section 3.3. Otherwise, growth enters the next phase.

3. **Phase III:** *Three-layer histology.* In this phase, $\eta < \zeta < \xi$, conditions (3.24) and (3.47) hold, and the growth dynamics are governed by the following ODE:

$$
3\xi^2 \frac{d\xi}{d\tau} = \xi^3 - \zeta^3 - 3\gamma\eta^3, \quad \tau \ge \tau_2, \tag{3.49}
$$

where τ_2 is the moment the quiescent layer begins to form, $\xi(\tau_2) = \xi_Q$ is the MCS radius at that moment, and $\eta(\tau_2) = \zeta(\tau_2)$.

So there are two possible asymptotic behaviors: either the inhibitor never has any effect and the MCS growth slows only due to necrosis as described in Section 3.3; or the inhibitor builds up sufficiently to cause a quiescent layer to form. Therefore, two questions arise: (1) under what conditions will the quiescent layer form, and (2) if it does form, how large will the MCS be when that happens? The following theorem answers the second question, and sets up the answer to the first.

PROPOSITION 3.1

The critical radius, ξ_Q, at which an MCS growing according to Greenspan's model develops a quiescent layer in the living region is

$$\xi_Q = \frac{Q}{\sqrt{2x_Q^2(1 - x_Q)}}, \tag{3.50}$$

where

$$x_Q = \frac{Q^2 + Q\sqrt{9Q^2 + 8}}{4(1 + Q^2)}, \tag{3.51}$$

and Q is defined by relation (3.47).

We leave the proof as an exercise. The following theorem provides the conditions for formation of the 3-layered MCS histology.

PROPOSITION 3.2

Consider an MCS growing according to Greenspan's model. Then a quiescent layer forms if and only if

$$\Gamma^{-1/3} > x_Q, \tag{3.52}$$

where Γ and ξ_Q are defined in (3.32) and (3.51), respectively.

Again, we leave the proof to the reader.

To make connection with his method for solving the MCS growth curve in Phase II (see differential equation (3.31)), Greenspan introduces the new variable,

$$y = \frac{\zeta}{\xi}, \tag{3.53}$$

which allows him to write the model in Phase III as follows:

$$\frac{9x}{(1 + 2x)(1 - x)} \frac{dx}{dt} = 1 - y^3 - 3\gamma x^3, \tag{3.54}$$

with

$$\xi = \frac{1}{\sqrt{(1-x^2)(1+2x)}}, \quad \frac{Q^2}{2\xi^2} = x^3 \left(\frac{1}{y} - 1\right).$$

and initial conditions,

$$x(\tau_2) = y(\tau_2) = x_Q, \quad \xi(\tau_2) = \xi_Q.$$

3.5 Testing Greenspan's model

Mathematical models are theoretical tools that describe and explain natural phenomena. As theory, however, they are useless unless they make some sort of experimentally testable prediction. In the case of the Greenspan model, as in most mathematical models, the obvious experimental test—grow spheroids in culture and compare their size and histological evolution with the model's prediction—suffers the crushing disadvantage of impossibility. Such a direct test would require accurate estimates of parameters that, practically, cannot be measured. For example, since we have no suspects for the inhibitory chemical, how can we measure λ and therefore γ?

Nevertheless, Greenspan's model can be tested quite rigorously and quantitatively. For example, we can take advantage of the definition of Q^2 (3.48) and its role in the model in the following way. Suppose we run two sets of experiments in which MCSs are grown in nearly identical cultures. In one set, however, the nutrient concentration is higher than in the other set, say it is doubled. Otherwise, all experimental units (individual MCSs) are treated identically.

Greenspan's model predicts that the ratio of Q^2 (defined in relation (3.47)) between the two sets, say Q_1^2 and Q_2^2, should be

$$\frac{Q_1}{Q_2} = \left(\frac{2\sigma_\infty - \sigma_l}{\sigma_\infty - \sigma_l}\right)^{\frac{1}{2}}, \tag{3.55}$$

where we have doubled the nutrient in set 1, σ_∞ is the nutrient concentration in the set 2 media, and σ_l is the necrotic threshold for nutrient (which can be measured in simple assays). Since σ_∞ and σ_l are measureable, we can use them, ξ_Q and x_Q to predict how this manipulation affects the size of the spheroids and their necrotic core when the quiescent layer first develops. Here is an outline of the procedure (which assumes an MCS system with reproducible behavior):

1. **Obtain an estimate of Q_1.** Grow spheroids and note the sizes of the MCS and its necrotic core when the quiescent layer first develops. (Quiescent and necrotic cells can be identified with fairly straightforward

assays; there would be some technical issues here since identifying the quiescent layer requires destruction of the spheroid, but this can be overcome by starting with a large number of MCSs treated identically, with sacrifice of some small fraction at regular time intervals.) These values give us ξ_Q and x_Q. Then it is straightforward to determine Q from relations (3.50) and (3.51) (numerically, if necessary).

2. **Predict Q_2.** Now that we have estimates of Q_1, σ_∞ and σ_l, solve equation (3.55) for Q_2 and take the positive root.

3. **Predict the outcome of the experimental increase in nutrient.** Use the estimate of Q_2 in equations (3.50) and (3.51) to predict ξ_Q and x_Q in the new experiment.

4. **Test this prediction** by running the same experiment as before, but now with double the nutrient concentration in the media. Measure the size of the MCS and necrotic core when quiescence first arises, and compare these measures to the predicted ξ_Q and x_Q.

Other tests may be devised. This simply illustrates how experimentally testable predictions can be generated from a complex model, the direct dynamics of which cannot be measured.

3.6   Sherratt-Chaplain model for avascular tumor growth

Most of the existing models of avascular tumor growth were influenced by the pioneer work of Greenspan published in 1972 [5], which often have a structure of a proliferating layer, a quiescent layer and a necrotic core. These compartments are viewed as discrete layers, separated by artificial and moving boundaries. In 2001, Sherratt and Chaplain [15] developed an alternative model formulated in terms of continuum densities of proliferating, quiescent and necrotic cells, together with a generic limiting nutrient. Their model solutions are capable of generating the usual three layer structure in earlier MCS models. Their model admits traveling wave solutions with a minimum speed that can be computed analytically.

We consider a situation that tumor density varies only in one spatial dimension such as in a thin disc shaped domain. Real tumors may differ from a MCS in its shape. They may initially develop in epithelia (known as carcinomas and represent majority of clinically observed tumors) and hence form approximately two-dimensional structures before invading surrounding tissues.

Let $p(x,t)$, $q(x,t)$ and $n(x,t)$ stand for the densities at time t and location x of proliferating, quiescent and necrotic cells, respectively. We assume that necrotic cells do not move. For simplicity, we also assume that cell movement

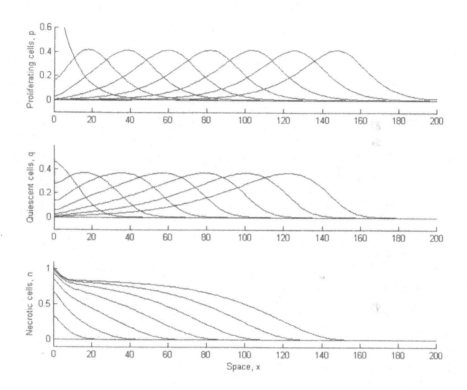

FIGURE 3.6: A numerical solution of the reduced model (3). The parameter values are $\gamma = 9$, $c_0 = 1$ and $\alpha = 0.8$. The rate functions are $f(c) = 0.5[1 - \tanh(4c - 2)]$, $g(c) = 1 + 0.1c$, and $h(c) = 0.5f(c)$. The initial conditions are $q = n = 0$, $p = e^{0.1x}$, and the boundary conditions used at $x = 0$ and $x = 200$ are $p_x = q_x = 0$.

is through diffusion. The growth dynamics of cells is determined by limiting nutrient density $c(x,t)$. The Sherratt-Chaplain model for avascular tumor growth takes the following form

$$
\begin{aligned}
\frac{\partial p}{\partial t} &= \frac{\partial}{\partial x}\left[\frac{p}{\partial p + q}\frac{\partial(p+q)}{\partial x}\right] + g(c)p(1 - p - q - n) - f(c)p,\\
\frac{\partial q}{\partial t} &= \frac{\partial}{\partial x}\left[\frac{q}{\partial p + q}\frac{\partial(p+q)}{\partial x}\right] + f(c)p - h(c)q,\\
\frac{\partial n}{\partial t} &= h(c)q,\\
\frac{\partial c}{\partial t} &= D_c\frac{\partial^2 c}{\partial x^2} + k_1c_0[1 - \alpha(p + q + n)] - k_1c - k_2pc.
\end{aligned}
\tag{3.56}
$$

Here we assume that a value of 1 corresponds to a packed population. We assume that the proliferating rate function $g(c)$ is increasing and $g(0) = 1$. The cell switching rate functions $f(c)$ and $h(c)$ are decreasing and tend to 0 when c tends to infinity. Since a cell can stay at the quiescent stage for a long time, it is conceivable that at a given limiting nutrient concentration level, a proliferating cell is more likely to switch to quiescent state than a quiescent cell is to become dead. Hence we assume that $f(c) > h(c)$. The availability of nutrient is represented by the source term $k_1c_0[1 - \alpha(p+q+n)]$. The parameter α is assumed to be in $(0, 1]$, and c_0 is the nutrient concentration outside of the tumor. k_1 is the nutrient degradation rate and k_2 is the nutrient uptake rate by proliferating cells.

Straightforward numerical simulations of the Sherratt-Chaplain model (3.56) by MATLAB®-supported pdepe-based programs (see such a code below) suggests that its typical solutions indeed reproduce the familiar structure of proliferating rim and a band of quiescent cells around a growing necrotic core (Figure 3.6).

3.6.1   MATLAB® file for Figure 3.6

```
------------------------------------------------------------------

function MCSwave
m = 0;
x = linspace(0,200,200);
t = linspace(0,14,141);

sol = pdepe(m,@mypde,@myic,@mybc,x,t);
u1 = sol(:,:,1);
u2 = sol(:,:,2);
u3 = sol(:,:,3);

subplot(3,1,1)
hold on

for i=0:2:14
```

```
        plot(x,u1(i*10+1,:));
end
ylabel('Proliferating cells, p');
xlim([0 200]); ylim([-.1 .6]);

subplot(3,1,2)
hold on

for i=0:2:14
    plot(x,u2(i*10+1,:));
end
ylabel('Quiescent cells, q');
xlim([0 200]); ylim([-.1 .6]);

subplot(3,1,3)
hold on

for i=0:2:14
    plot(x,u3(i*10+1,:));
end
ylabel('Necrotic cells, n');
xlabel('Space, x');
xlim([0 200]); ylim([-.1 1.1]);


% ------------------------------------------------------------------
function [c,f,s] = mypde(x,t,u,DuDx)
c = [1; 1; 1];

co = 1; alpha = .8; gamma = 9;
k = (co*gamma*(1-alpha*(u(1)+u(2)+u(3))))/(gamma+u(1));

fc = .5*(1-tanh(4*k-2));
gc = 1+.1*k;
hc = .5*fc;

f = [u(1)/(u(1)+u(2)) * (DuDx(1) + DuDx(2));
     u(2)/(u(1)+u(2)) * (DuDx(1) + DuDx(2));
     0];

s = [gc*(u(1)*(1-u(1)-u(2)-u(3))) - fc*u(1);
     fc*u(1) - hc*u(2);
     hc*u(2)];
% ------------------------------------------------------------------
function u0 = myic(x)
u0 = [exp(-.1*x); 0; 0];
% ------------------------------------------------------------------
function [pl,ql,pr,qr] = mybc(xl,ul,xr,ur,t)
pl = [0; 0; 0];
```

```
ql = [1; 1; 1];
pr = [0; 0; 0];
qr = [1; 1; 1];
```

Since nutrient movement and uptake are much faster processes than tumor cell growth and movement, it is practical to consider the nutrient kinetics are operating at a quasi-steady state. Hence we can replace the nutrient equation by the following

$$c = c_0\gamma[1 - \alpha(p + q + n)]/(\gamma + p), \quad \gamma = k_1/k_2. \tag{3.57}$$

3.6.2   Minimum wave speed

In the following, we would like to explore the possibility of the existence of wave-like solutions in the Sherratt-Chaplain model (3.56) and estimate their minimum speed. To this end, we introduce the wave variable $z = x - vt$ and $p(x, t) = P(z)$, $q(x, t) = Q(z)$ $n(x, t) = N(z)$, where v is the wave speed. Substituting these wave solution forms into (3.56) yields the following system of ordinary differential equations

$$\begin{aligned}
&\left(\frac{P(P' + Q')}{\partial P + Q}\right)' + vP' + g(C)P(1 - P - Q - N) - f(C)P = 0, \\
&\left(\frac{Q(P' + Q')}{\partial P + Q}\right)' + vQ' + f(C)P - h(C)Q = 0, \\
&vN' + h(C)Q = 0, \\
&C = c_0\gamma[1 - \alpha(P + Q + N)]/(\gamma + P).
\end{aligned} \tag{3.58}$$

We now consider the possible wave speed for solutions of (3.58). The conventional phase plane method to find minimum wave speed will not work in our situation here. Since near the wave fronts, population levels are low and hence they grow approximately exponentially. We therefore looking for admissible exponential solutions near the wave fronts. Linearizing (3.58) about $P = Q = N = 0$ and let $Pe^{-\xi z}, Qe^{-\xi z}, Ne^{-\xi z}$ be the leading terms of $P(z), Q(z), N(z)$, respectively. We obtain

$$\begin{aligned}
&\xi^2 P - v\xi P + [g(c_0) - f(c_0)]P = 0, \\
&\xi^2 Q - v\xi Q + f(c_0)P - h(c_0)Q = 0, \\
&-v\xi N + h(c_0)Q = 0.
\end{aligned} \tag{3.59}$$

This is equivalent to $Ax = 0$, where $x = (P, Q, N)^T$, $0 = (0, 0, 0)^T$, and

$$A = \begin{pmatrix} \xi^2 - v\xi + [g(c_0) - f(c_0)] & 0 & 0 \\ f(c_0) & \xi^2 - v\xi - h(c_0) & 0 \\ 0 & h(c_0) & -v\xi \end{pmatrix}.$$

Thus for nontrivial solutions, we must have $\det(A) = 0$. This yields

$$\xi = \frac{1}{2}\left(v \pm \sqrt{v^2 - 4[g(c_0) - f(c_0)]}\right).$$

Since P, Q and N must be positive, we require ξ to be real, so that

$$v^2 \geq 4[g(c_0) - f(c_0)].$$

Hence we have established the existence of a minimum possible wave speed

$$v_{min} = 2\sqrt{g(c_0) - f(c_0)}.$$

Many population models have traveling wave solutions traveling at their minimum speed and this is confirmed by numerical simulations for the Sherratt-Chaplain model (3.56) in Figure 3.6.

3.7 A model of in vitro glioblastoma growth

Glioblastoma multiforme (GBM) is an aggressive and fatal brain cancer. It is characterized by fast proliferation, stealthy infiltration and speedy migration, which contributes to the difficulty of treatment. Models of this type of cancer growth often include two separate equations to model proliferation or migration [12]. Stepien et al. [17] proposed a single equation which uses density-dependent diffusion to capture the behavior of both proliferation and migration. This single equation model is amenable for mathematical analysis, including the existence of traveling wave solutions and the minimum speed of such solutions. The solution of this model matches well with in vitro experimental data and the accuracy is comparable to a two population model in Stein et al. [16]. The material of this section is adapted from that of Stepien et al. [17].

3.7.1 Model formulation

The growth and diffusion of glioblastoma cells are governed by many processes including, but not limited to, random diffusion, chemotaxis, haptotaxis, cell-cell adhesion, cell-cell signaling, and microenvironmental cues such as oxygen and glucose. In Stein et al. [16], two human astrocytoma U87 cell lines are implanted into gels—one with a wild-type receptor (EGFRwt) and one with an over expression of the epidermal growth factor receptor gene (ΔEGFR). The resulting spheroids were left to grow over 7 days and imaged every day. One of the fundamental questions with tumor growth is its speed of advancing. The images allow us to approximate the tumor growth speed. However, it is highly desirable to quantify tumor spread, not just computationally, but also analytically.

The two population model proposed by Stein et al. [16] models the movement of the migratory cells (u_i) based on the experiment described above. The radius of the tumor core is modeled as increasing at a constant rate based on the experimental data. The model of Stein et al. [16] assumes that the tumor cells leave the tumor core and become invasive cells to invade the collagen gel at the edge of the tumor core. Their model takes the form of

$$\frac{\partial u_i(r,t)}{\partial t} = \underbrace{D\nabla^2 u_i}_{\text{diffusion}} + \underbrace{gu_i\left(1 - \frac{u_i}{u_{\max}}\right)}_{\text{logistic growth}} - \underbrace{\nu_i\nabla_r\cdot u_i}_{\text{taxis}} + \underbrace{s\delta(r - R(t))}_{\text{shed cells from core}}, \quad (3.60)$$

where u_i represents the invasive cells of the tumor at radius r and time t.

As the core of the tumor increases, cells are shed from the front of the expanding core to become invasive cells. Parameter D is the diffusion constant, g is the growth rate, $u_{\max}$ is the carrying capacity, ν_i is the degree at which cells migrate away from the core, s is the amount of cells shed per day, and δ is the Dirac delta function. The radius of the tumor core is modeled by $R(t) = R_0 + \nu_c t$, where R_0 is the initial radius of the tumor core and ν_c is the constant velocity at which the tumor core radius increases according to the data.

Stepien et al. [17] formulated a single equation model from a similar base. They considered a density-dependent convective-reaction-diffusion equation for the tumor cells $u(x,t)$ which takes the form of

$$\frac{\partial u}{\partial t} = \underbrace{\nabla \cdot \left(D\left(\frac{u}{u_{\max}}\right)\nabla u\right)}_{\text{density-dependent diffusion}} + \underbrace{gu\left(1 - \frac{u}{u_{\max}}\right)}_{\text{logistic growth}} - \underbrace{\text{sgn}(x)\nu_i\nabla \cdot u}_{\text{taxis}}, \quad (3.61)$$

where they consider the equation in Cartesian coordinates but assume there is radial symmetry. Parameters g and ν_i are as in equation (3.60), but now they are in relation to the entire tumor cell population instead of just the invasive cell population. The proliferating cells are assumed to diffuse slowly but grow in population more quickly. On the other hand, the migrating cells diffuse quickly, traveling far in a short amount of time, but do not grow in population as quickly. Due to slow movement, proliferating cells tend to be in a location with higher cell density. Experimental work from Stein et al. [16] suggests that diffusion is large for areas where the cell density is small (the migrating tumor cells), but diffusion is small where the cell density is large (the proliferating tumor cells). This relation could possibly be explained by cell–cell adhesion. To capture this, we let

$$D(u) = D_1 - \frac{D_2(u)^n}{a^n + (u)^n}. \quad (3.62)$$

For biologically relevant parameters, we assume that D_1, D_2, g, a, and ν_i are all positive, $n > 1$, and $D_2 \leq D_1$ to avoid "negative" diffusion, which can be a problem both biologically and numerically.

3.7.2 Traveling wave system properties

We would like to explore the existence of traveling wave solutions of (3.61) using the routine phase plane analysis.

Similar to Stein et al. [16], we consider (3.61) in one spatial dimension only. In this case, the governing equation is

$$\frac{\partial u}{\partial t} = D\left(\frac{u}{u_{\max}}\right) \frac{\partial^2 u}{\partial x^2} + \frac{1}{u_{\max}} D'\left(\frac{u}{u_{\max}}\right) \left(\frac{\partial u}{\partial x}\right)^2 - \nu_i \frac{\partial u}{\partial x} + gu\left(1 - \frac{u}{u_{\max}}\right).$$
(3.63)

To simplify our analysis, we rescale the above equation with the following transformations

$$t^* = gt, \qquad x^* = x\sqrt{g}, \qquad u^* = \frac{u}{u_{\max}}.$$
(3.64)

Let

$$v = \frac{\nu_i}{\sqrt{g}},$$
(3.65)

and, dropping the asterisks and dividing through by $gu_{\max}$, we obtain

$$\frac{\partial u}{\partial t} = D(u)\frac{\partial^2 u}{\partial x^2} + D'(u)\left(\frac{\partial u}{\partial x}\right)^2 - v\frac{\partial u}{\partial x} + u(1-u).$$
(3.66)

A traveling wave solution of (3.66) is a solution of the form

$$u(x,t) = w(x - kt),$$
(3.67)

where $k \geq 0$ is the speed of the traveling wave and the function $w(z) = w(x - kt)$ is defined on $(-\infty, \infty)$ and satisfies

$$\lim_{z \to -\infty} w(z) = 1, \qquad \lim_{z \to \infty} w(z) = 0.$$
(3.68)

Substituting (3.67) into (3.66) results in the following equation

$$w''(z) + \frac{1}{D(w(z))}\Big((k - v)w'(z) + D'(w(z))(w'(z))^2 + w(z)(1 - w(z))\Big) = 0.$$
(3.69)

As usual, we convert (3.69) to a phase plane system of first-order ordinary differential equations,

$$w' = y,$$
(3.70a)

$$y' = \frac{-1}{D(w)}\Big((k - v)y + D'(w)y^2 + w(1 - w)\Big).$$
(3.70b)

This system has exactly two steady states, $(w, y) = (0, 0)$ and $(w, y) = (1, 0)$. The Jacobi matrix evaluated at $(1, 0)$ is

$$J(1, 0) = \begin{pmatrix} 0 & 1 \\ \frac{1}{D(1)} & \frac{-(k-v)}{D(1)} \end{pmatrix}, \tag{3.71}$$

from which we have $\det J(1, 0) = \frac{-1}{D(1)} < 0$, implying $(1, 0)$ is a saddle point. The Jacobian matrix evaluated at $(0, 0)$ is

$$J(0, 0) = \begin{pmatrix} 0 & 1 \\ \frac{-1}{D(0)} & \frac{-(k-v)}{D(0)} \end{pmatrix}, \tag{3.72}$$

from which we have $\det J(0, 0) = \frac{1}{D(0)} = \frac{1}{D_1} > 0$ and assuming that $k > v$, then $\text{tr}\, J(0, 0) = \frac{-(k-v)}{D_1} < 0$ and $(0, 0)$ is a stable node or spiral. Since a stable spiral cannot result in physiologically relevant solutions, we must have the origin as a stable node. In other words, we must have the eigenvalues of $J(0, 0)$ as negative numbers. This implies that

$$k \geq k_{\min} = 2\sqrt{D_1} + v, \tag{3.73}$$

which in terms of the original dimensional equation (3.63) is

$$k \geq k_{\min} = 2\sqrt{D_1 g} + \nu_i. \tag{3.74}$$

3.7.3   Existence of traveling wave solutions

We would like to establish the following result on the existence of traveling wave solutions in the partial differential equation (3.66).

THEOREM 3.1
There exists a traveling wave solution (3.67) of the partial differential equation (3.66) with boundary conditions $u(x, t) \to 1$ as $x \to -\infty$ and $u(x, t) \to 0$ as $x \to \infty$ with $0 < u(x, t) < 1$, whose orbit connects the steady states $u \equiv 0$ and $u \equiv 1$ if and only if (3.73) is satisfied.

To this end, we would like to construct a positively invariant region in which to trap the unstable manifold of the saddle point. Figure 3.7 illustrates such a trapping region and the heteroclinic orbit connecting the two steady states for a set of parameters generated by fitting the model (3.61) to the data reported in Stein et al. [16].

Observe that the solution of the system (3.70) across the line segment

$$\mathcal{T}_1 = \{(w, y) : 0 \leq w \leq 1, \ y = 0\} \tag{3.75}$$

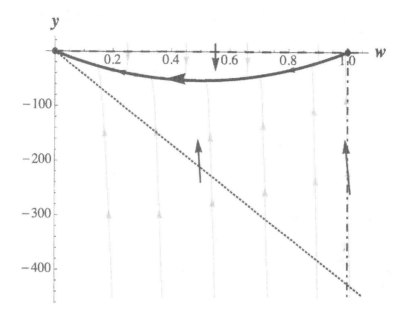

FIGURE 3.7: Phase portrait of the system (3.70) with parameter values $D_1 = 5.5408 \times 10^{-6}, \mathrm{cm}^2/\mathrm{day}$, $D_2 = 5.3910 \times 10^{-6}, \mathrm{cm}^2/\mathrm{day}$, $a = 0.021188 \,\mathrm{cells/cm}^3$, $n = 1.2848$, $g = 0.49120/\mathrm{day}$, $\nu_i = 4.6801 \times 10^{-5}\mathrm{cm/day}$. The solid curve is the unstable manifold, the dashed horizontal line segment is $\mathcal{T}_1$, the dotted slanted line is $\mathcal{T}_2$ (3.78) parallel to the eigenvector associated to the most negative eigenvalue of $J(0,0)$, and the dash-dotted vertical line segment is $\mathcal{T}_3$ (3.79). Arrows show direction of flow. The nondimensional wave speed $k = 2\sqrt{D_1} + v = 2\sqrt{D_1} + \nu_i/\sqrt{g} \approx 0.0047746$.

perpendicularly downwards. Consider also the line parallel to the eigenvector corresponding to the more negative eigenvalue of the linearized system at $(0,0)$, i.e.,

$$y(w) = \alpha_0 w, \tag{3.76}$$

with

$$\alpha_0 = \frac{1}{2D(0)}\left(-(k-v) - \sqrt{(k-v)^2 - 4D(0)}\right) < 0. \tag{3.77}$$

We define

$$\mathcal{T}_2 = \{(w,y) : 0 \leq w \leq 1, \ y = \alpha_0 w\} \tag{3.78}$$

and

$$\mathcal{T}_3 = \{(w,y) : w = 1, \ \alpha_0 \leq y \leq 0\}. \tag{3.79}$$

We would like to show that there is a value k such that the triangle $\mathcal{T}$ formed by $\mathcal{T}_1$, $\mathcal{T}_2$ and $\mathcal{T}_3$ is positively invariant with respect to the solutions of the system (3.70). To this end, we would like to first establish the following key statement.

LEMMA 3.1
The flow at any point along the line $y(w) = \alpha_0 w$, for $w \in (0,1]$, crosses that line in the positive y direction for k sufficiently small.

PROOF A normal vector to the graph of $(w, \alpha_0 w)$ pointing in the positive y direction is $(-\alpha_0, 1)$. Along the line $y(w) = \alpha_0 w$, the system (3.70) becomes

$$w' = \alpha_0 w, \tag{3.80a}$$

$$y' = \frac{-1}{D(w)}\Big((k-v)\alpha_0 w + D'(w)\alpha_0^2 w^2 + w(1-w)\Big). \tag{3.80b}$$

We would like to choose k so that the solution of the phase plane system (3.70) cross the line $y(w) = \alpha_0 w$ and into the triangle region $\mathcal{T}$. This is equivalent to requiring that the inner product $(-\alpha_0, 1) \cdot (w', y') \geq 0$ along the graph of $(w, \alpha_0 w)$ which implies

$$-\alpha_0^2 w - \frac{1}{D(w)}\Big((k-v)\alpha_0 w + D'(w)\alpha_0^2 w^2 + w(1-w)\Big) \geq 0, \tag{3.81}$$

which can be rearranged as

$$-\frac{\alpha_0 w}{D(w)}\left(\alpha_0 D(w) + (k-v) + D'(w)\alpha_0 w + \frac{1-w}{\alpha_0}\right) \geq 0. \tag{3.82}$$

Since $\alpha_0 < 0$ and $D(w) > 0$ for $w \in (0,1]$, we have

$$\alpha_0 D(w) + (k-v) + D'(w)\alpha_0 w + \frac{1-w}{\alpha_0} \geq 0. \tag{3.83}$$

Since (3.83) must hold for all $w \in [0,1]$, then we obtain the condition

$$k \geq v - \max_{w \in [0,1]}\left\{\frac{d}{dw}\big(D(w)(\alpha_0 w)\big) - \frac{w(1-w)}{\alpha_0 w}\right\}. \tag{3.84}$$

However, since α_0 depends on k, substitute (3.77) into (3.84), and after tedious algebraic manipulation the inequality becomes (we leave the details as an exercise)

$$k \leq v + \min_{w \in [0,1]}\left\{\frac{D(w) + wD'(w) - D_1(1-w)}{\sqrt{w(D(w) + wD'(w) - D_1)}}\right\}. \tag{3.85}$$

Consider the function

$$h(w) = \frac{D(w) + wD'(w) - D_1(1-w)}{\sqrt{w(D(w) + wD'(w) - D_1)}}. \tag{3.86}$$

The minimum of $h(w)$ is attained in the interior of the domain $[0,1]$ if $h'(w) = 0$ for some $w \in (0,1)$. This means that either (i) $D_1(1+w) - \frac{d}{dw}(wD(w)) = 0$ or (ii) $D_1 - D(w) + w\frac{d}{dw}(wD'(w)) = 0$. In case (i), this means that the diffusion function must be of the form $D(w) = D_1(1+w) + \frac{C_1}{w}$, where C_1 is an arbitrary constant. In case (ii), the diffusion function must be of the form $D(w) = D_1 + C_1\frac{w^2-1}{w} + iC_2\frac{w^2-1}{2}$, where C_1 and C_2 are arbitrary constants. Since our diffusion function (3.62) is of neither of these forms, the minimum cannot be attained in the interior of $[0,1]$ and must be attained at either of the endpoints $w = 0$ or $w = 1$.

Since $h(w)$ tends to infinity as $w \to 0$ (assuming that parameter $n > 1$), then the minimum occurs at $w = 1$, and thus if the condition

$$k \leq v + \frac{D(1) + D'(1)}{\sqrt{D(1) + D'(1) - D_1}} = v + \frac{(a+1)^2 D_1 + (1+a+an)D_2}{(a+1)\sqrt{(1+a+an)D_2}} \quad (3.87)$$

is satisfied, then the lemma holds. □

Since

$$2\sqrt{D_1} \leq \frac{(a+1)^2 D_1 + (1+a+an)D_2}{(a+1)\sqrt{(1+a+an)D_2}}, \quad (3.88)$$

the flow across the line $\mathcal{T}_2$ is in the positive y direction for at least the minimum wave speed k. For larger speeds, the nonlinearities of the system require that $\mathcal{T}_2$ be nonlinear such that

$$k \leq v - \max_{w \in [0,1]} \left\{ \frac{d}{dw}\big(D(w)f(w)\big) - \frac{w(1-w)}{f(w)} \right\}, \quad (3.89)$$

is satisfied for some function f where $f(0) = 0$ and $f(w) \leq 0$ for $w \in (0,1]$. The minimum wave speed is determined by taking the infimum on the set of functions f (Sánchez-Garduño et al. [14]).

The flow across $\mathcal{T}_3$ is in the negative w direction. Hence $\mathcal{T}$ is a positively invariant set for system (3.70).

We are now ready to prove Theorem 3.1.

PROOF We first show that the unstable manifold of the saddle at $(1,0)$ remains in the region $\mathcal{T}$ all the time.

The vertical ($y' = 0$) nullclines are the solutions to the quadratic equation

$$D'(w)y^2 + (k-v)y + w(1-w) = 0, \quad (3.90)$$

which are

$$y_+(w) = \frac{1}{2D'(w)}\Big(-(k-v) + \sqrt{(k-v)^2 - 4D'(w)w(1-w)} \Big), \quad (3.91a)$$

$$y_-(w) = \frac{1}{2D'(w)}\Big(-(k-v) - \sqrt{(k-v)^2 - 4D'(w)w(1-w)} \Big). \quad (3.91b)$$

Since $k > v$, the slope of the y_+ nullcline at $w = 1$ is

$$y'_+(1) = \frac{1}{k-v} > 0, \tag{3.92}$$

and the eigenvector of the linearized system corresponding to the positive eigenvalue at the saddle point $(1,0)$ is

$$\vec{\eta} = \left(1, \quad \frac{1}{2}\left(\frac{-(k-v)}{D(1)} + \sqrt{\left(\frac{k-v}{D(1)}\right)^2 + \frac{4}{D(1)}}\right)\right)^T. \tag{3.93}$$

All trajectories that leave the point $(1,0)$ in the region $\mathcal{R} = \{(w,y) : 0 \leq w \leq 1, \ y \leq 0\}$ have the tangent vector $\vec{\eta}$ at $(1,0)$. Comparing the slope of eigenvector $\vec{\eta}$ and the slope of the y_+ nullcline at the point $(1,0)$, we find that the slope of $\vec{\eta}$ is less than the slope of y_+. Therefore, trajectories leaving point $(1,0)$ leave above y_+. Since the flow across the nullcline y_+ is horizontal in the negative w direction, and y_+ is contained in $\mathcal{T}$ near $w = 1$, the unstable manifold of the saddle at $(1,0)$ has nonempty intersection with $\mathcal{T}$. For the same reason (the flow across the nullcline y_+ is horizontal in the negative w direction), y_+ can not intersect $\mathcal{T}_\in$ for $w \in (0,1)$ since otherwise, the flow will cross $\mathcal{T}_\in$ and leaves

$$\mathcal{T}.$$

Therefore, the unstable manifold of the saddle at $(1,0)$ remains in the region $\mathcal{T}$ for all time, and furthermore, the ω-limit set of the corresponding orbit is also in $\mathcal{T}$. Since $w' = y \leq 0$ within $\mathcal{T}$, by the Poincaré–Bendixson theorem, there are no periodic orbits or equilibrium points in the interior of $\mathcal{T}$. The ω-limit set must be contained in the boundary of $\mathcal{T}$, and therefore, the ω-limit set is $(0,0)$.

Hence, there exists a heteroclinic orbit connecting the equilibrium points $(w,y) = (0,0)$ and $(w,y) = (1,0)$ as long as the condition (3.73) is satisfied, which implies that a traveling wave solution exists, at least for the minimum speed $k_{\min} = 2\sqrt{D_1} + v$. □

Numerical simulation results confirm that the wave speed of the invasive cells is constant and satisfies condition (3.74) [17].

3.8   Derivation of one-dimensional conservation equation

A starting point of formulating many partial differential equations is the conservation equation, which is also referred as the balance equation. In this section, we provide a rigorous derivation of the one-dimensional conservation

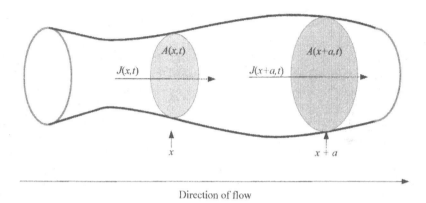

FIGURE 3.8: An illustration for the derivation of the one dimensional conservation equation.

equation. The word of "cells" can be replaced by "populations" or "particles". To this end, we define the following notations:

$u(x,t)$ = concentration of cells at (x,t),

$A(x,t)$ = the area of cross section at (x,t),

$J(x,t)$ = cell flux at (x,t) which is the number of cells crossing a unit area at x in the positive direction in a unit of time,

$p(x,t)$ = rate of cells produced or died in a unit volume at (x,t).

With the above notations, we can derive the conservation equation in a tubular flow. Let $N(x,a,t)$ be the total number of cells located within the tube in the interval of $(x, x+a)$ at time t, then

$$N(x,a,t) = \int_x^{x+a} u(s,t)A(s,t)ds. \qquad (3.94)$$

Let $P(x,a,t)$ be the net cell production or death within the tube in the interval of $(x, x+a)$ at time t, then

$$P(x,a,t) = \int_x^{x+a} p(s,t)A(s,t)ds. \qquad (3.95)$$

The law of conservation of cells implies that

$$\frac{\partial N(x,a,t)}{\partial t} = J(x,t)A(x,t) - J(x+a,t)A(x+a,t) + P(x,a,t). \quad (3.96)$$

Dividing both sides of the above equation by a and letting a tend to zero yields the following general one dimensional conservation equation

$$\frac{\partial(u(x,t)A(x,t))}{\partial t} = -\frac{\partial(J(x,t)A(x,t))}{\partial x} + p(x,t)A(x,t). \quad (3.97)$$

The most common cell movement is diffusion. It is the result of random motion and can be described by the so-called Fick's law, which amounts to say that the cell flux due to random motion is proportional to the cell gradient

$$J(x,t) = -D\nabla u(x,t). \quad (3.98)$$

Assume that the cell flux is purely driven by random motion and the area of cross section is constant, then equation (3.97) is reduced to the following one dimensional reaction diffusion equation which is often seen in the literature

$$\frac{\partial u(x,t)}{\partial t} = D\frac{\partial^2 u(x,t)}{\partial x^2} + p(x,t). \quad (3.99)$$

3.9   Exercises

Exercise 3.1: Consider the equilibrium nutrient profile for model (3.11).

1. Calculate the general solution of model (3.11) for $\partial\sigma/\partial t = 0$. By general, we mean do not impose the boundary conditions.

   Solution:

   $$\sigma(r,\cdot) = \frac{C_1 \sinh\left(\sqrt{\frac{A}{D}}r\right) + C_2 \cosh\left(\sqrt{\frac{A}{D}}r\right)}{r},$$

   where C_1 and C_2 are integration constants.

2. Now suppose σ is bounded for all $r \in (0, R_0)$, and set $\sigma(R_0,\cdot) = \sigma_\infty$. We impose no other conditions. What is the equilibrium nutrient profile under these conditions?

   Solution:

   $$\sigma(r,\cdot) = \frac{\sigma_\infty R_0 \sinh\left(\sqrt{\frac{A}{D}}r\right)}{r \sinh\left(\sqrt{\frac{A}{D}}R_0\right)}.$$

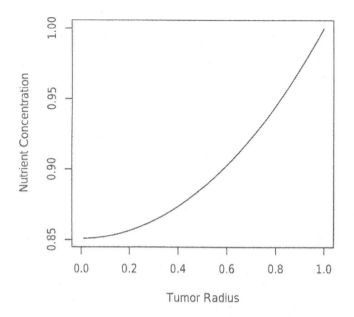

FIGURE 3.9: Equilibrium nutrient profile of model (3.11).

3. Graph this solution on $r \in (0, 1)$ assuming all parameters are unity. Ignore the no-flux (Neumann) boundary condition, but impose the outer boundary (Dirichlet) and boundedness conditions.

Solution: See Fig. 3.9.

Exercise 3.2: Make the substitution $u = r\sigma$ into model (3.11); what now is the model's form? What insight do you gain from this change in form?

Solution:

$$\frac{\partial u}{\partial t} = D \frac{\partial^2 u}{\partial r^2} - Au.$$

Exercise 3.3: Show that solutions to model (3.11) cannot become negative for any initial distribution $\sigma(r, 0) = \sigma_0(r) \geq 0$ for all $r \in (0, R_0)$.

Exercise 3.4: Assume that $R_i > 0$ in equation (3.13). Let $h = R_0 - R_i$ and $\gamma = R_i / R_0$. Show that equation (3.13) implies

$$h^2 = \frac{6D(\sigma_\infty - \sigma_l)}{A(1 + 2\gamma)} \leq \frac{6D}{A}(\sigma_\infty - \sigma_l). \tag{3.100}$$

If γ tends to 1, then h^2 tends to $\frac{2D}{A}(\sigma_\infty - \sigma_l)$.

Exercise 3.5: (The next four exercises are adapted from material in Chapter 8 of the book of Britton [1].) Assume that the living cells in a multicell spheroid (MCS) proliferate at a rate of P and die at a rate of L. Assume the MCS has a volume of $V(t)$ and its necrotic core has a volume of $V_i(t)$. Assume further that the mass density of the MCS is the same everywhere. Show that

$$\frac{dV}{dt} = P(V - V_i) - LV_i. \tag{3.101}$$

Assume the MCS has a radius R and its necrotic core has a radius R_i. Show that if the MCS has a finite eventual size (i.e., the limit of $V(t)$ is finite), then we have

$$\lim_{t \to \infty} \frac{R_i^3}{R^3} = \frac{P}{P + L}. \tag{3.102}$$

Exercise 3.6: As in the previous exercise, assume that the living cells in a MCS proliferate at a rate of P and die at a rate of L. Assume that the mass density of this MCS is $\rho(r, t)$ and the velocity field of the cells in this MCS is $\mathbf{v}(r, t) = \nabla v$. Assume the MCS has a radius R and its necrotic core has a radius R_i.

1. Use the mass conservation law to show that the density function satisfies the following equations

$$\begin{aligned} \frac{\partial \rho}{\partial t} &= -\rho L - \nabla \cdot (\rho \mathbf{v}), \quad r \in (0, R_i), \\ \frac{\partial \rho}{\partial t} &= \rho P - \nabla \cdot (\rho \mathbf{v}), \quad r \in (R_i, R). \end{aligned} \tag{3.103}$$

2. Assume that the densities of the tumor in the necrotic core and the proliferative layer are constants. Show that the conservation of mass equation become (Hint: Use the gradient formula in spherical polar coordinates which can be found in Section C.2.3 in [1].)

$$\begin{aligned} \nabla \cdot \mathbf{v} &= \frac{1}{r^2} \frac{\partial(r^2 v)}{\partial r} = -L, \quad r \in (0, R_i), \\ \nabla \cdot \mathbf{v} &= \frac{1}{r^2} \frac{\partial(r^2 v)}{\partial r} = P, \quad r \in (R_i, R). \end{aligned} \tag{3.104}$$

Show that $v(r) = \frac{1}{3} Pr - \frac{1}{3}(P + L)\frac{R_i^3}{R^2}$.

3. Observe that

$$\frac{dR(t)}{dt} = v(R(t)).$$

Show that

$$\frac{dR(t)}{dt} = \frac{1}{3} PR \left(1 - \frac{P + L}{P} \frac{R_i^3}{R^3} \right). \tag{3.105}$$

Show that if L is much smaller than P and $h = R - R_i$, then R can be approximated by $3h(P + L)/L$.

Exercise 3.7: In many situations, tumors grow around microvessels which supply them nutrient and hence take the shape of circular cylinders. This form of tumor is referred as tumor cord. Tumor cords are one of the fundamental forms of solid tumors.

Let $C(r, t)$ be the concentration of the most limiting nutrient for the growth of the tumor cells at time at radius r from the microvessel. Assume $C(r, t)$ satisfy the following growth equation

$$\frac{\partial C}{\partial t} = D\nabla^2 C - k, \tag{3.106}$$

where D is the constant diffusion coefficient and k is a constant standing for the limiting nutrient uptake rate.

1. Using cylindrical coordinate, convert equation (3.106) into an equation in terms of the radius. Let $c(r)$ be the steady state solution of the resulting equation. Show that

$$c(r) = \frac{1}{4}\frac{k}{D}r^2 + A\ln r + B.$$

2. Find the maximum outer radius of a tumor cord without necrotic cylindrical shell.

3. Find the limit of the band width of the necrotic cylindrical shell when the outer radius is large relative to it.

Exercise 3.8: Intermediate levels of nutrient may sustain tumor cells but often not enough for them to proliferate. This is often viewed as a mechanism explaining the existence of a quiescent layer of tumor cell sandwiched between the proliferative layer and the necrotic core. Formulate such a model with appropriate boundary conditions. Carry out a systematic mathematical analysis similar to the previous exercise.

Exercise 3.9: Substitution (3.26) leading to the differential equation (3.31) is due to Greenspan (see Section 3.4). Use differential equations (3.22) and (3.31) to reproduce Greenspan's figure 2, from $\tau = 0$ to $\tau = 20$. Why, biologically, do $\xi(\tau)$ and $\eta(\tau)$ vary as they do with changes in γ?

Exercise 3.10: Solve equation (3.31), with initial conditions $\xi_0 = 1$, $x_0 = 0$.

Exercise 3.11: Prove propositions (3.1) and (3.2).

Exercise 3.12: Reproduce Figure 6 in Greenspan [5].

Exercise 3.13: Reproduce Figure 2 in [15].

Exercise 3.14: Assume that the growth dynamics of an avascular cancer cell population can be described by the following reaction-diffusion equation with logistic growth form, also known as Fisher equation.

$$\frac{\partial P}{\partial t} = D\frac{\partial^2 P}{\partial x^2} + rP(1 - P). \tag{3.107}$$

1. Show that traveling solutions to (3.107) must satisfy

$$dP/dz = -S, \qquad dS/dz = \frac{r}{D}P(1 - P) - \frac{v}{D}S. \tag{3.108}$$

2. Show that if $v > 2(rD)^{1/2}$, then (3.107) has a biologically meaningful traveling solution of the form $p(x, t) = P(x - vt)$ (Hint: you need to show that the heteroclinic orbit is positive. A good reference on this is the section 18 in Kot (2001) [8]).

3. Linearizing (3.108) ahead of the wave (about $P = \theta$, $S = 0$) and assuming $P(z) - \theta = Pe^{-bz}$, $S(z) = Se^{-bz}$ to leading order. Show that in order to have $b > 0$, we must have $v > 2(aD)^{1/2}$.

Exercise 3.15: Assume that the growth dynamics of an avascular cancer cell population can be described by the following equation

$$\frac{\partial P}{\partial t} = D\frac{\partial^2 P}{\partial x^2} + rP(P - \theta)(1 - P), \quad \theta \in (0, 1). \tag{3.109}$$

1. Show that traveling solutions to (3.109) must satisfy

$$dP/dz = -S, \qquad dS/dz = \frac{r}{D}P(P - \theta)(1 - P) - \frac{v}{D}S. \tag{3.110}$$

2. Show that if $v > (rD)^{1/2}$, then (3.109) has a biologically meaningful traveling solution of the form $p(x, t) = P(x - vt)$. (Hint: you need to show that the heteroclinic orbit is positive.)

3. Linearizing (3.110) ahead of the wave (about $P = \theta$, $S = 0$) and assuming $P(z) - \theta = Pe^{-bz}$, $S(z) = Se^{-bz}$ to leading order. Show that in order to have $b > 0$, we must have $v > 2(r\theta(1 - \theta)D)^{1/2}$.

Exercise 3.16: Show that the condition

$$k \geq v - \max_{w \in [0,1]} \left\{ \frac{d}{dw}(D(w)(\alpha_0 w)) - \frac{w(1 - w)}{\alpha_0 w} \right\}$$

and

$$\alpha_0 = \frac{1}{2D(0)}\left(-(k - v) - \sqrt{(k - v)^2 - 4D(0)}\right)$$

implies that

$$k \leq v + \min_{w \in [0,1]} \left\{ \frac{D(w) + wD'(w) - D_1(1 - w)}{\sqrt{w(D(w) + wD'(w) - D_1)}} \right\}.$$

3.10 Projects

We propose some natural alternative approaches to Greenspan's [5] modeling efforts. These ideas may be pursued by students as a class project or pursued in an extensive fashion by seasoned researchers leading to possible publications.

3.10.1   Nutrient limitation induced quiescence

Develop a model in which there is only nutrient (no inhibitor), and quiescence and necrosis arise from lack of nutrient. In particular, there are two thresholds: σ_q and σ_n, with $\sigma_q > \sigma_n$. Cells become necrotic wherever $\sigma \le \sigma_n$, and living cells become quiescent wherever $\sigma \le \sigma_q$. Compare the dynamics and implications of this model with those of Greenspan's. Devise an experimental test to differentiate between these two hypotheses.

3.10.2   Inhibitor generated by living cells

Rederive and reanalyze Greenspan's model with the assumption that the inhibitor is some sort of chemical produced by living cells instead of necrotic ones. All other assumptions remain unchanged.

3.10.3   Glioblastoma growth in a Petri dish or in vivo

Study the dynamics and the traveling wave solution for the Stepien et al. model (3.61) in a disc domain. One can transform their model into an one-spatial-dimensional equation using radial symmetry.

One can also study the dynamics and the traveling wave solution for the Stepien et al. model (3.61) in vivo and assume the tumor takes the shape of a sphere. One can again transform their model into an one-spatial-dimensional equation using radial symmetry.

Moreover, one can formulate various plausible two population reaction diffusion models to probably capture additional properties of the go or grow population interaction.

3.10.4   A simple model of tumor-host interface

Gatenby et al. hypothesized that the tumor-host interface of an invasive cancer is similar to a traveling wave in which the tumor edge represents the wave front propagating into the surrounding normal tissue [4]. To see if this can be confirmed by some plausible mathematical model, they considered the following system of reaction-diffusion equations of Lotka-Volterra type. For simplicity, they considered one dominant tumor population, T, interacting with one dominant native (normal) cell population, N:

$$\frac{\partial N}{\partial t} = D_N \frac{\partial^2 N}{\partial x^2} + r_N N\left(1 - \frac{N}{K_N} - \frac{b_{NT} T}{K_N}\right),$$
$$\frac{\partial T}{\partial t} = D_T \frac{\partial^2 T}{\partial x^2} + r_T T\left(1 - \frac{T}{K_T} - \frac{b_{TN} N}{K_T}\right). \tag{3.111}$$

where r_N and r_T are maximum growth rates of normal cells and tumor cells; K_N and K_T denote the maximal normal and tumor cell densities; b_{NT}

and b_{TN} are the lumped competition terms; D_N and D_T are cellular diffusion constants.

It can be shown that the trivial steady state $(0,0)$ is unstable state and hence is biologically irrelevant.

The steady state $(K_N, 0)$ corresponds to the tumor free state. One can show that solutions with positive initial values will tend to this state if both $b_{TN}K_N/K_T > 1$ and $b_{NT}K_T/K_N < 1$. If the starting point is close to $N = K_N$, $T = 0$, only the first condition is needed.

The steady state $N = 0$, $T = K_T$ corresponds to the situation that tumor invasion with total destruction of nearby normal cells. It can be shown that solutions with positive initial values will tend to this state if both $b_{TN}K_N/K_T < 1$ and $b_{NT}K_T/K_N > 1$. If the starting point is close to $N = 0$, $T = K_T$, only the second condition is needed.

The positive steady state $(N^*, T^*) = \left(\frac{K_N - b_{NT}K_T}{1 - b_{NT}b_{TN}}, \frac{K_T - b_{TN}K_N}{1 - b_{NT}b_{TN}} \right)$ corresponds to coexistence of tumor and normal cells. The solutions tend this state if both $b_{NT}K_T/K_N < 1$ and $b_{TN}K_N/K_T < 1$. One limitation of this model is that if the carrying capacities K_N and K_T are limited only by space, this state of coexistence is biologically impossible since if both b_{TN} and b_{NT} are very small, N^* will be approximately equals to K_N and T^* will be approximately equals to K_T and hence $N^*/K_N + T^*/K_T > 1$, violating the spatial constraint rule.

It can be shown that if the tumor invasion is possible (equivalent to say that the steady state $N = 0$, $T = K_T$ is stable which in turn requires $b_{TN}K_N/K_T < 1$ and $b_{NT}K_T/K_N > 1$), the propagation speed of total tumor invasion into the tumor free state is given by

$$v_{TN} \geq 2\sqrt{r_T D_T (1 - b_{TN}K_N/K_T)}. \tag{3.112}$$

Since K_T is approximately the same as K_N, the above inequality implies that tumor invasion is likely if b_{NT} is large, and b_{TN} is small. This amounts to say that the presence of tumor has a significantly adverse effect on the normal cell population but not the other way around. On the other hand, if both $b_{TN}K_N/K_T > 1$ and $b_{NT}K_T/K_N < 1$ are true, then the normal cells can recover from tumor invasion with a speed

$$v_{NT} \geq 2\sqrt{r_N D_N (1 - b_{NT}K_T/K_N)}. \tag{3.113}$$

A key finding of this modeling effort is that tumor treatment must find ways to change some of the crucial parameters related to the tumor-host interaction in order to be successful. The usual slash and burn approach will not change the tumor-host interaction and hence will do little to reverse the tumor invasion.

There are several alternative ways this interesting research topic on tumor-host interaction can be considered. For examples, the linear diffusion terms can be replaced by the cross diffusion terms such as in the work of Sherratt and Chaplain [15] or the density dependent diffusion in Stepien et al. [17].

There are also many other possibly more realistic ways to describe the tumor and host interaction functions. It will be especially interesting to also include explicitly a limiting nutrient which is required by both tumor and normal cells.

References

[1] Britton NF: *Essential Mathematical Biology.* London: Springer, 2003.

[2] Burton AC: Rate of growth of solid tumours as a problem of diffusion. *Growth* 1966, 30:157–176.

[3] Edelstein-Keshet L: *Mathematical Models in Biology.* Philadelphia: SIAM, 2005.

[4] Gatenby RA, Maini PK, Gawlinski ET: Analysis of tumor as an inverse problem provides a novel theoretical framework for understanding tumor biology and therapy. *Appl Math Lett* 2002, 15:339–345.

[5] Greenspan HP: Models for the growth of a solid tumor by diffusion. *Stud Appl Math* 1972, 52:317–340.

[6] Greenspan HP: On the growth and stability of cell cultures and solid tumors. *J Theor Biol* 1976, 56:229–242.

[7] Haddow A: The biological characters of spontaneous tumours of the mouse, with special reference to the rate of growth. *J Path Bact* 1938, 47:553–565.

[8] Kot M: *Elements of Mathematical Ecology.* Cambridge UK: Cambridge University Press, 2001.

[9] Laird AK: Dynamics of tumor growth. *Brit J Cancer* 1964, 18:490–502.

[10] Laird AK: Dynamics of tumour growth: Comparison of growth rates and extrapolation of growth curve to one cell. *Brit J Cancer* 1965, 19:278–291.

[11] Logan JD: *An Introduction to Nonlinear Partial Differential Equations.* Hoboken, New Jersey: Wiley, 2008.

[12] Martirosyan NL, Rutter EM, Ramey WL, Kostelich EJ, Kuang Y, Preul MC: Mathematically modeling the biological properties of gliomas: A review. *Math Biosc Eng* 2015, 12:879–905.

[13] Mayneord WV: On a law of growth of Jensen's rat sarcoma. Am. J. Cancer 1932, 16:841–846.

[14] Sánchez-Garduño F, Maini PK, Pérez-Velásquez J: A non-linear degenerate equation for direct aggregation and traveling wave dynamics. *Discrete Cont Dyn-B* 2010, 138:455–487.

[15] Sherratt JA, Chaplain MAJ: A new mathematical model for avascular tumour growth. *J Math Biol* 2001, 43:291–312.

[16] Stein AM, Demuth T, Mobley D, Berens M and Sander LM: A mathematical model of glioblastoma tumor spheroid invasion in a three-dimensional in vitro experiment. *Biophy J* 2007, 92:356–365.

[17] Stepien TL, Rutter EM, Kuang K: 2015. A data-motivated density-dependent diffusion model of in vitro glioblastoma growth. *Math Biosc Eng* 2015, 12:1157–1172.

[18] Thomlinson RH, Gray LH: The histological structure of some human lung cancers and possible implications for radiotherapy. *Br J Cancer* 1955, 9:539–549.

Chapter 4

Physiologically Structured Tumor Growth

4.1 Introduction

In this chapter we further explore the hypothesis explaining Gompertz-like tumor growth kinetics proposed by Gyllenberg and Webb. They extended their original quiescence model (section 2.4) to a more sophisticated framework that increases the realism beyond what simple models from Chapter 2 can handle. This new framework also introduces an important modeling feature—physiological stage structure (see [9] for an introduction to such models; detailed theory can be found in [4, 5]). These models, like the MCS models of Chapter 3, are based on partial differential equations.

Recall that Gyllenberg and Webb's original model (2.64, section 2.4) assumes that, besides being proliferative or quiescent, all cells are otherwise identical [7]. In a follow-up paper [8] they relax this assumption and allow tumor cells to vary in size and cell cycle progression.

In the type of model Gyllenberg and Webb used, one studies an object of the form,

$$P(x, y, t) = \int_x^y p(s, t) \, \mathrm{d}s, \qquad (4.1)$$

where $P(x, y, t)$ is interpreted as the number of cells between sizes (alternatively stages or ages) x and y at time t. The density function $p(s, t)$ is defined on some physiological domain, S, of possible sizes or stages. Intuitively, we tend to equate the biological and mathematical concepts of "density" by calling $p(s, t)$ cell density. However, at a point $s \in S$, "cell density" $p(s, t)$ has no intuitive meaning. It simply generates the population size via integration. Also, at this point we avoid the temptation to make the common claim that $p(s, t) \, \mathrm{d}s$ is the number of cells on the interval $(s, s + \mathrm{d}s)$, which is really correct only approximately. That approximation will be made explicit below.

Gyllenberg and Webb interpret $P(x, y, t)$ as the number of physiologically active (growing or proliferating) cells with a size between x and y in a tumor at time t. Similarly, they define the number of quiescent cells (neither growing nor proliferating) between u and v at time t to be

$$Q(u, v, t) = \int_u^v q(s, t) \, \mathrm{d}s. \qquad (4.2)$$

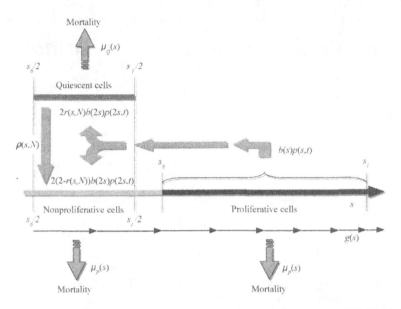

FIGURE 4.1:   Schematic representation of the size structured Gyllenberg-Webb model. Function definitions and explanations are given in the text.

The total number of active and quiescent cells is therefore

$$P(t) = \int_0^\infty p(s,t)\,\mathrm{d}s \quad \text{and} \quad Q(t) = \int_0^\infty q(s,t)\,\mathrm{d}s, \qquad (4.3)$$

respectively, and the total cell population is

$$N(t) := P(t) + Q(t). \qquad (4.4)$$

4.2　Construction of the cell-size structured model

Our goal here is to motivate Gyllenberg and Webb's size structured model in an intuitive way. The theory of physiologically structured models like this one is explored in detail in a number of excellent sources [2, 3, 9, 11]. We develop the theory in some detail, focusing on a biological way of thinking, because this very important and useful class of models has largely gone unnoticed by the medical modeling community.

In biology, population-level models are almost always built from assumptions on the behavior of individuals composing them, although often these assumptions are implicit. The population level model therefore represents some sort of average, either an average across individuals in a population or across repeated runs of an implicit stochastic process, what is referred to by physicists as the ensemble average. Here we make these assumptions explicit. In particular, each individual cell is assigned a particular value of an attribute we will call "size," $s(t)$, at time t. Although we call it size, s could represent biomass, amount of a critical nutrient, critical macromolecule, position in the cell cycle or any other continuous measure of maturity. Whatever it represents, s is the *structuring variable*, making this a *physiologically structured model*. In this setting we construct two models: one for cell growth and another for proliferation. The model is presented schematically in Fig. 4.1.

Cell size, s, is limited to a subset of $\mathbb{R}^+$ as a consequence of two assumptions: (1) new cells arise only from division of existing cells, and (2) cells can divide only if they are large enough; that is, the probability of division is positive only for $s \in (s_0, s_1]$, where s_1 is finite. Cells may grow larger than s_1, but if they do the model loses track of them because it is further assumed that cells cannot shrink and thereby reenter the proliferative class. Furthermore, cells cannot divide twice without a growth phase in between; in other words, we assume that $s_1/2 < s_0$. Finally, cell division is symmetrical. Therefore, the natural domain of the cell growth model is $s \in [(1/2)s_0, s_1]$, and we can limit

$$N(t) = \int_{\frac{1}{2}s_0}^{s_1} p(s,t)\,\mathrm{d}s + \int_{\frac{1}{2}s_0}^{\frac{1}{2}s_1} q(s,t)\,\mathrm{d}s = P(t) + Q(t). \qquad (4.5)$$

Active cell growth is modeled by the continuously differentiable function $g(s)$, $s \in [(1/2)s_0, s_1]$. Function $g(s)$ defines the per capita growth rate of active cells of size s, so

$$\frac{\mathrm{d}s(t)}{\mathrm{d}t} = g(s). \qquad (4.6)$$

We assume that $g(s)$ is almost always positive (active cells always grow).

Active cell proliferation is governed by a function, $b(s)$, $s \in [(1/2)s_0, s_1]$, assumed to be continuous, positive on the interval (s_0, s_1), but 0 elsewhere on its domain; therefore, $b(s_0) = b(s_1) = 0$.

When a cell divides, its daughters have a certain probability of entering quiescence instead of the active class. This probability depends on the mother cell's size and the total number of cells in the tumor. In particular, transition to quiescence is governed by a continuous function $r(s, N)$, $s \in [(1/2)s_0, (1/2)s_1]$, $N \in [0, \infty)$. The function $r(s, N)$ is interpreted as the average number of daughter cells that become quiescent. Therefore,

$$r : [(1/2)s_0, (1/2)s_1] \times [0, \infty) \mapsto [0, 2).$$

Note that by assumption the probability of entering quiescence is never 1.

Quiescent cells neither grow nor divide. They may, however, reactivate their growth programs, which is assumed to occur at rate $\rho(s, N)$. The reactivation function ρ is continuous on the interval $[(1/2)s_0, (1/2)s_1]$. Both active and quiescent cells die at rates $\mu_P(s)$ and $\mu_Q(s)$, respectively. Both μ_P and μ_Q are continuous and nonnegative on $[(1/2)s_0, s_1]$ and $[(1/2)s_0, (1/2)s_1]$, respectively.

Consider $\hat{s} \in (s_0/2, s_1/2)$ and a tiny interval $[\hat{s} - \Delta s/2, \hat{s} + \Delta s/2]$, $\Delta s \ll 1$. To model the proliferative cell time dynamics on this interval of length Δs centered on

$\hat{s}$ we start with the cell conservation law on a tiny time interval, $\Delta t \ll 1$:

$$\begin{pmatrix} \text{Cells at time} \\ t + \Delta t \end{pmatrix} - \begin{pmatrix} \text{Cells at} \\ \text{time } t \end{pmatrix} = \begin{pmatrix} \text{Cells entering} \\ \text{at age } \hat{s} - \Delta s/2 \end{pmatrix} - \begin{pmatrix} \text{Cells leaving} \\ \hat{s} + \Delta s/2 \end{pmatrix} -$$

$$\begin{pmatrix} \text{Cells leaving} \\ \text{for division} \end{pmatrix} + \begin{pmatrix} \text{reactivation} \\ \text{from quiescence} \end{pmatrix} + \text{Births} - \text{Deaths}. \tag{4.7}$$

It is easy to see that

$$\begin{pmatrix} \text{Cells at time} \\ t + \Delta t \end{pmatrix} - \begin{pmatrix} \text{Cells at} \\ \text{time } t \end{pmatrix} = \int_{\hat{s}-\frac{1}{2}\Delta s}^{\hat{s}+\frac{1}{2}\Delta s} [p(s, t + \Delta t) - p(s, t)] \, ds. \tag{4.8}$$

From the assumptions above one can see that cells enter and leave our size interval in four ways and their quantities in this tiny time interval can be expressed accordingly:

1. *Cells entering at age $\hat{s} - \Delta s/2$:*

$$\int_t^{t+\Delta t} g(\hat{s} - \Delta s/2)p(\hat{s} - \Delta s/2, \tau) \, d\tau. \tag{4.9}$$

2. *Cells leaving at age $\hat{s} + \Delta s/2$:*

$$\int_t^{t+\Delta t} g(\hat{s} + \Delta s/2)p(\hat{s} + \Delta s/2, \tau) \, d\tau. \tag{4.10}$$

3. *Cells "leaving" via division:*

$$\int_t^{t+\Delta t} \int_{\hat{s}-\frac{1}{2}\Delta s}^{\hat{s}+\frac{1}{2}\Delta s} b(s)p(s, \tau) \, ds d\tau \tag{4.11}$$

4. *Reactivation from quiescence:*

$$\int_t^{t+\Delta t} \int_{\hat{s}-\frac{1}{2}\Delta s}^{\hat{s}+\frac{1}{2}\Delta s} \rho(s, N)q(s, \tau) \, ds d\tau. \tag{4.12}$$

5. *Birth or proliferation (Cell division).* Because cell division is symmetrical, new cells are precisely half their mothers' size. Therefore, the amount of new cells "born" into the interval $s \in [\hat{s} - \Delta s/2, \hat{s} + \Delta s/2]$ from the interval $\bar{s} \in [2\hat{s} - \Delta s, 2\hat{s} + \Delta s]$ in the period of $[t, t + \Delta t]$ is

$$\int_t^{t+\Delta t} \int_{2\hat{s}-\Delta s}^{2\hat{s}+\Delta s} b(\bar{s})p(\bar{s}, \tau) \, d\bar{s}d\tau$$

$$= 2 \int_t^{t+\Delta \tau} \int_{\hat{s}-\frac{1}{2}\Delta s}^{\hat{s}+\frac{1}{2}\Delta s} b(2s)p(2s, \tau) \, ds d\tau. \tag{4.13}$$

Intuitively, the leading factor of 2 represents a "compression" effect; mother cells in a given size interval contribute daughters to a size interval exactly half as large. Alternatively, think of a monolayer of cells completely covering the surface of a culture dish and imagine all the cells dividing at the same moment. Then instantly the cell density (number of cells per unit surface area) would double. Upon division, new cells immediately "choose" whether or not to enter quiescence. The number that return to the active class is, by definition of $r(s, N)$,

$$\int_t^{t+\Delta \tau} \int_{\hat{s}-\frac{1}{2}\Delta s}^{\hat{s}+\frac{1}{2}\Delta s} 2(2 - r(s, N))b(2s)p(2s, \tau)\, ds d\tau, \tag{4.14}$$

while

$$\int_t^{t+\Delta \tau} \int_{\hat{s}-\frac{1}{2}\Delta s}^{\hat{s}+\frac{1}{2}\Delta s} 2r(s, N)b(2s)p(2s, \tau)\, ds d\tau \tag{4.15}$$

become quiescent. Note that equations (4.9)–(4.15) hold for all $\hat{s} \in ((1/2)s_0, s_1)$.

6. *Death.* Dying cells leave the interval as follows:

$$\int_t^{t+\Delta t} \int_{\hat{s}-\frac{1}{2}\Delta s}^{\hat{s}+\frac{1}{2}\Delta s} \mu_p(s)p(s, \tau)\, ds d\tau. \tag{4.16}$$

Using the above expressions, we can restate the conservation law (4.8) as follows:

$$\int_{\hat{s}-\frac{1}{2}\Delta s}^{\hat{s}+\frac{1}{2}\Delta s} [p(s, t + \Delta t) - p(s, t)]\, ds = \int_t^{t+\Delta t} g(\hat{s} - \Delta s/2)p(\hat{s} - \Delta s/2, \tau)\, d\tau$$

$$- \int_t^{t+\Delta t} g(\hat{s} + \Delta s/2)p(\hat{s} + \Delta s/2, \tau)\, d\tau - \int_t^{t+\Delta t} \int_{\hat{s}-\frac{1}{2}\Delta s}^{\hat{s}+\frac{1}{2}\Delta s} b(s)p(s, \tau)\, ds d\tau$$

$$+ \int_t^{t+\Delta t} \int_{\hat{s}-\frac{1}{2}\Delta s}^{\hat{s}+\frac{1}{2}\Delta s} \rho(s, N)q(s, \tau)\, ds d\tau$$

$$+ \int_t^{t+\Delta \tau} \int_{\hat{s}-\frac{1}{2}\Delta s}^{\hat{s}+\frac{1}{2}\Delta s} 2(2 - r(s, N))b(2s)p(2s, \tau)\, ds d\tau$$

$$- \int_t^{t+\Delta t} \int_{\hat{s}-\frac{1}{2}\Delta s}^{\hat{s}+\frac{1}{2}\Delta s} \mu_p(s)p(s, \tau)\, ds d\tau. \tag{4.17}$$

Dividing both sides by $\Delta s \Delta t$ and taking the limit as $\Delta s, \Delta t \to 0$ yields the model for proliferative cells:

$$\frac{\partial}{\partial t}p(s, t) + \frac{\partial}{\partial s}[g(s)p(s, t)] = 2(2 - r(s, N(t)))b(2s)p(2s, t)$$

$$+ \rho(s, N(t))q(s, t) - b(s)p(s, t) \tag{4.18}$$

$$- \mu_P(s)p(s, t),$$

which holds for all $s \in (s_0/2, s_1)$ and $t > 0$. A similar argument leads to the following model for quiescence:

$$\frac{\partial}{\partial t}q(s, t) = 2r(s, N(t))b(2s)p(2s, t) - \mu_Q(s)q(s, t) - \rho(s, N(t))q(s, t), \tag{4.19}$$

for $s \in (s_0/2, s_1/2)$ and $t > 0$. To preserve cell conservation, this model adopts the left-hand boundary condition that

$$p\left(\frac{s_0}{2}, t\right) = 0 \qquad (4.20)$$

for all $t > 0$. Finally we specify the initial distributions,

$$p(s, 0) = \phi(s), \ s \in (s_0/2, s_1), \qquad (4.21)$$

and

$$q(s, 0) = \psi(s), \ s \in (s_0/2, s_1/2). \qquad (4.22)$$

4.3 No quiescence, some intuition

Model (4.18)–(4.22) is a natural extension of models originally proposed independently by a number of other authors [1, 10]. These foundational models differ only in having no quiescence, so they are recovered by setting ρ and r identically to zero. Metz and Diekmann [9] analyze this simplification in an intuitively clear way, so the following development is largely due to them.

A concept similar to R_0 in epidemiology and pathology models can be developed for this type of model. In this context, R_0 is the number of viable new pathogens (or infections) produced by an existing pathogen (or infected person). An infection is therefore viable—the disease-free equilibrium is unstable—when $R_0 > 1$. The analogous concept here is the number of daughter cells an existing cancer cell produces that survive to reproduce. We use π_0 to represent this analogue to R_0.

One obtains π_0 with the following intuitive argument. Imagine a cell of size s_0, the minimum size at division. We ask, how many of this cell's daughters will themselves reach s_0? Intuitively, this number depends on the probability of three events: event (A), the mother cell survives long enough to grow from size s_0 to some $s \in (s_0, s_1]$; event (B), the mother cell divides at this size s; and event (C), the daughters survive long enough to grow from size $s/2$ to s_0. Therefore, π_0 is the sum (integral) over all $s \in (s_0, s_1]$ of

$$2(\Pr\{C|A, B\} \Pr\{B|A\} \Pr\{A\}),$$

where $\Pr\{X|Y\}$ is the probability of event X given event Y. The factor of 2 represents the fact that each mother splits into two daughters.

1. $\Pr\{A\}$. Consider a cohort of cells all of size s_0 at time t_0. Let $n(t)$ be the number of cells in this cohort at time $t \geq t_0$. No new cells are born into this cohort, and cells can leave in only two ways: death, at rate $\mu_P(s)$; and division, at rate $b(s)$. Therefore, this cohort's dynamics is described by the following system of differential equations:

$$\frac{dn}{dt} = -(\mu_P(s(t)) + b(s(t)))n(t), \quad \frac{ds}{dt} = g(s(t)), \qquad (4.23)$$

with $g(t_0) = s_0$ and $n(t_0) = n_0$. Solving for $n(t)$ yields

$$n(t) = n_0 \exp\left(-\int_{t_0}^{t} [\mu_P(s(\tau)) + b(s(\tau))] \, d\tau\right). \qquad (4.24)$$

Substituting $\xi = s(\tau)$ in the right-hand-side of (4.24) yields

$$n_0 \exp\left(-\int_{s_0}^{s} \frac{\mu_P(\xi) + b(\xi)}{g(\xi)} \, d\xi\right). \tag{4.25}$$

Formula (4.25) gives us the number of cells in the cohort that reach size s before dying or dividing, regardless of when they get there. Therefore,

$$\exp\left(-\int_{s_0}^{s} \frac{\mu_P(\xi) + b(\xi)}{g(\xi)} \, d\xi\right) \tag{4.26}$$

can be interpreted as the probability that a mother cell neither dies nor divides between sizes s_0 and $s \in (s_0, s_1]$.

2. $\Pr\{B|A\}$. Let

$$\int_{a}^{b} \chi(s) \, ds, \qquad a \geq s_0, \tag{4.27}$$

be the *fraction* of cells that disappear (either die or divide) between sizes a and b. Therefore,

$$n_0 \exp\left(-\int_{s_0}^{s} \frac{\mu_P(\xi) + b(\xi)}{g(\xi)} \, d\xi\right) = n_0 \left(1 - \int_{s_0}^{s} \chi(s') \, ds'\right). \tag{4.28}$$

Differentiating both sides gives us

$$\chi(s) = \frac{\mu_P(s) + b(s)}{g(s)} \exp\left(-\int_{s_0}^{s} \frac{\mu_P(\xi) + b(\xi)}{g(\xi)} \, d\xi\right). \tag{4.29}$$

We can roughly interpret $\chi(s)\Delta s$, $\Delta s \ll 1$, as the joint probability per unit size that a cell in our cohort successfully grows from size s_0 to size s, and having done so, either divides or dies essentially "at s." Notice that, by formula (4.26), the exponential term in (4.29) is the former probability. Therefore, we can interpret

$$\frac{\mu_P(s) + b(s)}{g(s)}$$

as the conditional probability that a cell disappears at size s. Since "disappearing" includes two disjoint events—a cell cannot both die and divide—we can partition $\chi(s)$ into

$$\chi(s) = \chi_\mu(s) + \chi_b(s)$$
$$= \frac{\mu_P(s)}{g(s)} \exp\left(-\int_{s_0}^{s} \frac{\mu_P(\xi) + b(\xi)}{g(\xi)} \, d\xi\right) +$$
$$\frac{b(s)}{g(s)} \exp\left(-\int_{s_0}^{s} \frac{\mu_P(\xi) + b(\xi)}{g(\xi)} \, d\xi\right),$$

where $\chi_\mu(s)$ and $\chi_b(s)$ are the probabilities of growing from size s_0 to s and then dying or dividing at s, respectively. Therefore, the conditional probability we seek is

$$\frac{b(s)}{g(s)} \Delta s.$$

3. $\Pr\{C|A, B\}$. Using an argument similar to the one for $\Pr\{A\}$ above, we find that a newly minted cell of size $s/2$ survives to the reproductive threshold s_0 with probability

$$\exp\left(-\int_{\xi/2}^{s_0} \frac{\mu_P(\eta)}{g(\eta)}\, d\eta\right). \tag{4.30}$$

Putting all these arguments together and integrating for all $\xi \in (s_0, s_1)$ yields

$$\pi_0 = 2\int_{s_0}^{s_1} \frac{b(\xi)}{g(\xi)} \exp\left(-\int_{\xi/2}^{\xi} \frac{\mu_P(\eta) + b(\eta)}{g(\eta)}\, d\eta\right) d\xi. \tag{4.31}$$

We now have π_0, our analog of R_0 for tumor growth, which bears the same relation to viability its analog does; namely, $\pi_0 > 1$ implies tumor growth, and the reverse inequality implies tumor regression.

We now ask what perhaps may seem a surprising question. Suppose we wanted to stabilize this tumor so that it neither grew nor regressed. At what (constant) rate would we have to add or remove cells on the interval $((1/2)s_0, s_1)$ to achieve such a stabilization? In other words, if the tumor is growing, how fast do we have to kill cells to reach the "break-even" point in which every cell is replaced by exactly one of its daughters (or how fast would we have to add cells if the tumor is regressing)?

To this end, we define $\sigma \in (-\infty, \infty)$ as this fictitious alteration rate of the reproductive cell population. In particular, we define σ such that $\sigma > 0$ is the rate at which we remove cells, and $\sigma < 0$ is the rate at which cells are added. Therefore, σ is added as a constant per capita loss (if positive) or source (if negative) term on the right-hand-side of equation (4.18). Repeating the computations above with σ added alters equation (4.31) to

$$\pi(\sigma) = 2\int_{s_0}^{s_1} \frac{b(\xi)}{g(\xi)} \exp\left(-\int_{\xi/2}^{\xi} \frac{\sigma + \mu_P(\eta) + b(\eta)}{g(\eta)}\, d\eta\right) d\xi, \tag{4.32}$$

and so $\pi(0) = \pi_0$. The alteration rate we seek therefore satisfies the characteristic equation

$$\pi(k) = 1; \tag{4.33}$$

that is, when $\sigma = k$, we have achieved the "break-even" alteration rate, assuming such a (real) solution exists (see exercises). Here is the payoff: under certain conditions, k is the **Malthusian parameter**, or what ecologists call the **intrinsic growth rate** of the tumor cell population. In particular,

$$k > 0 \Leftrightarrow \pi_0 > 1 \Rightarrow \text{the tumor is growing.}$$

(See Theorem 4.1 below for a restriction of this interpretation.)

4.4 Basic behavior of the model

Ultimately, we want an expression for $p(s, t)$. Experience with the theory of differential equations leads us to suggest a test solution of the form,

$$p(s, t) = \phi(s)e^{kt}, \tag{4.34}$$

where k solves the characteristic equation (4.33). If this form is correct, then

$$\frac{\partial}{\partial t} p(s, t) = k\phi(s)e^{kt}. \qquad (4.35)$$

Therefore, our model (equation (4.18), still assuming $r = \rho = 0$) becomes

$$\frac{\partial}{\partial s} g(s)\phi(s) = 4b(2s)\phi(2s) - b(s)\phi(s) - \mu_P(s)\phi(s) - k\phi(s). \qquad (4.36)$$

On the interval $(s_1/2, s_1]$ there is no production of new cells, so $4b(2s)\phi(2s) \equiv 0$ here. On this interval we can solve the differential equation (4.36) to obtain

$$\phi(s) = \frac{C}{g(s)} \exp\left(-\int_{\frac{1}{2}s_1}^{s} \frac{k + \mu_P(\xi) + b(\xi)}{g(\xi)} \, d\xi\right), \qquad (4.37)$$

where C is an integration constant which from here on we set to 1 without loss of generality.

On the remainder of the domain, $(s_0/2, s_1/2]$, we have to account for production of new cells. But now we have an expression for $\phi(s) \in (s_1/2, s_1)$. From boundary condition (4.20), we also have $\phi(s_0/2) = 0$. So we can solve model (4.36) as an initial value problem on $(s_0/2, s_1/2]$, which yields

$$\phi(s) = \frac{\lambda(s)}{g(s)} \exp\left(-\int_{\frac{1}{2}s_0}^{s} \frac{k + \mu_P(\xi) + b(\xi)}{g(\xi)} \, d\xi\right), \qquad (4.38)$$

$$\lambda(s) = 2\int_{s_0}^{2s} \frac{b(\xi)}{g(\xi)} \exp\left(-\int_{\frac{1}{2}\xi}^{\xi} \frac{k + \mu_P(\eta) + b(\eta)}{g(\eta)} \, d\eta\right) d\xi. \qquad (4.39)$$

Now we see mathematically why $\pi(k) = 1$; with the boundary condition $\phi(s_0/2) = 0$, the condition $\pi(k) = 1$ is needed to make the equation for $\phi(s)$ continuous at $s = (1/2)s_1$. Also, equation (4.53) represents the solution on the entire interval $(s_0/2, s_1)$ as long as we replace relation (4.54) with

$$\lambda(s) = \begin{cases} 2\int_{s_0}^{2s} \frac{b(\xi)}{g(\xi)} \exp\left(-\int_{\frac{1}{2}\xi}^{\xi} \frac{k + \mu_P(\eta) + b(\eta)}{g(\eta)} \, d\eta\right) d\xi \; ; \; s \in \left[\frac{s_0}{2}, \frac{s_1}{2}\right], \\ 1 \qquad\qquad\qquad\qquad\qquad\qquad ; \; s \in \left(\frac{s_1}{2}, s_1\right]. \end{cases} \qquad (4.40)$$

Example 4.1
Suppose $g(s) \equiv 1$, $\mu_P(s) \equiv 0$ and $s_0 = 3$ and $s_1 = 5$. Let

$$b(s) = \begin{cases} 2(s-3)(5-s) \; ; \; 3 \le s \le 5, \\ 0 \qquad\qquad ; \; \text{elsewhere}, \end{cases} \qquad (4.41)$$

and suppose there is no quiescence. Find π_0.

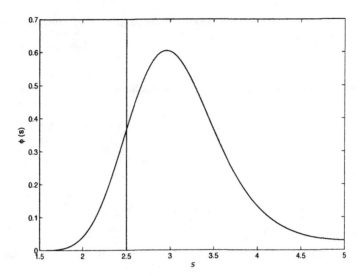

FIGURE 4.2:   Cell size distribution $\phi(s)$ satisfying equation (4.34) for example 4.1. Vertical line marks $s_1/2$; to the right, $\lambda(s) \equiv 1$, and to the left, $\lambda(s)$ is calculated with expression (4.52).

SOLUTION: From equation (4.31),

$$\pi_0 = 2 \int_3^5 2(s-3)(5-s) \exp\left(-\int_{s/2}^s b(\eta)\,d\eta\right) ds. \tag{4.42}$$

$$= 2 \int_3^5 2(s-3)(5-s) \exp\left(-\int_3^s 2(\eta-3)(5-\eta)\,d\eta\right) ds, \tag{4.43}$$

because cells divide only on the interval $s \in [3,5]$. Therefore,

$$\pi_0 = 2 \int_3^5 2(s-3)(5-s) \exp\left(\frac{2}{3}s^3 - 8s^2 + 30s - 36\right) ds \tag{4.44}$$

$$= 2(1 - e^{-8/3}) \tag{4.45}$$

$$\approx 1.8610. \tag{4.46}$$

Since $\pi_0 > 1$, the tumor must be growing. Of course, this result is in no way surprising given that there is no cell death.   ☐

Example 4.2
Once again let $g(s) \equiv 1$, $s_0 = 3$, $s_1 = 5$ and $b(s)$ be equation (4.41), but now suppose cell death rate $\mu_P \equiv 0.2$. Find the tumor cell intrinsic rate of increase (Malthusian parameter) k.

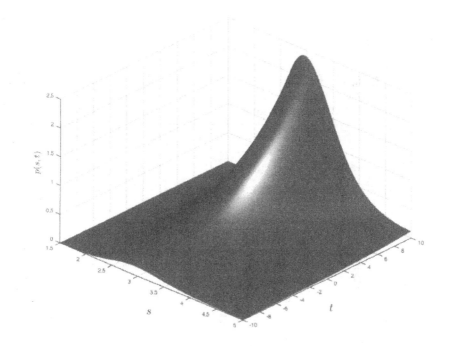

FIGURE 4.3:   The solution (4.34) in example 4.1 on the time interval $[-10, 10]$, using $\phi(s)$ as graphed in Figure 4.2.

SOLUTION: From the definition of k and equations (4.32) and (4.33), we have

$$1 = 2 \int_3^5 2(s-3)(5-s) \exp\left(-\int_{s/2}^s k + 0.2 + b(\eta)\, d\eta\right)\, ds. \qquad (4.47)$$

As before, since cells only divide on the interval $s \in [3,5]$, the integral in the exponential function is naturally partitioned as follows:

$$\int_{s/2}^s k + 0.2 + b(\eta)\, d\eta = \int_{s/2}^s k + 0.2\, d\eta + \int_{s_0}^s 2(\eta-3)(5-\eta)\, d\eta \qquad (4.48)$$

$$= -\frac{2}{3}s^3 + 8s^2 + \left(\frac{k}{2} - 29.9\right)s + 36. \qquad (4.49)$$

Therefore, k satisfies

$$1 = 2 \int_3^5 2(s-3)(5-s) \exp\left[\frac{2}{3}s^3 - 8s^2 - \left(\frac{k}{2} - 29.9\right)s - 36\right]\, ds. \qquad (4.50)$$

From here one can obtain a numerical estimate of k using equation (4.50), as can be done directly in Maple, for example. Such a numerical solution places $k \approx 0.1378$. □

Continuing with this example, we can now find $\phi(s)$. From (4.53) and the value of k just obtained we have that, on the interval $(s_1/2, s_1]$,

$$\phi(s) = \exp\left[\frac{2}{3}s^3 - 8s^2 + 29.8311s - 36\right], \qquad (4.51)$$

because on this interval $\lambda(s) = 1$. To the left, on the interval $[s_0/2, s_1/2]$,

$$\lambda(s) = 2 \int_3^{2s} 2(\xi-3)(5-\xi) \exp\left(0.5833\,\xi^3 - 6\,\xi^2 + 14.8311\,\xi\right)\, d\xi, \qquad (4.52)$$

which can be evaluated numerically, for example using the trap function in MATLAB®. Therefore, $\phi(s)$ on the interval $[s_0/2, s_1/2]$ is recovered by multiplying the r.h.s. of (4.51) by the value calculated for $\lambda(s)$. This distribution is shown in Figure 4.2, and the solution (4.34) for $t \in [-10, 10]$ is shown in Figure 4.3.

In a simple linear ODE model with no feedback between growth rate and population size, exponential growth or decay is the rule. Note that $\mu_p(s)$ and $b(s)$ do not depend on N; therefore, this structured model is linear in that sense—death and birth rates may be nonlinear functions of s, but they are linear functions of total population size. So, intuitively we expect exponential growth or decay, and that is precisely what we see except for one surprising exception. We state this exception in the following theorem. Details can be found in [6].

THEOREM 4.1
In model (4.18) − −(4.22), let $\rho, r, \psi(s) \equiv 0$ and $\phi(s) > 0$. If cell growth is not exponential—that is, $2g(s) \neq g(2s)$ for some $s \in [s_0/2, s_1]$—then the total population growth $N(t)$ is exponential; that is, $N(t) = N_0 e^{kt}$, where k satisfies

the characteristic equation (4.33). If $2g(s) = g(2s)$ for all $s \in [s_0/2, s_1]$, then total population growth is not exponential.

In other words, in this simple case *tumors grow exponentially if and only if their constituent cells do not grow (physically get larger) exponentially.* If cells do grow exponentially, then tumor growth rate oscillates as cohorts of cells all grow together in the same clump, and we lose touch with k as the "intrinsic" growth rate. (See Fig. 3 in reference [8].)

This "cohort effect" causing oscillating growth rate under exponential cell growth is obliterated by quiescence. In the simplest case in which quiescence is linear— i.e., the rate at which cells enter quiescence does not depend on tumor size N— growth synchrony is deranged as some members of the cohort enter quiescence at every cell division, and then reenter at some later time independent of the size of their proliferating peers. So, growth rate oscillations decay, and tumors approach exponential growth asymptotically. (See e.g., Fig. 4 in reference [8].)

If the rate at which cells become quiescent does depend on tumor size, then a variety of behaviors are possible. Simple exponential growth is still possible, as is a Gompertzian (sigmoidal) growth form with a constant fraction of proliferating cells. Gompertzian growth can also be recovered in a nonlinear version of the necrosis model, in which "quiescence" is a permanent state. In this case, however, the Gompertzian form is artificial—the tumor mass levels off when all cells become necrotic. As a final example, if cells tend to leave quiescence when they are large or the tumor is small, then a transient sigmoidal growth form precedes a slow, but unbounded, growth phase. Details of these examples are presented by Gyllenberg and Webb [8], as are the conditions leading to each. These conditions are closely related to the concepts presented in this chapter, although the techniques applied are more advanced.

4.5 Exercises

Exercise 4.1: Solve model (4.36) as an initial value problem on $(s_0/2, s_1/2]$, and show that

$$\phi(s) = \frac{\lambda(s)}{g(s)} \exp\left(-\int_{\frac{1}{2}s_0}^{s} \frac{k + \mu_P(\xi) + b(\xi)}{g(\xi)} \, d\xi\right), \qquad (4.53)$$

$$\lambda(s) = 2\int_{s_0}^{2s} \frac{b(\xi)}{g(\xi)} \exp\left(-\int_{\frac{1}{2}\xi}^{\xi} \frac{k + \mu_P(\eta) + b(\eta)}{g(\eta)} \, d\eta\right) d\xi. \qquad (4.54)$$

Exercise 4.2: Use MATLAB to reproduce Figures 4.2 and 4.3.

Exercise 4.3: Gyllenberg and Webb suggest the following example for the model with quiescence [equations (4.18) and (4.19)]: Let $s_0 = 3$, $s_1 = 5$, $g(s) \equiv 1$, $\mu_p(s) \equiv 0$, $\mu_Q = 0$ and $b(s)$ be defined by equation 4.41. Further, suppose cells enter the quiescent state at some constant positive rate ($r(s, N) \equiv r > 0$), but if so, they never return to a proliferative state ($\rho(s, N) \equiv 0$). In this example, we interpret "quiescence" as necrosis because cells that enter that state never leave. Derive the correct characteristic equation for π_0, analogous to equation (4.31), and use it to confirm that if $r = 1.1$, then $k \approx 0.096$ and if $r = 0.7$ that $k \approx 0.103$.

Exercise 4.4: Let
$$\phi(s) = 100(s - 1.5)(5 - s).$$
Use this and the results from the previous question to reconstruct Figs. 1 and 2 in [8].

Exercise 4.5: Prove that there exists a unique solution to the characteristic equation (4.33).

References

[1] Bell GI, Anderson EC: Cell growth and division I. A mathematical model with applications to cell volume distributions in mammalian suspension cultures. *Biophys J* 1967, 7:329–351.

[2] de Roos AM: A gentle introduction to physiologically structured population models. In: Tuljapurkar SD, Caswell H, editors. *Structured Population Models in Marine, Terrestrial and Freshwater Ecosystems*. New York: Chapman and Hall, 1997. p. 119–204.

[3] Diekmann O, Gyllenberg M, Metz JAJ, Thieme HR: The 'cumulative' formulation of (physiologically) structured population models. In: Clément Ph, Lumer G, editors. *Evolution Equations, Control Theory and Biomathematics, Lecture Notes in Pure and Applied Mathematics*. Berlin: Springer Verlag, 1994. p. 145–154.

[4] Diekmann O, Gyllenberg M, Metz JAJ, Thieme HR: On the formulation and analysis of general deterministic structured population models. I. Linear theory. *J Math Biol* 1998, 36:349–388.

[5] Diekmann O, Gyllenberg M, Huang H, Kirkilionis M, Metz, JAJ, Thieme HR: On the formulation and analysis of general deterministic structured population models. II. Nonlinear theory. *J Math Biol* 2001, 43:157–189.

[6] Greiner G, Nagel R: Growth of cell populations via one-parameter semigroups of positive operators. In: *Mathematics Applied to Science*. Boston: Academic Press, 1988. p. 79–105

[7] Gyllenberg M, Webb GF: Quiescence as an explanation of Gompertzian tumor growth. *Growth Dev Aging* 1989, 53:25–33.

[8] Gyllenberg M, Webb GF: A nonlinear structured population model of tumor growth with quiescence. *J Math Biol* 1990, 28:671–694.

[9] Metz JAJ, Diekmann O: *The dynamics of physiologically structured populations*. Berlin: Springer, 1986.

[10] Sinko JW, Streifer W: A new model of age-size structure of a population. *Ecol* 1967, 48:810–918.

[11] Thieme H: *Mathematical models in population biology*. Princeton: Princeton University Press, 2003.

Chapter 5

Prostate Cancer: PSA, AR, and ADT Dynamics

5.1 Introduction

The prostate is a walnut-shaped and sized organ that envelops the urethra. Accounting for at most 0.1% (and as little as 0.01%, depending upon the individual) of the male body mass, it remains something of a mystery why neoplasms of this tiny organ contribute so greatly to cancer burden. Prostate cancer (PC) accounts for 25% of new cancer diagnoses and 10% of cancer deaths in American men [28].

Prostate cancer incidence and mortality vary widely worldwide, with those in the West far more likely to die of the disease. The natural history of prostate cancer spans decades, and can begin in the third or fourth decade of life, as shown by autopsies of men who died of other causes. Most men develop prostate enlargement at some point in life, and by age 90, 90% of men worldwide have developed some form of preclinical prostate cancer [59]. While it is diagnosed more often than any other cancer (in men), cancer aggressiveness varies widely between individual patients, and a majority of those diagnosed will not actually die of the disease. Over 90% of prostate cancer cases are diagnosed at local or regional stages for which survival approaches 100% [28].

The long preclinical phase of this neoplasm suggests a typically slow evolutionary progression toward the malignant phenotype; how ecological factors drive this selective process is not known. Indeed, the role of androgens in the prostate cancer etiology is controversial; we discuss this in some depth in Section 5.4.

Since the late 1980s, prostate cancer screening is typically done by measuring the serum prostate specific antigen (PSA). PSA is a proteinase produced by healthy prostate epithelium, and it is not an intrinsic marker of cancer. Normally, only a very small amount of PSA leaks through the prostate interstitium into the blood. Thus, elevated levels of blood PSA often correlate with prostatic hyperplasia.

While PSA levels have been generally understood to correlate with prostate neoplasia mass, the efficacy of PSA screening is unclear and remains very controversial. A recent study by the U.S. Preventive Services Task Force [33] concluded that the natural history of PSA-detected cancer is poorly understood, and screening can cause significant psychological harm. The task force recommended that men over 75 not be screened, as potential harm outweighed the benefits, while the evidence for those under 75 was insufficient to recommend for or against testing. In light of this controversy, we open this chapter with several simple clinical [9, 66] and dynamical models [61, 67] relating serum PSA and tumor growth dynamics.

Here, as elsewhere throughout this text, we invoke Theodosius Dobzhansky's rule that "nothing in biology makes sense except in the light of evolution" [13] (see also Chapter 1) to help explain the biology and clinical behavior of prostate cancer. This neoplasm's physiology is intimately linked to androgens (e.g., testosterone), the male sex hormones. Androgens are essential resources for survival and development of healthy prostate tissue. These hormones mediate proliferation, apoptosis, oxidative stress, and perhaps inflammation in the prostate. Thus they play a central role in modulating selection for cancerous cells.

In this chapter we investigate a multi-scale model of prostate cancer that includes dynamics of the concentrations of serum testosterone, intracellular testosterone and dihydrotestosterone (DHT), the androgen receptor (RA) and its complexes with the two androgens. The primary observable variable in both model and clinic is serum PSA. We will then apply this model to study androgen ablation therapy, which is a treatment of last resort for advanced tumors. Since healthy prostate epithelial cells rely on androgen, it is not surprising that prostatic epithelial tumors—the main type of prostate cancer—rely on androgens for growth and survival, or that androgen blockade (via castration) essentially invariably causes tumor regression. This type of hormone therapy is called androgen deprivation therapy (ADT) or androgen suppression therapy. The androgen suppression can be applied continuously or intermittently. Unfortunately, in either case, resistance to the treatment inevitably evolves as androgen insensitive tumor clones arise and become the dominant phenotypes. The mechanism appears to be natural selection, as clones that have less need for androgen become favored in the competition for this hormonal "resource."

Unfortunately, due to the lack of clinical data and the complexity of our comprehensive multi-scale model, we are unable to perform any serious model validation. Nevertheless, some very limited clinical data sets can be employed to estimate a subset of the parameters in our model. In the next chapter, we explicitly and mechanistically address nutrient-limited cell growth and competition in the context of prostate cancer dynamics subject to intermittent ADT. By focusing on just the cell and population scales, we are able to formulate more tractable models and validate them with clinical data. Understanding the evolution of aggressive prostate cancer in response to androgen ablation therapy has also been the focus of several mathematical models. If a model can accurately describe this evolution, then it could serve as a guide for rational treatment strategies that minimize evolution to an androgen-independent phenotype.

5.2   Models of PSA kinetics

The value of PSA in screening for prostate cancer is controversial, but in cases of confirmed cancer, serum PSA correlates with tumor volume across patients. However, in individual cases, serum PSA correlates poorly with tumor volume and prognosis. For example, grouping cancers into volume groups can yield excellent correlation between volume group and serum PSA, but the variance is so large that serum PSA cannot be used to predict individual tumor volumes [39].

Several other metrics have diagnostic and prognostic value, and other PSA kinetic parameters have been studied, including PSA velocity (the rate of PSA change), PSA density (the ratio of PSA to prostate volume), and the PSA doubling time (or equivalently, the relative PSA velocity). There has been a great deal of interest in predicting outcomes in response to therapy on the basis of pre-treatment findings. To this end, a number of statistical models have been employed that use metrics such as serum PSA, PSA density, PSA velocity, positive digital rectal examination (DRE), prostate volume in transrectal ultrasound, Gleason grade of biopsy samples, and patient age. Such predictive aids are referred to as *nomograms*, and a great number have been published [53].

Response to therapy, survival and quality of life are what is ultimately of clinical interest, and nomograms designed to predict tumor volume or pathological grade are only of secondary clinical interest, as such metrics must then themselves be related to survival and therapeutic response. However, we restrict our consideration to how PSA kinetics predict tumor volume and changes in tumor volume.

5.2.1 Vollmer et al. model

As an example, we begin with the study by Vollmer et al. [66], who present a model that tracks serum PSA in untreated prostate cancer (i.e., *watchful waiting*). These researchers describe serial measurements of PSA with a log-linear model:

$$\log(y(t)) = a + bt, \tag{5.1}$$

where $y(t)$ represents serum PSA. This is easily transformed to the familiar case of exponential growth,

$$y(t) = y_0 e^{bt}, \tag{5.2}$$

where $y_0 = e^a$. Parameter a is referred to as PSA amplitude, and

$$b = \ln(2)/(\text{PSA doubling time})$$

is referred to as the *relative PSA velocity*. If one wishes to relate it to clinically measured PSA velocity and serum PSA, it can also be expressed as

$$b = \frac{\mathrm{d}y/\mathrm{d}t}{y}. \tag{5.3}$$

Vollmer et al. concluded that PSA amplitude together with the relative PSA velocity best predicted outcomes. The key result of this work is that, at least in slower growing, untreated cancers, the underlying dynamic describing the change in serum PSA is exponential growth. Moreover, other authors have found that PSA relative velocity (or doubling time) is a strong marker of disease progression [38], consistent with the notion that serum PSA changes parallel changes in tumor growth, which is typically characterized by a doubling time.

5.2.2 Prostate cancer volume

Serum PSA is a marker for PSA production by both healthy and cancerous prostate cells. A number of simple mathematical and empirical relationships can be derived relating PSA to prostate cancer volume. For example, D'Amico et al.

[9] derived the "calculated prostate cancer volume" using simple mathematics to relate prostate volume, as measured by ultrasound (which includes tumor and healthy cells), cancer Gleason grade in biopsy cores, and serum PSA to estimate volume of the actual tumor. Let serum PSA concentration be y, the volume of benign prostate epithelium be V_b, the volume of cancerous prostate tissue be V_c, and represent the per-unit volume contribution of PSA to the serum for benign and cancerous tissue by c_1 and c_2, respectively. The fundamental relationship between these is

$$y = c_1 V_b + c_2 V_c, \tag{5.4}$$

which implies that

$$V_c = \frac{y - c_1 V_b}{c_2}. \tag{5.5}$$

D'Amico et al. refer to the numerator as the "*cancer-specific PSA*," as it is the total PSA minus the PSA contribution from benign tissues; the quantity $c_1 V_b$ is called the "*PSA from benign epithelial tissue*." The latter quantity is itself determined as a function of the prostate volume as determined by transrectal ultrasound and the epithelial fraction of the prostate. Letting V_p represent the total prostate volume and ϕ be the epithelial fraction gives

$$V_b = \phi V_p. \tag{5.6}$$

Finally, cancer volume is calculated in terms of serum PSA, prostate volume on ultrasound, prostate epithelial fraction, and PSA leak from benign and tumor tissue as follows:

$$V_c = \frac{y - c_1 \phi V_p}{c_2}. \tag{5.7}$$

Lepor et al. [32] measured a PSA increase of 0.33 ng/ml for every additional cm^3 of prostate epithelial tissue, a value D'Amico et al. used to calculate c_1. A more recent measurement by Fukatsu et al. [17] places this value at 1.27 ng/ml per cm^3 of epithelium.

The epithelial fraction was measured by Marks et al. [36] in 20 prostate samples from men with benign prostatic hyperplasia (BPH) and ranged from 0.117 to 0.308 with an average of 0.199, implying that $\phi \approx 0.20$. Thus, from the simple relationship (5.4) it follows that prostate cancer volume can be estimated using several empirical measurements and parameters. However, these parameters themselves are functions of lower-level dynamical parameters, and variation between them may still limit the predictive ability of the "calculated prostate cancer volume" in individuals. Although in their original work, D'Amico et al. found the calculated prostate cancer volume to be a much better predictor of actual volume than was serum PSA, a subsequent study failed to validate this finding [7].

Using newer estimates of PSA per cm^3 of epithelial tissue, or assuming different epithelial fractions depending upon whether BPH is present, are possible modifications that could improve the ability of the calculated prostate cancer volume to predict the actual cancer volume. But ultimately, it is not surprising that such a model cannot predict individual tumor volumes, as the parameters c_1 and c_2 are in reality functions of individually variable dynamical parameters, as we shall see in the following section.

5.3 Dynamical models

Serum PSA concentration is determined by the balance between input from the prostate into the serum compartment and clearance from this compartment. The parameters c_1 and c_2 in equation 5.4 then represent the aggregate result of this flux balance.

Two simple dynamical models have been proposed to account for the dynamics underlying serum PSA kinetics. The first, by Swanson et al. [61], considers serum PSA dynamics as a function of tumor growth. A later model by Vollmer and Humphrey [67] focuses on production of PSA in the prostate compartment, PSA transfer to the serum, and its subsequent clearance from this compartment. Below we discuss these models, their construction, and the insight they give into PSA dynamics.

5.3.1 Swanson et al. model

Because serum PSA level and prostate cancer volume correlate poorly at individual patient level, Swanson et al. [61] proposed a simple dynamical model to examine the relationship between these two quantities. In their model, serum PSA is produced by healthy and cancerous cells at different rates and is eliminated from the serum by first-order kinetics. Serum PSA is represented by $y(t)$ (in ng/ml), and V_h and $V_c(t)$ (both in mm^3) give the volume of healthy and cancerous prostatic cells, respectively. Note that V_h is assumed to be constant. These considerations yield the following linear model:

$$\frac{dy}{dt} = \beta_h V_h + \beta_c V_c(t) - ky(t). \tag{5.8}$$

The tumor is assumed to grow exponentially, so

$$V_c(t) = V_0 e^{\rho t}. \tag{5.9}$$

We have β_h as the rate at which PSA is produced by healthy prostate, β_c the rate of PSA production for cancerous prostate, and k the rate at which PSA is cleared, while ρ is the per-unit-volume rate of tumor volume increase, which is related to the doubling time, t_d, by $t_d = \ln(2)/k$.

This model can be criticized for its simplicity; the assumption of exponential tumor growth clearly cannot hold for all time, and any number of background biological processes are neglected. However, it is important to keep the issue that motivates the model in mind, which is the relationship between tumor volume and PSA serum dynamics. Thus, the simple model can still generate important insight in this area, and can potentially be more insightful than a more complicated model. Furthermore, this model takes into account the underlying biological process (i.e., tumor growth) determining PSA dynamics. This contrasts with the widely employed statistical nomograms, although the scope of the two approaches differ.

Due to its importance for the scientific predictions, we devote significant space to the parameter estimation process. This model was parameterized using in vivo data from a nude (athymic) mouse xenograft model of prostate cancer by Ellis et al. [15]. Because it is a xenograft (human cancer implanted into an immune-deficient

TABLE 5.1:   Parameters (P) obtained from 3 human-derived
mouse xenograft sublines (LuCaP 23.1, 23.8 and 23.12) used for
Swanson's PSA model.

P	LuCaP 23.1	LuCaP 23.8	LuCaP 23.12	Units
ρ	0.0655	0.0504	0.0487	day^{-1}
β_h	0	0	0	ng ml^{-1} mm^{-3} day^{-1}
β_c	1.7210	2.1841	6.9722	ng ml^{-1} mm^{-3} day^{-1}
k	1.2896	1.2896	1.2896	day^{-1}
V_0	20–25	20–25	20–25	mm^3
V_h	0	0	0	mm^3

mouse), $V_h = 0$. Parameters V_0 and k (from PSA half-life) are given by Ellis et al., while Swanson et al. estimated the tumor growth rate, ρ, from time-series data for tumor volume (see Table 5.1).

To determine PSA production rate, Swanson et al. used the analytical solution for $y(t)$ in conjunction with PSA:volume ratios reported in [15]. Here is the procedure. First, note that

$$y(t) = \frac{\beta_c}{\rho + k}(V_c - V_0 e^{-kt}) \qquad (5.10)$$

satisfies $dy/dt = \beta_c V_c - ky$, $y(0) = 0$. Suppose that asymptotically, y and V_c approach constant values. As $t \to \infty$, $e^{-kt} \to 0$, and we have that

$$\frac{y}{V_c} = \frac{\beta_c}{\rho + k}. \qquad (5.11)$$

Since the ratio y/V_c and parameters ρ and k are known for all cell lines, it is now a simple matter to calculate estimates for β_c. Ellis et al. reported PSA indices (PSA:tumor volume ratios) for cell sublines of human prostate cancer derived from three different metastatic sites in a single patient (LuCaP 23.1, LuCaP 23.8 and LuCaP 23.12). These ratios were 1.27, 1.63, and 5.21 ng ml^{-1} mm^{-3} for LuCaP 23.1, 23.8, and 23.12, respectively, yielding $\beta_c = 1.7210$, 2.1841, and 6.9722 ng ml^{-1} mm^{-3} day^{-1}. Table 5.1 gives all parameter values for these mouse xenografts, while Table 5.2 gives likely values for human prostate cancer as derived below.

Swanson et al. suggested that the dimensionless ratio $\mu = k/\rho$ determines the usefulness of the PSA level in predicting tumor volume, and therefore differences in tumor growth rates can explain the disconnect between measured PSA level and tumor volume. However, on closer examination, while the data supports the notion that normal variations in parameter value between individuals can explain the poor correlation between PSA and tumor volume, the aggregate parameter μ is not the key parameter. Rather, we argue that k, the PSA clearance rate, and β_c, the PSA production rate, are much more important.

Parametrization of human prostate cancer. To determine if and how parameter differences impact the PSA:volume ratio, we must first de-

TABLE 5.2: Likely parameter values in man under Swanson's PSA model framework.

Parameter	Value	Reference
ρ	0.0012–0.0045 day^{-1} (non-metastatic)	[4]
	$\leq$ 0.021 day^{-1} (metastatic)	
β_h	2.870×10^{-5} to 1.354×10^{-4}	see text
	ng ml^{-1} mm^{-3} day^{-1}	
β_c	varies	
k	0.1754 to 0.4030 day^{-1}	[34]
V_0	varies	
V_h	3333 to 52333 mm^3 epithelium	[65, 60]
	(10000–157000 mm^3 total)	

termine the biologically reasonable parameter range for human prostate cancers. Berges et al. [4] determined tumor doubling times for prostate cancers at various stages in their evolution. For non-metastatic cancer, the smallest doubling time was 154 ± 22 days ($\rho = 0.0045$ day^{-1}). The smallest doubling time overall was for lymph node metastases in hormonally untreated cancer, at 33 ± 4 days ($\rho = 0.021$ day^{-1}). The largest doubling time was 577 ± 68 ($\rho = 0.0012$ day^{-1}) for low-grade localized cancer.

In [4], it is reported that athymic (nude) mice clear PSA seven times faster than humans. Since nude mice were used in [15], this implies a human half-life of 3.76 days. This corresponds to estimates in the literature that all range between 1.72 days and 3.95 days [19, 34], giving $k = 0.1754$ to 0.4030 day^{-1}.

For the sake of completeness, we note that normal prostate volume in younger men is around 20 cm^3. Vesely et al. [65] measured non-cancerous prostate volumes and serum PSA mainly in older men. Prostate volume ranged from 10 to 157 cm^3, with an overall average of 40.1 ± 23.9. These values correlate with age: mean prostate size for men under 54 years was 27.5 cm^3, while the mean for men under 80 was 48.2 cm^3. Mean serum PSA follows a similar pattern. For all men in the study, serum PSA averaged 3.9 ± 4.2 ng/ml, with means of 1.5 ng/ml for men under 54 and 5.4 ng/ml for men under 80. Assuming a stroma to epithelium ratio of 2:1 for a healthy prostate [60], this gives $V_h = 3333$ to 52333 mm^3. Setting $V_c = 0$, and assuming a steady state yields the simple relationship,

$$\beta_h = \frac{ky}{V_h}. \tag{5.12}$$

Using $V_h = 9167$ mm^3, $y = 1.5$ ng/ml (values for men under 54), and $k \in [0.1754, 0.4030]$ yields $\beta_h = 2.870 \times 10^{-5}$ to 6.595×10^{-5} ng ml^{-1} mm^{-3} day^{-1}. Using $V_h = 16067$ mm^3, $y = 5.4$ ng/ml (values for men under 80), and $k \in [0.1754, 0.4030]$ yields $\beta_h = 5.895 \times 10^{-5}$ to 1.354×10^{-4} ng ml^{-1} mm^{-3} day^{-1}. These parameter values suggest that older men produce PSA at a higher rate per volume of prostate tissue than do younger men. Furthermore, benign prostatic hyperplasia, common in older men, results in a greater stroma:epithelium ratio [60]. Increasing this ratio yields an even larger β_h value, suggesting that our estimate is a lower bound.

Predictions. Using the parameter estimates just derived, it is a simple matter to show that while varying the tumor growth rate does have a small effect on the PSA:tumor volume ratio, this effect is very minor. We leave it as an exercise to confirm this.

On the other hand, we find that the PSA clearance rate, k, is a key parameter. Longer PSA half-lives (i.e., smaller ks) result in a greater PSA:volume ratio, as we expect intuitively, and over the realistic parameter range the PSA:volume ratio changes more than two-fold. Moreover, the per-capita production rate of PSA by cancer cells varies widely. If we examine the PSA:volume ratio under biologically reasonable values of β_c (at least for the mouse), we see that changes in β_c have a profound effect. Thus, the data and model together imply that, while the raw serum PSA concentrations may have poor predictive value, the *rate* at which cancer cells produce PSA is very important. Our results also indicate that the relative *change* in serum PSA parallels the relative change in tumor volume, assuming β_c remains constant.

This interpretation of the model and data presented in [61] suggests that variation in tumor growth rates is unlikely to affect how well serum PSA predicts tumor volume. Instead, variation in the rates at which PSA is produced by cancer cells and cleared in the serum appear to be much more important in determining the relationship between PSA and tumor volume. From these results we can also conclude that, in individual cases, while the absolute PSA level cannot reliably predict tumor volume, it may be valuable to track *relative* changes in PSA level as a marker for the relative change in tumor volume.

5.3.2 Vollmer and Humphrey model

In 2003, Vollmer and Humphrey [67] developed a simple, two-compartment model for serum PSA kinetics that includes production of PSA in the prostate and leak into the serum. This model did not explicitly consider tumor growth. Letting $f(t)$ represent tissue PSA (ng/ml) and $y(t)$ represent serum PSA (ng/ml), the basic model becomes

$$\frac{df}{dt} = \alpha - \beta f, \tag{5.13}$$

$$\frac{dy}{dt} = \beta f \frac{V_p}{V_s} - ky. \tag{5.14}$$

PSA is produced in the tissue compartment at rate α and leaks into serum at rate β, by first order kinetics. PSA concentration is diluted upon entry into the serum. Therefore the influx βf is modified by the ratio V_p/V_s, where V_p is the volume of the prostate compartment, and V_s is the serum volume. Serum PSA degrades by first-order kinetics with rate constant k.

The production of tissue PSA is, according to our basic assumption, a function of both benign and cancerous prostate epithelium. We introduce the flux constants Q_b and Q_c (in units ng ml^{-1} day^{-1}) which give the production of tissue PSA per unit volume of benign and cancerous tissue, respectively. Letting V_b and V_c represent the volumes of benign and cancerous tissue, respectively, the rate of PSA production in ng/day is given by

$$Q_b V_b + Q_c V_c. \tag{5.15}$$

Normalizing this to tissue concentration gives α (in ng/ml/day) as

$$\alpha = \frac{Q_b V_b + Q_c V_c}{V_p}. \tag{5.16}$$

We first look at the model steady states; other than the trivial one, $(0,0)$, a single steady state, (f_s, y_s), exists and is given as

$$f_s = \frac{\alpha}{\beta} = \frac{Q_b V_b + Q_c V_c}{\beta V_p}, \tag{5.17}$$

$$y_s = \frac{\alpha V_p}{k V_s} = \frac{Q_b V_b + Q_c V_c}{k V_s}. \tag{5.18}$$

The dependencies upon k, α, V_s, and V_p in y_s are biologically expected, but surprisingly, y_s does not depend upon β. That is, the model predicts that serum PSA does not depend upon the rate of PSA leak from tissue to serum, at least at steady state.

With some clever manipulations, Vollmer and Humphrey used this dynamical model to obtain the basic equation relating PSA to benign and prostatic tissue given in equation (5.4). Recall this fundamental relationship,

$$y = c_1 V_b + c_2 V_c. \tag{5.19}$$

To derive this statement from the dynamical model, Vollmer and Humphrey first substitute the right-hand side of equation (5.16) into equation (5.13) and rearrange, giving

$$\frac{Q_b V_b + Q_c V_c}{V_p} = \frac{df}{dt} + \beta f. \tag{5.20}$$

Since serum PSA (y) is what is measured and related in equation (5.4), we need f and df/dt in terms of y. Using equation (5.14) allows f to be determined in terms of y and dy/dt as

$$f = \left(\frac{dy/dt + ky}{\beta V_p}\right) V_s. \tag{5.21}$$

Differentiating gives df/dt:

$$\frac{df}{dt} = \left(\frac{d^2y/dt^2 + kdy/dt}{\beta V_p}\right) V_s. \tag{5.22}$$

Finally, plugging equations (5.21) and (5.22) into equation (5.20) yields the following relationship:

$$Q_b V_b + Q_c V_c = V_s \left(\frac{d^2 y/dt^2 + (\beta + k)dy/dt + \beta k y}{\beta} \right). \tag{5.23}$$

This relationship is then related to the simple exponential model for serum PSA, which, as Vollmer et al. [66] previously had shown, describes serum PSA dynamics in untreated cancer. That is,

$$y(t) = y_0 e^{\gamma t}. \tag{5.24}$$

From this, it follows that $dy/dt = \gamma y$ and $d^2 y/dt^2 = \gamma^2 y$. Plugging these into equation (5.23) and rearranging yields, finally,

$$y = c_1^* V_b + c_2^* V_c, \tag{5.25}$$

where

$$c_1^* = \frac{Q_b \beta}{V_s(\gamma^2 + (k + \beta)\gamma + \beta k)}, \tag{5.26}$$

$$c_2^* = \frac{Q_c \beta}{V_s(\gamma^2 + (k + \beta)\gamma + \beta k)}. \tag{5.27}$$

Thus, from a simple dynamical model, it can be shown that the constants c_1 and c_2 are functions of PSA production by tissue (Q_b, Q_c), plasma volume (V_s), the rate at which PSA leaks into serum (β), PSA serum degradation rate (k), and the PSA relative velocity (γ). However, our analysis of Swanson et al.'s model, supported by biological studies, indicates that the relative change in serum PSA directly tracks tumor growth. Thus, γ reflects the growth rate of both benign and malignant tissues.

Vollmer and Humphrey [67] used data from 100 men with prostate cancer who underwent prostatectomy to obtain estimates for c_1 or c_2. The volume of benign and cancerous tissue was measured, and serum PSA at prostatectomy was known. Using equation (5.4) gave average values of $c_1 = 0.117$ and $c_2 = 1.30$ ng/(ml · cm^3). Vollmer and Humphrey also estimated k and β using data obtained after either biopsy or radical prostatectomy; these values are reported in Table 5.3 for both free and total PSA. In reality, some PSA is free, while much is complexed to large serum proteins. However, we restrict our consideration to total PSA. Using a median $\gamma = 4.4 \times 10^{-4}$ day^{-1} from the same data set and an estimated serum volume of $V_s = 3,360$ ml, Vollmer and Humphrey estimated $Q_b = 100$ and $Q_c = 1,070$ ng/(ml · day).

Using these and our previous estimates for parameter values (in the context of Swanson et al.'s model) we examine how changing each parameter within biologically reasonable parameter space affects serum PSA and the PSA:tumor volume ratio. Initially, we restrict our attention to a single tissue type influencing PSA, and arbitrarily choose cancerous tissue; all results directly translate to the case when only benign tissue is present. Since we have set $V_b = 0$, we also have that

$$c_2 = \frac{y}{V_c}. \tag{5.28}$$

In other words, c_2 is precisely the PSA:tumor volume ratio, which we also examined in Swanson et al.'s model. From that model, we have argued that the PSA clearance and production rates are the primary parameters in serum PSA variance. Now, the PSA production rate by cancer cells has been effectively expanded from a single parameter to two: the actual tissue production, Q_c, and the rate of PSA leak, β. From the equation for c_2 it is apparent that serum PSA will increase in direct proportion to Q_c. The influence of β is less clear—rearranging gives

$$c_2 = \frac{Q_c \beta}{V_s \gamma^2 + V_s k \gamma + V_s \beta(k + \gamma)}. \tag{5.29}$$

Thus, for sufficiently large β, c_2 will become unaffected by changes in β. Using numerical values, we find that the effect of β on c_2 is generally insignificant, except in the case of a large γ—i.e., a fast growing tumor. Assuming γ is identical to the tumor growth rate, the largest biologically feasible is $\gamma = 0.021$. For this value of γ, varying β from 0.03 to 4.0 (the range of β determined after biopsy) results in a 69% increase in c_2.

Interestingly, the effect of the tumor growth rate, γ, on c_2 is most pronounced when β is small. For $\beta = 0.03$, ranging γ from 4.4×10^{-4} to 0.021 reduces c_2 by nearly one half. It is also interesting that these two effects on c_2 are likely competing. That is, as the tumor growth rate increases, microvessel density and permeability are also likely to increase, and their effects on PSA:tumor volume ratio may largely cancel each other out.

The two parameters that have by far the greatest effect on c_2 are Q_c and k. Therefore, the conclusion of this model, when restricting our attention to a single tissue type, like that of Swanson et al., is that while increasing the tumor growth rate can reduce the PSA:volume ratio modestly, it is the actual production of PSA by cancerous cells and the serum PSA half-life that likely are the most important factors in inter-individual PSA:tumor volume variation.

5.3.3 PSA kinetic parameters: Conclusions from dynamical models

Using the relatively simple dynamical models we have examined so far, we can predict the prognostic value and relation to the underlying dynamics of four widely used PSA kinetic parameters: serum PSA, PSA velocity, relative PSA velocity, and PSA density. We have already extensively studied serum PSA, finding that it correlates with tumor volume, but PSA half-life and cellular PSA production cause significant variation.

TABLE 5.3:   Parameters β and k in Vollmer and Humphrey's 2003 model [67].

Parameter	After Biopsy	After Prostatectomy
Total PSA		
β (day^{-1})	0.216 (0.03 − 4.0)	5.6 (0.2 − 21.9)
k (day^{-1})	0.067 (0.0034 − 1.29)	.216 (0.031 − 0.57)
Free PSA		
β (day^{-1})	3.74 (0.309 − 4.09)	16.3 (2.66 − 24.6)
k (day^{-1})	0.406 (0.375 − 0.437)	.909 (0.279 − 3.34)

First, we examine PSA velocity, the rate of change of PSA (i.e., dy/dt). The simple exponential model for serum PSA implies that PSA velocity alone gives no information that is not given by serum PSA. Recall the model,

$$y(t) = y_0 e^{bt}, \tag{5.30}$$

Which of course implies the ODE:

$$\frac{dy}{dt} = by. \tag{5.31}$$

Thus, PSA velocity (dy/dt) is simply a linear scaling of serum PSA and is therefore not a marker of the prostate (or tumor) growth rate. Rather, b is the meaningful parameter, which can be simply calculated in a clinical setting as

$$b = \frac{dy/dt}{y}. \tag{5.32}$$

This is simply the relative PSA velocity. This prediction holds under the more complex model of Swanson et al. [61]. For $V_h = 0$, we have

$$\frac{dy/dt}{y} = \frac{\beta_c V_c - ky}{y} = \frac{\beta_c V_c}{y} - k, \tag{5.33}$$

where

$$y = \frac{\beta_c}{\rho + k}(V_c - V_0 e^{-kt}). \tag{5.34}$$

As $t \to \infty$, $y \to \beta_c V_c/(\rho + k)$, implying that, as $t \to \infty$,

$$\frac{dy/dt}{y} = \frac{\beta_c V_c}{(\beta_c V_c)/(\rho + k)} - k = \rho. \tag{5.35}$$

Here, ρ is the tumor growth rate, but we have concluded that the PSA growth rate tracks the tumor growth rate very well. Thus, we can conclude that PSA velocity alone provides no new information, but a clinical measurement for PSA velocity could be used along with PSA level to estimate the more meaningful relative PSA velocity.

PSA density—i.e., serum PSA/prostate volume—is another common kinetic parameter. Vollmer and Humphrey [67], on the basis of their model results, claimed that corrections made for prostate volume simply by dividing by prostate volume are invalid, since

$$\text{PSA Density} = \frac{y}{V_p} = \frac{c_1 V_b + c_2 V_c}{V_p}. \tag{5.36}$$

This clearly remains of function of c_1 and c_2, which are complex functions of highly variable parameters. However, before abandoning the PSA density as useless, consider the scenario of a slow growing prostate—implying small γ—and assume $\gamma \ll k$, which is universally true for feasible parameter values. Then,

$$c_1 = \frac{Q_c \beta}{V_s(\gamma^2 + (k+\beta)\gamma + \beta k)} \approx \frac{Q_b \beta}{V_s k \beta} = \frac{Q_b}{V_s k}. \tag{5.37}$$

Similarly, $c_2 \approx Q_c/(V_s k)$. Plugging these into (5.36) yields

$$\frac{y}{V_p} = \frac{Q_b V_b + Q_c V_c}{k V_s V_p} = \frac{\alpha}{k V_s}. \tag{5.38}$$

Thus, PSA density reflects the rate of PSA production by both healthy and cancerous tissues as well as serum volume and PSA half-life. If we divide PSA density by V_p again, we have

$$\frac{y}{V_p^2} = \frac{\alpha}{V_p} \frac{1}{k V_s}. \tag{5.39}$$

This new metric reflects PSA production per unit of prostate tissue, which is expected to be higher if cancer is present (as $Q_c \gg Q_b$). We propose that this modified PSA density may have greater prognostic value than the standard one.

Vollmer et al. [66] argued on the basis of their clinical data that relative PSA velocity was an essential parameter in predicting outcomes in prostate cancer. We argue this too, but *on the basis of the underlying system dynamics*, as the dynamical models of both Swanson et al. [61] and Vollmer and Humphrey [67] imply that relative PSA velocity tracks tumor growth.

In conclusion, a brief analysis of simple dynamical models implies that PSA velocity gives no more information than serum PSA. PSA density, in the case of slowly growing prostate tumors, primarily reflects the ratio $\alpha/(kV_s)$. Therefore, dividing serum PSA by prostate volume squared (y/V_p^2) would reflect α/V_p; larger values of this index may imply the presence of cancer. Relative PSA velocity is the best marker for underlying disease progression.

5.4   Androgens and the evolution of prostate cancer

Androgens, the male sex hormones, have long been central to the study and treatment of prostate cancer. Androgens are essential survival factors for prostate secretory epithelial cells and act by binding with the androgen receptor (AR). Androgens are steroid hormones and freely cross cell membranes to bind with cytoplasmic AR. Testosterone, the primary androgen in the serum, is converted to dihydrotestosterone (DHT) by the enzyme 5-α-reductase in the prostate [20]. Testosterone and DHT both bind to AR, but DHT is more active, displaying greater binding affinity and stabilization of the AR complex [70]. Upon binding, androgen:AR complexes are phosphorylated, dimerize, and translocate to the nucleus, where they bind to androgen-response elements in the promoter regions of target genes [30] to modulate the transcriptional activity of at least several hundred target genes.

The importance of androgens is readily demonstrated by rat castration models. Following castration, over 70% of androgen sensitive cells undergo apoptosis [20], and the prostate epithelial mass decreases dramatically to only 7% of its original mass at 21 days [50]. Exogenous androgens induce prostate regrowth [46, 70], but high levels of androgen alone do not generally induce the prostate to grow beyond its normal size; androgen induced proliferation is apparently regulated by the normal prostate cell count, although the mechanism for this is unclear [46].

Most clinical prostate cancers are AR-dependent, and this observation has motivated androgen ablation therapy. Such therapy consists of chemical or surgical castration, which reduces serum testosterone by up to 95%, but reduces intraprostatic DHT levels by only 50% [20]. More complete androgen blockade can be achieved by supplementing castration with anti-androgens such as flutamide, nilutamide, and bicalutamide, and such therapy is referred to as maximal androgen blockade (MAB) [30]. However, the benefit to MAB over castration is uncertain, and a large meta-analysis suggested that any additional benefit to MAB is only slight [47].

Most men respond initially to androgen ablation, and often experience dramatic cancer regression. However, most cancers progress to a hormone refractory (HR) state even with near total androgen ablation. While time to progression can vary greatly [30], patients with metastatic prostate cancer eventually experience recurrence on average between 12 and 18 months following treatment [20]. Most cancers are more aggressive following HR recurrence, there are no effective treatments for such cancers, and median survival following progression does not exceed 15 months [30]. These cancers are often referred to as androgen independent, but most retain at least some dependence on the AR for survival.

5.4.1 Evolutionary role

Because of their role in protecting against apoptosis and promoting proliferation and the (transient) efficacy of androgen ablation therapy, it has long been thought that high levels of androgens play a causal role in prostate cancer development. The fact that eunuchs and men with genetic deficiencies in 5-α-reductase do not typically experience prostate cancer, along with the fact that androgen deprivation causes cancer regression have long been cited in support of this notion. But as Raynaud recently pointed out [48], such scenarios have little if anything to do with cancer development under the normal physiologic androgen range. However, in support of the high androgen hypothesis, in several animal models androgens were capable of inducing cancer, and some clinical studies have suggested a link between high testosterone and cancer incidence [46, 48].

In 1999, Prehn [46] proposed an alternate hypothesis: that low levels of androgen creates selective pressure for prostate cells that are less dependent upon androgen for growth. Declining levels of androgen could result in hyperplastic foci that resist atrophy but remain susceptible to further neoplastic transformation. In indirect support of this hypothesis, a number of clinical studies have failed to support the notion that high androgen levels increase the risk of prostate cancer [48, 58], and some data suggests that low serum testosterone is associated with aggressive, therapy-resistant tumors. In a prospective study including 17,049 men, high serum testosterone did not increase risk of prostate cancer and lowered the risk of aggressive tumors [55], and Sofikerim et al. recently found a significantly increased risk of cancer detection in men with low versus high serum testosterone [58]. Such data has led many authors to conclude that normal or high androgen promotes normal differentiation and function in epithelial cells, protecting against rather than promoting carcinogenesis [48, 55].

Although the role of androgens in predicting the incidence of prostate cancer has not been definitively settled, a broad literature dating from at least 1981 has consistently demonstrated poorer response to hormonal therapy in men with low pre-treatment serum testosterone (see [14] for a review of these studies).

In the following sections, we build a multi-scale framework, first presented by Eikenberry et al. [14], for the role of androgens in prostate growth and cancer evolution. We first present a model of the intracellular kinetics of the AR and androgens. We then use this model of the AR androgen binding kinetics to inform a higher level model of prostate epithelial growth in response to androgens. This model is finally used to study the evolution of prostate cells toward a malignant phenotype in early cancer etiology.

We focus upon the evolution of AR expression because of its deep importance in hormone therapy resistance and the fact that higher AR expression has been correlated with higher grade tumors [20]. We find that low serum testosterone strongly selects for greater AR expression. We also find that treatment with a 5-α-reductase inhibitor (e.g., finasteride) similarly selects for increased AR expression. Together, these results suggest that low androgen environments select more strongly for hormone therapy resistance and possibly more aggressive cancer clones than do normal or elevated androgen environments.

5.4.2 Intracellular AR kinetics model

The intracellular chemical kinetics model is founded on the following assumptions:

1. Free testosterone influx into the prostate is an empirical function of serum testosterone concentration, and this hormone is uniformly distributed to the intracellular compartment of all prostate cells.

2. Free intracellular testosterone is converted to free DHT by the enzyme 5-α-reductase. The intraprostatic 5-α-reductase level is assumed to be constant.

3. Free testosterone and DHT both degrade according to first-order kinetics.

4. Free testosterone and DHT bind to AR to form T:AR and DHT:AR complexes according to mass action kinetics. These complexes do not degrade.

5. Intracellular free AR binds to testosterone and DHT according to mass action kinetics, degrades by first order kinetics, and is produced at a rate that depends upon the homeostatic AR concentration set-point and current free AR concentration.

The model tracks the following concentrations:

1. $T_S(t)$ = Total serum testosterone concentration (nM),

2. $R(t)$ = Free intracellular androgen receptor concentration (nM),

3. $T(t)$ = Free intracellular testosterone concentration (nM),

4. $D(t)$ = Free intracellular DHT concentration (nM),

5. $C_{T:R}(t)$ = T:AR complex concentration (nM),

6. $C_{D:R}(t)$ = DHT:AR complex concentration (nM).

The basic mass action binding between T and DHT with AR and the conversion from T to DHT by 5-α-reductase is illustrated schematically as follows:

$$T + R \underset{k_a^T}{\overset{k_d^T}{\rightleftharpoons}} C_{T:R},$$

$$D + R \underset{k_a^D}{\overset{k_d^D}{\longleftrightarrow}} C_{D:R},$$

$$T \xrightarrow{5\alpha-\text{reductase}} D.$$

Translating this scheme into an ODE, and also taking into account T influx, AR production, and free T, DHT, and AR degradation yields the following chemical kinetics model:

$$\frac{dR}{dt} = \lambda - k_a^T TR + k_d^T C_{T:R} - k_a^D DR + k_d^D C_{D:R} - \beta_R R, \quad (5.40)$$

$$\frac{dT}{dt} = U - k_a^T TR + k_d^T C_{T:R} - \alpha k_{cat} \frac{T}{K_M + T} - \beta_T T, \quad (5.41)$$

$$\frac{dD}{dt} = \alpha k_{cat} \frac{T}{K_M + T} - k_a^D DR + k_d^D C_{D:R} - \beta_D D, \quad (5.42)$$

$$\frac{dC_{T:R}}{dt} = k_a^T TR - k_d^T C_{T:R}, \quad (5.43)$$

$$\frac{dC_{D:R}}{dt} = k_a^D DR - k_d^D C_{D:R}. \quad (5.44)$$

The rate of AR production is denoted λ, and AR, T, and DHT degrade at rates β_R, β_T, and β_D, respectively. 5-α-reductase converts T to DHT by Michaelis-Menten enzyme kinetics, where α is the concentration of 5-α-reductase, k_{cat} is the turnover number—i.e., the maximum rate at which T is converted to DHT by each unit of enzyme—and K_M is the Michaelis constant. Parameters k_a^T, k_d^T, k_a^D, k_d^D are the mass action rate constants for T and DHT binding to AR.

The rate of T influx, U, was estimated by Eikenberry et al. [14] as an empirical function of serum T, T_S, viz.

$$U(T_S) = \begin{cases} 0.02938 T_S^3 - 0.006729 T_S^2 + 0.05514 T_S + 0.0004823 & T_S \leq 1.375, \\ 0.01138 T_S^3 - 0.06751 T_S^2 + 0.203 T_S - 0.05441 & 1.375 < T_S \leq 7, \\ -0.0012 T_S^3 + 0.048 T_S^2 + 0.42 T_S - 2.9 & T_S > 7. \end{cases}$$
$$(5.45)$$

This is an empirical form. An alternative, based on first principles, is given in Section 6.5.4. We also write the total AR concentration, R_t, as

$$R_t = R + C_{T:R} + C_{D:R}. \quad (5.46)$$

We assume that homeostatic mechanisms keep R_t constant, giving the AR production rate as follows:

$$\lambda = \beta_R^*(R_t - C_{TR} - C_{DR}), \quad (5.47)$$

where β_R^* is the normal AR turnover rate. This assumption is necessary for the model to match data in [71], as discussed in Eikenberry et al. [14].

TABLE 5.4: Parameters and baseline values for the AR kinetics model.

Parameter	Value	Reference
R_t	16–45 nM	[14]
k_a^T	0.14 nM^{-1} hr^{-1}	[69]
k_d^T	0.069 hr^{-1}	[69]
k_a^D	0.053 nM^{-1} hr^{-1}	[69]
k_d^D	0.018 hr^{-1}	[69]
β_R	$\ln(2)/3$ hr^{-1}	[18]
β_T	$\ln(2)/3$ hr^{-1}	
β_D	$\ln(2)/9$ hr^{-1}	[14, 68]
α	5.0 mg L^{-1}	[14]
k_{cat}	18 ± 15 nmol hr^{-1} mg^{-1}	[40]
K_M	75 ± 33 nM	[40]
K_I	0.46 ± 0.21 nM	[40]

Serum testosterone (T_S), while in reality a function of time, is always imposed in this model and does not vary according to a governing ODE. Significantly, we have not modeled dimerization of androgen:AR complexes or their nuclear localization and binding to gene promoter regions under the assumption that the concentrations of androgen:AR complexes can be taken as surrogates for such activities. We have also assumed that all prostate androgens are intracellular and uniformly distributed among the epithelial cells. Prostate testosterone concentration can be much higher than serum concentration [37, 71], and DHT prostate concentration can be over 50 times that of serum concentration. This suggests that most is intracellular, as extracellular androgens would presumably equilibrate with serum androgens. We ignore all the details of transport between serum, extracellular, and intracellular compartments, and instead have T transported directly into the intracellular compartment.

5.4.3 Basic dynamics of the AR kinetics model

The likely normal physiologic range for serum T (T_S) in rat is 3 to 6 nM [1, 2]. Values for most of the basic kinetic parameters, k_a^T, k_d^T, k_a^D, k_d^D, β_R, β_T, k_cat, and K_M are available directly from empirical biological data; these values with references are given in Table 5.4. The other parameters, λ, U, α, β_D, were estimated by Eikenberry et al. [14] using a combination of empirical data and steady state analysis, relying heavily on data from [71].

We briefly characterize the basic dynamics of the AR kinetics model. The time-dependent dynamics of the model are demonstrated by initially setting R_t to a constant and all other variables to 0. Serum T concentration is prescribed, and the model is run to steady state, as shown in Figure 5.1. For baseline parameter values, the free T concentration is always small, as

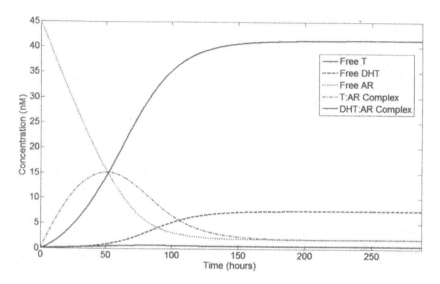

FIGURE 5.1: Time-series for the AR kinetics model. Free AR is set to 45 nM as an initial condition; all other variables are initially zero and baseline parameter values are used. Serum T is prescribed at 5 nM, inducing an influx of testosterone, and the model runs to a steady state.

most T is rapidly converted to DHT. There is a transient peak in T:AR complex concentration early in time, but the DHT:AR complex dominates within several days; this pattern is a consequence of the time it takes 5-α-reductase to produce DHT. For physiologic values of serum T and prostate AR, once steady state is reached most androgen is bound to its receptor, and nearly all intraprostatic androgen is DHT. These dynamics are biologically expected.

5.5 Prostate growth mediated by androgens

We now link intracellular androgen concentrations to the proliferation and apoptosis of prostate epithelial cells. While we generally refer to low or high androgen levels causing a behavior, it is really the concentrations of AR:T and AR:DHT complexes that mediate these androgen-related activities. We introduce the variable C_t to represent the "effective" androgen:AR concentration. In [71], DHT was 2.4 times as potent as T in maintaining prostate weight and duct lumen mass, and these quantities varied linearly with either androgen. Therefore, we take C_t to be a simple linear combination of $C_{T:R}$

and $C_{D:R}$:

$$C_t = C_{T:R} + 2.4 C_{D:R}. \tag{5.48}$$

This approach allows the previously studied androgen kinetics model to be coupled directly to a model of prostate growth mediated by androgens. We let $P(t)$ represent the number of prostate epithelial cells. We assume that the change in P is governed by two distinct death and proliferation signals; the per-capita proliferation rate is $M(C_t, S)$ and the per-capita death rate is $N(C_t, S)$, yielding the basic model framework:

$$\frac{dP}{dt} = PM(C_t, S) - PN(C_t, S). \tag{5.49}$$

We now determine the formal forms for $M(C_t, S)$ and $N(C_t, S)$. Prostate epithelial proliferation and death are regulated by androgens in several ways.

1. Androgens induce stroma to produce factors, mainly bFGF and FGF-7, that support epithelial growth in a paracrine manner by supporting the prostate vasculature, induce epithelial proliferation, protect the epithelium from apoptosis, and regulate AR protein levels.

2. Androgens may have a direct mitogenic effect upon epithelial cells through upregulation of proteins required for cell cycle progression.

3. Androgens directly protect against apoptosis by negatively regulating TGF-β and increasing bcl-2 levels.

4. Androgens mediate oxidative stress and the production of reactive oxygen species (ROS) within epithelial cells, which can induce proliferation, stasis, or death, depending upon the concentration.

Proliferation. It is generally accepted that androgens induce epithelial proliferation in vivo when the cell count is below normal, and androgen administration following castration induces rapid prostate regrowth in the rat [68, 70]. However, proliferation is thought to be limited, at least to some degree, by the homeostatic size of the prostate [46]. In our model, we assume that high levels of androgen directly induce proliferation while low levels cause apoptosis.

Redox state. The prostate redox state is also influenced by androgens, and this may be deeply important in epithelial death, proliferation, and carcinogenesis. Androgen blockade induces the production of ROS and subsequent oxidative stress. Several rat models have demonstrated that castration [62, 42] and treatment by either finasteride (5-α-reductase inhibitor) or flutamide (an anti-androgen) are strongly pro-oxidant [6]. Androgen withdrawal also causes vascular regression and prostate hypoxia [56], which in turn can induce ROS and increase expression of hypoxia inducible factor-1α (HIF-1α).

Androgen administration has also been shown to induce oxidative stress. Tam et al. [63] found that administration of testosterone with 17β-estradiol resulted in oxidative and nitrosative stress in the lateral lobe of the Noble rat. Ripple et al. [49] found that physiologic levels of DHT induced ROS in the LNCaP carcinoma cell line, and ROS generation preceded DHT induced proliferation.

Thus, a normal androgen environment likely promotes a balance between antioxidant and pro-oxidant activity [62], but both low and high androgen environments are pro-oxidant. Therefore, we assume that both low and high levels of androgen induce the formation of ROS.

ROS effects. At low levels, ROS act as important intracellular signalling molecules. A number of transcription factors, including NF-κB and AP-1 are redox sensitive, and modest levels of ROS are mitogenic. Higher levels of ROS can induce growth arrest and apoptosis, while very high levels can cause necrosis [10].

We assume that both low and high concentrations of C_t induce ROS and that there is some background level of ROS independent of androgens. We choose S to represent ROS and formally take

$$S = \mu + \frac{\theta_1^n}{C_t^n + \theta_1^n} + \frac{C_t^m}{C_t^m + \theta_2^m}. \tag{5.50}$$

Here, μ is the background ROS level, and ROS is induced by low C_t and high C_t according to the first and second Hill functions, respectively. The half-maximal C_t for ROS induction by low androgen is θ_1, while θ_2 is the half-maximal C_t for high androgen induced ROS.

We assume that prostate proliferation is induced by increasing concentrations of such complexes, and a high cell count inhibits proliferation. Low AR:ligand complex concentration induces apoptosis, and there is always some small baseline turnover rate (1–2% of cells turnover daily in the healthy prostate [20]). Modest levels of ROS induce proliferation, while higher level cause growth arrest and apoptosis. These assumptions lead to our formal choices for $M(C_t, S)$ and $N(C_t, S)$:

$$M(C_t, S) = \frac{r}{2} \left(\underbrace{\left(\frac{C_t^2}{\varphi_1^2 + C_t^2} \right)}_{\text{Androgen signal}} + \underbrace{\phi S e^{1 - \phi S}}_{\text{ROS signal}} \right) - \underbrace{\sigma P}_{\text{crowding inhibition}}, \tag{5.51}$$

$$N(C_t, S) = \frac{\delta}{2} \left(\underbrace{\left(\frac{\varphi_2^2}{C_t^2 + \varphi_2^2} \right)}_{\text{Androgen signal}} + \underbrace{\left(\frac{S^q}{\omega^q + S^q} \right)}_{\text{ROS signal}} \right) + \underbrace{\delta_0}_{\text{normal turnover}}. \tag{5.52}$$

Our incorporation of S as a function of C_t into the equation for dP/dt allows both direct and indirect ROS mediated effects of androgens on prostate growth to be incorporated into a single differential equation. This construction is

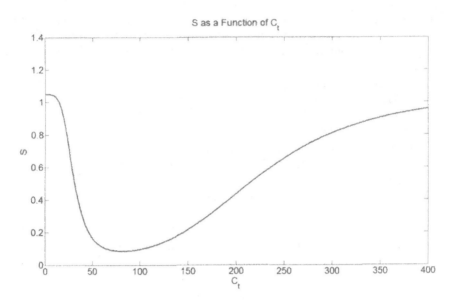

FIGURE 5.2:   Oxidative stress, S, as a function of C_t.

somewhat similar to the ecological model of planktonic algae interaction with vegetation in shallow lakes proposed by Scheffer et al. [54].

The maximum per-capita proliferation rate is r, and the maximum death rate is $\delta + \delta_0$. The direct proliferation signal due to androgens is modeled by a Hill function, with the strength of the signal increasing with C_t. The signal due to ROS is strong for low S and attenuates as S becomes large. Hill functions are used to model the death signals due to both androgen and ROS, with the signals strong for low C_t and high S, respectively. Cells also die at the background rate δ_0, and the $-\sigma P$ term prevents unbounded growth.

The growth of the prostate epithelium in our model is governed solely by the intraprostatic weighted AR:ligand complex concentration, C_t, and the prostate cell count, $P(t)$. There are two independent proliferation signals, one mediated by C_t and the other by S, and two similar death signals. The shapes of these signals are determined by the parameters θ_1, θ_2, φ_1, φ_2, ϕ, ω, n, m, and q.

Oxidative stress, S, is shown as a function of C_t in Figure 5.2. Figure 5.3 shows the proliferation and death signals due to C_t and S and their sums; these figures disregard the $-\sigma P$ crowding term as this always depends upon prostate size. The qualitative form of these curves is preserved over most of the parameter space.

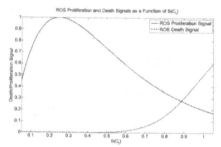

(a) Signal due to ROS as a function of S.

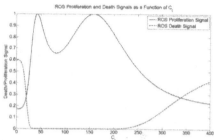

(b) Signal due to ROS as a function of C_t.

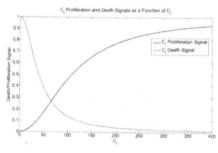

(c) Signal due to C_t as a function of C_t.

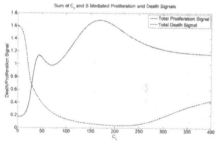

(d) Sum of ROS and C_t signals as a function of C_t.

FIGURE 5.3: Curves for the different growth signals. Attenuation of proliferation by crowding is disregarded. Baseline parameter values are $\theta_1 = 30$ nM, $\theta_2 = 225$ nM, $n = 4$, $m = 4$, $\varphi_1 = 110$ nM, $\varphi_2 = 40$ nM, $\phi = 4$, $\omega = 1$, $q = 8$, $\sigma = 1.5 \times 10^{-10}$ cell^{-1} hour^{-1}, $\delta = 0.004$ hour^{-1}, $r = \ln(2)/24$ hour^{-1}, $\delta = \ln(2)/24$ hour^{-1}.

5.6 Evolution and selection for elevated AR expression

In advanced cancers, upregulation of the AR protein is perhaps the single most important pathway by which cancers become hormone refractory. Chen et al. [8] found that in seven prostate cancer xenograft models, increased androgen receptor expression was the only change consistently associated with HR cancer progression. Rapid HR cancer recurrence in a xenograft model by Rocchi et al. was always associated with increased AR expression [51].

Because of the importance of the AR in prostate cancer progression, we investigate how different androgen environments select for cell lines expressing different levels of the AR; i.e., we examine selection upon R_t. To do this, we use the coupled AR kinetics and prostate growth model described in the previous section.

We have performed a series of numerical experiments where two or more epithelial cell strains, each having different values of R_t, compete with each other. Notably, these experiments suggest that selective pressure for increased AR expression varies with R_t and T_S, but this nonuniform behavior makes it hard to gain insight using this simple approach. We suggest that implementing such a two-strain competition model numerically would be a useful exercise.

5.6.1 Model

To model competition between a large number of cell lines that evolve in time, Eikenberry et al. [14] proposed a state-transition model where cells transition between states that each represent a different level of R_t expression. We define a set of states

$$\{P_i\}_{i=1}^{L}. \tag{5.53}$$

Each state represents a strain of prostate cells with a different AR concentration (i.e., R_t), and R_t varies linearly with i. Cells transition from P_i to P_{i-1} and P_{i+1} at rate γ, representing mutation. Populations of each strain grow according to the following coupled kinetics-growth model:

$$\frac{\mathrm{d}P_i}{\mathrm{d}t} = \begin{cases} P_i M(C_{t_i}, S_i) - P_i N(C_{t_i}, S_i) - 2\gamma P_i + \gamma P_{i-1} + \gamma P_{i+1}, & 2 \leq i < L, \\ P_i M(C_{t_i}, S_i) - P_i N(C_{t_i}, S_i) - \gamma P_i + \gamma P_{i+1}, & i = 1, \\ P_i M(C_{t_i}, S_i) - P_i N(C_{t_i}, S_i) - \gamma P_i + \gamma P_{i-1}, & i = L. \end{cases} \tag{5.54}$$

We choose our model to have 100 states representing R_t from 15 to 114 nM. That is, we set $L = 100$ and R_{t_i}, the total AR level at state i, to be $14 + i$ nM, $i \in \{1, 2, ..., 100\}$. The total number of cells in each state is tracked through time, and the average R_t at all time steps is calculated.

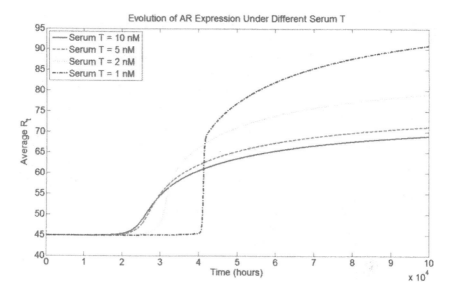

FIGURE 5.4: Evolution of average R_t in the state-transition model under different serum T. AR expression is presumably a marker for the malignant potential of a strain. Low serum T selects for higher R_t than the normal environment (serum T = 5 nM), but takes longer to do so. High serum T selects for a slightly lower R_t.

5.6.2 Results

Figure 5.4 shows the evolution of average R_t under different serum T levels. In general, results indicate that physiologic serum T selects for increased AR expression, as do all androgen environments. Thus, the model predicts that even in healthy men, prostate epithelial cells will increase their potential for malignancy with time. Under high serum T, selection for an elevated R_t is actually slightly weaker. In comparison, low androgen environments (i.e., low serum T) demonstrate rapid, late selection for increased AR expression. While this selection occurs later in time, a higher average R_t is ultimately obtained.

We performed a sensitivity analysis to determine whether this behavior is preserved under other growth model parameter values. We found that the prediction that low androgen selects for a greater final R_t is always preserved. The parameters θ_1, θ_2, and μ, which govern the level of ROS, have the greatest effect on the dynamics. Overall, low androgen appears to select for a greater final R_t in all parameter space, but this selection becomes apparent later in time than under normal androgen for most parameter space.

These results suggest that a low androgen environment may delay the development of a malignant phenotype, but result in a more malignant or therapy-resistant strain later in time. This result could also be interpreted to mean that low androgen reduces the overall incidence of cancer, as the expected time to the development of a malignant strain is increased, but those cancers that do arise may be more aggressive. This notion is consistent with the results of finasteride treatment in men [64].

Having established one possible model framework by which the role of androgens in early cancer evolution may be studied, we now turn our attention to several models examining cancer recurrence driven by competition between androgen dependent (AD) and androgen independent (AI) cancer cell lines following androgen deprivation therapy (ADT). The focus of these models is evolution in a clinically meaningful, aggressive cancer, rather than long-term evolution predisposing cell lines toward malignancy.

5.7 Jackson ADT model

Jackson proposed a model [24, 25] describing the growth of a tumor spheroid consisting of two populations of cells—one that depends on androgen for survival and proliferation, and one that does not. Because the mathematical model is similar in form to that of Ideta et al. [23], discussed in the next section, we restrict our attention to some of Jackson's key modeling assumptions: (1) the proliferation of androgen dependent (AD) cells increases in the presence of androgen, (2) the proliferation of androgen independent (AI) cells is unaffected by androgen levels, (3) the rate at which AD cells undergo apoptosis decreases in the presence of androgen, and (4) the rate at which AI cells undergo apoptosis increases in the presence of androgen. (As we discuss below, assumption (4) is debatable.)

In [24], Jackson predicted that androgen deprivation therapy would successfully control tumor growth for all time only in a small region in parameter space. This region is larger for total androgen blockade compared to partial androgen deprivation, but is still small. However, it is likely that this prediction is a consequence of the assumption that androgens harm AI cells. In [25], Jackson predicted that, following androgen ablation therapy, cancer could recur either through androgen-independent mechanisms such as upregulation of the anti-apoptotic protein bcl-2 in AI cells or through AR upregulation in AD cells. These two mechanisms corresponded to two qualitatively different patterns of recurrence, and the reader may consult the original paper for details.

The essence of the Jackson model is the tumor-wide balance between proliferation and death, which is expressed by the following ODE system:

$$\frac{dp(t)}{dt} = \alpha_p \theta_p(a)p - \delta_p \omega_p(a)p, \tag{5.55}$$

$$\frac{dq(t)}{dt} = \alpha_q \theta_q(a)q - \delta_q \omega_q(a)q. \tag{5.56}$$

Here, a is the concentration of androgen, assumed to be uniform throughout the tumor. Maximum per-capita growth rates are given by constants α_p and α_q, while constants δ_p and $\delta_q(a)$ represent the maximum per capita death rates for AD and AI cells, respectively. Parameters $\theta_p(a) \in [0,1]$, $\theta_q(a) \in [0,1]$, $\omega_p(a)$ and $\omega_q(a)$ mediate cell proliferation and death according to androgen levels.

Androgen-dependent cells are assumed to proliferate at rate α_p in the presence of sufficient androgen. This rate decreases as androgen levels decrease, becoming $\theta_1 \alpha_p$ in the complete absence of androgen. AI cells always proliferate at the maximum rate α_q. These assumptions can be satisfied by the following specific expressions:

$$\theta_p(a) = \theta_1 + (1 - \theta_1)\frac{a}{a + K}, \tag{5.57}$$

$$\theta_q(a) = 1. \tag{5.58}$$

Since $0 \le \theta_1 \le 1$, then $\theta_p(0) = \theta_1$ and $\theta_p(a) \to 1$ as $a \to \infty$. The term $a/(a + K)$ can be thought to represent saturation of a receptor governing the proliferative response to androgen.

Similarly, apoptosis rate in AD cells is assumed to be a decreasing function of androgen concentration. However, androgens are assumed to increase the the death rate of AI cells, which again is probably debatable. The functions ω_p and ω_q take on the same form as that used for θ_p:

$$\omega_p(a) = \omega_1 + (1 - \omega_1)\frac{a}{a + K}, \quad \omega_1 > 1, \tag{5.59}$$

$$\omega_q(a) = \omega_2 + (1 - \omega_2)\frac{a}{a + K}, \quad 0 \le \omega_2 \le 1. \tag{5.60}$$

We require $\omega_1 > 1$ to make $\omega_p(a)$ a decreasing function of androgen. Also we have $0 \le \omega_2 \le 1$, implying that $\omega_q(a)$ decreases with increasing androgen. Parameters δ_p and δ_q represent the respective rates of apoptosis for AD and AI cells in a normal (high) androgen environment. As the androgen level approaches 0, the death rates go to $\omega_1 \delta_p > \delta_p$ and $\omega_2 \delta_q < \delta_q$, respectively. Thus, ω_1 and ω_2 are the factors by which AD and AI cell death rates are modified in the complete absence of androgen.

To monitor the spatial distributions of the different cell types, one can keep track of the fluxes of the two classes of cancer cell, say $J_p(r,t)$ and $J_q(r,t)$. The net rate of collective cellular motion is determined by the balance between cell growth and death and the diffusive flux. The resulting partial differential equation model will have a moving tumor boundary. The rate of radial tumor expansion can also be tracked.

Jackson's model considered both diffusive and advective fluxes. Diffusive flux is due to random cellular motion. Advective flux is often determined by some vector field. However, Jackson's model "reverses" this—cells are not transported according to some vector field, but the net rate of collective cellular motion is given by the vector $\mathbf{u}$, which is itself determined from the balance between cell growth and death and the diffusive flux. If we let $p(r,t)$ and $q(r,t)$ represent the tumor volume fraction of each class of cell (rather than some other metric such as absolute cell count), then Jackson's model takes the form of the following conservation equations:

$$\frac{\partial p}{\partial t} + \nabla \cdot (\mathbf{u}p) = D_p \nabla^2 p + \alpha_p \theta_p(a)p - \delta_p \omega_p(a)p, \tag{5.61}$$

$$\frac{\partial q}{\partial t} + \nabla \cdot (\mathbf{u}q) = D_q \nabla^2 q + \alpha_q \theta_q(a)q - \delta_q \omega_q(a)q. \tag{5.62}$$

Noting that $p+q = k$ is constant, adding equations (5.61) to (5.62) yields the following expression for $\nabla \cdot \mathbf{u}$:

$$k\nabla \cdot \mathbf{u} = (D_p - D_q)\nabla^2 p + \alpha_p \theta_p(a)p + \alpha_q \theta_q(a)(k-p) - \delta_p \omega_p(a)p - \delta_q \omega_q(a)(k-p). \tag{5.63}$$

Jackson assumes a spherically symmetric geometry, so the only component of the velocity is the radial component. This gives the rate of radial expansion for the tumor spheroid, which in turn allows tumor volume to be tracked. Using $R(t)$ to represent the outer tumor radius, one arrives at an ODE for the rate of radial expansion, which is simply

$$\frac{dR}{dt} = u(R(t), t). \tag{5.64}$$

To complete the system, no flux boundary conditions are imposed at the inner $(r = 0)$ and outer boundaries $(r = R)$ of the spheroid. That is, at $r = 0$ and $r = R$

$$D_p \frac{\partial p}{\partial r}(r,t) - u(R,t)p(r,t) = 0, \quad D_p \frac{\partial p}{\partial r}(r,t) - u(R,t)p(r,t) = 0. \tag{5.65}$$

By symmetry, at $r = 0$,

$$\frac{\partial p}{\partial r}(0,t) = \frac{\partial q}{\partial r}(0,t) = 0. \tag{5.66}$$

These, together with no flux boundary condition at $r = 0$, yield

$$u(0,t) = 0. \tag{5.67}$$

The initial tumor radius is assumed to be $R_0 > 0$, and cell conditions are uniform in space, with $p(r,0) = p_0$ and $q(r,0) = q_0$.

TABLE 5.5: Parameter values for Jackson's model, as reported in [24, 25].

Parameter	Value
α_p	0.4621 day^{-1}
α_q	0.4621 day^{-1}
δ_p	0.3812 day^{-1}
δ_q	0.4765 day^{-1}
θ_1	0.8
ω_1	1.18—1.35
ω_2	0.25—1.0
p_0	0.995

Finally, androgen levels are modified by treatment. Jackson considered two classes of treatment: castration through either surgical or chemical means (ADT); and total androgen deprivation (TAD), meaning castration in addition to anti-androgen drugs. The former reduces testosterone serum levels by up to 95%, but only reduces intraprostatic DHT by 50%, while the latter can reduce prostate DHT levels by 90% [20]. It is assumed that the androgen level, $a(r, t)$, is at a steady-state until treatment is initiated at time T, at which point it declines exponentially to a new steady-state, either non-zero in the case of ADT or zero for TAB. Mathematically,

$$a(r, t) = \begin{cases} a_0 & ; t < T, \\ a_0 e^{-bt} + a_s & ; t \geq T, \end{cases} \tag{5.68}$$

where $a_s > 0$ for ADT and identically 0 for TAB. We note that this model does not consider transitions between the cell classes, and the presence of androgen independent cells must be imposed in the initial conditions.

Jackson derived reasonable parameter values for growth and death rates of AD cells using biological data. Berges et al. [4] reported cell cycle times varying between 48 ± 5 hours for cancer cells. Based on this, a doubling time of 36 hours was assumed, still well within what is biologically reasonable (tumor cell doubling times typically vary between 1 and 4 days), yielding $\alpha_p = \alpha_q = 0.4621$ day^{-1}. Using the data of Ellis et al. [15], the same data used to parameterize Swanson et al.'s PSA model [61], Jackson estimated $\delta_p = 0.3812$, implying an overall growth rate of $\rho = 0.0798$ day^{-1}. This is somewhat larger than Swanson et al.'s estimates. The parameter estimates for AI cells are more ad hoc. Table 5.5 shows all the parameters governing growth and death used by Jackson. No values or estimates for K, b, or a_s were published.

5.8   The Ideta et al. ADT model

Building on Jackson's work, in 2008 Ideta et al. [23] proposed a model of intermittent androgen deprivation therapy and evolution toward androgen-independent cancer recurrence. Like Jackson's, this model considers androgen levels, represented by $a(t)$, androgen-dependent cells, $x_1(t)$, and androgen-independent cells, $x_2(t)$. Androgen levels approach some homeostatic set-point that can be modified by androgen deprivation therapy (ADT). Androgen-mediated growth and death reflect Jackson's assumptions, but three different hypotheses concerning how androgens affect AI cell proliferation are considered:

1. Normal androgen levels have no effect on AI cell proliferation.

2. AI cell proliferation is negatively regulated by androgens, and at a defined normal androgen level (a_0) proliferation and apoptosis completely balance each other, giving a net growth rate of zero.

3. AI proliferation is negatively regulated by androgen, and the net growth rate is negative at the normal androgen level.

Androgen-dependent cells mutate to become AI cells at a rate that increases with decreasing androgen levels. This assumption apparently stems from the notion that low androgen induces transformation into an androgen-independent phenotype. These considerations yield the following model:

$$\frac{da}{dt} = -\gamma a + \gamma a_0 (1 - u), \tag{5.69}$$

$$\frac{dx_1}{dt} = \alpha_1 p_1(a) x_1 - \beta_1 q_1(a) x_1 - m(a) x_1, \tag{5.70}$$

$$\frac{dx_2}{dt} = \alpha_2 p_2(a) x_2 - \beta_2 q_2(a) x_2 + m(a) x_1, \tag{5.71}$$

where, like in Jackson's model,

$$p_1(a) = k_1 + (1 - k_1)\frac{a}{a + k_2}, \quad 0 \le k_1 \le 1, \tag{5.72}$$

$$q_1(a) = k_3 + (1 - k_3)\frac{a}{a + k_4}, \quad k_3 \ge 1, \tag{5.73}$$

$$p_2(a) = \begin{cases} 1 & ; \text{ for hypothesis (1)}, \\ 1 - \left(1 - \frac{\beta_2}{\alpha_2}\right)\frac{a}{a_0} & ; \text{ for hypothesis (2)}, \\ 1 - \frac{a}{a_0} & ; \text{ for hypothesis (3)}, \end{cases} \tag{5.74}$$

$$q_2(a) = 1, \tag{5.75}$$

$$m(a) = m_1 \left(1 - \frac{a}{a_0}\right). \tag{5.76}$$

Treatment is represented by the effect of u, which is either 1 (treatment on) or 0 (treatment off). Similar to Jackson's model, AD cells proliferate at the baseline rate $\alpha_1 k_1$. As $a \to \infty$, this rate approaches α_1 as the proliferative response saturates. For $k_3 > 1$, $q_1(a)$ causes the death rate to increase with decreasing androgen, equaling $\beta_1 k_3$ when $a = 0$. The Ideta et al. model departs from Jackson's in considering three hypotheses for AI proliferation in response to androgen and in assuming a constant AI apoptosis rate regardless of androgen concentration ($q_2(a) \equiv 1$).

The rate of transition from AD to AI cells decreases linearly from m_1 at $a = 0$ to 0 when androgen concentration is normal; i.e., $a = a_0$. As discussed previously, this is controversial. The dependence on a_0 in many of the functions causes AI proliferation to become negative if $a > a_0$ for hypotheses (2) and (3), while $m(a)$ also becomes negative, implying that AI cells begin transitioning into AD cells in a high androgen environment.

Ideta et al. proposed an algorithm for governing treatment in which androgen deprivation is turned on or off depending on the level of system androgen and the rate of change in tumor growth. It is assumed that, clinically, only the serum PSA level can be tracked, rather than the actual tumor volume. Represented by $y(t)$, PSA concentration is assumed to be a simple function of AD and AI cells with no lag in production; therefore,

$$y(t) = c_1 x_1(t) + c_2 x_2(t). \tag{5.77}$$

This equation is the same fundamental relationship that we studied earlier in the context of PSA dynamics (see equation (5.4)). Ideta et al. set $c_1 = c_2 = 1$, implying that PSA tracks tumor growth precisely. In our earlier exploration of PSA dynamical models, we determined that the PSA level in individuals correlates poorly with absolute tumor volume, but likely tracks relative changes in tumor volume quite well. Thus, we suggest that this assumption is only a reasonable initial approximation.

The algorithm for cycling treatment turns treatment on when PSA is both sufficiently high and increasing, turns it off when PSA is both sufficiently low and decreasing; this is formally modeled as follows:

$$u(t) = \begin{cases} 0 \to 1 & \text{when } y(t) = r_1 \text{ and } dy/dt > 0, \\ 1 \to 0 & \text{when } y(t) = r_0 \text{ and } dy/dt < 0, \end{cases} \tag{5.78}$$

where r_1 is the threshold PSA for activating treatment, and r_0 is the threshold PSA for ceasing treatment.

Ideta et al. examined the efficacy of cycling androgen deprivation therapy under the three hypotheses for the growth of AI cells in an androgen rich environment. Importantly, they found that only under hypotheses (2) and (3), where androgens inhibit AI cell proliferation, does intermittent therapy delay androgen-independent relapse. Since we have argued on biological grounds that these hypotheses are unlikely in an in vivo tumor, or at most account for only a small fraction of AI tumors, the results under hypothesis (1) are the most clinically relevant.

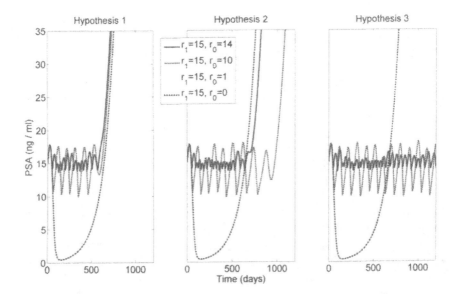

FIGURE 5.5: Intermittent androgen deprivation therapy in Ideta et al.'s model [23] under the three hypotheses for the effect of androgens on AI cell proliferation. Parameter values are $\alpha_1 = 0.0204$ (bone mets) to 0.0290 (lymph node mets), $\alpha_2 = 0.0242$ to 0.0277, $\beta_1 = 0.0076$ to 0.0085, $\beta_2 = 0.0168$ to 0.0222, $k_1 = 0$, $k_2 = 2$, $k_3 = 8$, $k_4 = 0.5$, $\gamma = 0.08$, $a_0 = 30$, and $m_1 = 0.00005$ to 0.0002.

Figure 5.5 shows how different schedules of intermittent therapy (i.e., different r_0s) affect time to AI relapse. Under hypothesis (3), any treatment cycling at all permanently prevents relapse, while continuous androgen deprivation results in aggressive recurrence. Under hypothesis (2), relapse is only delayed by intermittent therapy, while under hypothesis (1), intermittent therapy slightly reduces the time to relapse. However, because it can reduce the side effects of therapy and improve quality of life, a slightly shorter time to relapse may be an acceptable trade-off.

Interestingly, the Ideta et al. treatment algorithm suggests that modifying the one governing parameter, r_0, can greatly affect the time to relapse, with a lower r_0 significantly increasing the time to relapse. As can be seen from Figure 5.5, a lower r_0 roughly translates to a longer period of treatment cycling. Moreover, the simplicity of this algorithm makes its clinical use feasible.

5.9   Predictions and limitations of current ADT models

Both the Jackson and Ideta et al. model formulations assume that AI cells are either out-competed or do not arise in normal androgen environments. As the authors note, this assumption is reasonably justified from the biological observation that such cells are not routinely observed in untreated prostate cancers. It remains an open question, however, how selection operates on AI cells in an androgen-rich environment. It is certainly possible that selection disfavors AI cells in such an environment. Alternatively, the AI phenotype may be neutral—neither favored by selection nor disfavored. In the latter case, one could expect at least a small population of AI cells to exist in prostate cancers before androgen ablation begins. In a few cases, the population may become quite sizable prior to androgen ablation, since neutral traits can spread through populations by genetic drift.

The Jackson model specifically hypothesizes that selection acts against androgen-independent clones in environments with high androgen concentrations. There is some evidence, primarily from in vitro cell line data, that in some cases physiologic androgen levels can induce apoptosis in androgen independent prostate cancer cells. However, administration of androgens in hormone-refractory cancer nearly always results in disease flare [20]. Furthermore, AI cells generally retain some dependence upon the AR and usually evolve to a so-called androgen independent state by activating AR dependent genes through other means. These observations support the alternative hypothesis that androgens promote proliferation in both AI and AD cells. The degree to which this assumption affects the dynamics and predictions of Jackson's model is unclear. If this alternative hypothesis were true, then the Ideta et al. model would also be affected since it explores the possibilities of androgen having either no effect on AI cell growth or a negative effect. However, the Ideta et al. model could be extended to explore this possibility and its effects on treatment cycling.

These models highlight the difficulty of examining evolutionary dynamics when competition between a limited number of pre-defined clones is considered. However, despite their limitations, the Jackson and Ideta et al. models do make several interesting predictions. Specifically, they predict that only a small region of parameter space will allow androgen ablation therapy to completely control prostate cancer. Furthermore, in the case of AI cells whose growth is inhibited by androgens, intermittent androgen deprivation can delay or prevent AI relapse. If the proliferation of AI cells is unaffected by androgens, then intermittent therapy can slightly reduce the time to relapse. Also, more infrequent cycling of therapy delays relapse.

5.10   An immunotherapy model for advanced prostate cancer

Cancer immunotherapy has been studied with a mathematical model by Kirschner and Panetta [29]. Their model examines the dynamics between the adaptive immune system, tumor cells, and the cytokine interleukin-2 (IL-2). The model shows that the immune system can control tumors with average to high antigenicity at a dormant state. However, their results suggest that treatment with IL-2 alone may not clear the tumor without administering toxic levels of the cytokine. Partially motivated by the work of Kirschner and Panetta [29], Porta and Kuang [44] formulated a mathematical model of advanced prostate cancer treatment to examine the combined effects of ADT and immunotherapy.

From previous sections we know that ADT, while initially successful, eventually results in a relapse after two to three years in the form of androgen-independent prostate cancer. Intermittent androgen deprivation therapy attempts to enhance quality of life and occasionally prevent relapse by cycling the patient on and off treatment. Over the past decade, dendritic cell (DC) vaccines have been used with some success in clinical studies for the immunotherapy of prostate cancer. Although these studies found that the DC vaccines could slow the progression of the disease in hormone refractory patients, they did not show how the vaccine would affect patients actively undergoing hormone therapy.

In this section we present the Portz-Kuang model, which studies the efficacy of dendritic cell vaccines when used with continuous or intermittent ADT schedules. Their model may determine if intermittent therapy has any benefits over continuous therapy, other than improved quality of life, when

TABLE 5.6: Variables in the Portz-Kuang model.

Variable	Meaning	Unit
X_1	number of androgen-dependent cancer cells	cells
X_2	number of androgen-independent cancer cells	cells
T	number of activated T cells	cells
I_L	concentration of cytokines	ng/ml
A	concentration of androgen	nmol/ml
D	number of dendritic cells	cells

combined with the use of DC vaccines. The necessary conditions for disease elimination or stabilization using such combined treatments can also be examined by this model. Numerical simulations of the Portz-Kuang model suggest that immunotherapy can indeed successfully stabilize the disease using both continuous and intermittent androgen deprivation. This section is adapted from the work contained in Portz and Kuang [44].

In Portz and Kuang [44], the prostate cancer treatment by immunotherapy and androgen deprivation therapy is modeled by a system of ordinary differential equations which takes the form,

$$\frac{dX_1}{dt} = \underbrace{r_1(A)X_1}_{\text{proliferation and death}} - \underbrace{m(A)X_1}_{\text{mutation to AI}} - \underbrace{\frac{e_1 X_1 T}{g_1 + X_1}}_{\text{killed by T cells}}, \tag{5.79}$$

$$\frac{dX_2}{dt} = \underbrace{r_2 X_2}_{\text{proliferation and death}} + \underbrace{m(A)X_1}_{\text{mutation from AD}} - \underbrace{\frac{e_1 X_2 T}{g_1 + X_2}}_{\text{killed by T cells}}, \tag{5.80}$$

$$\frac{dT}{dt} = \underbrace{\frac{e_2 D}{g_2 + D}}_{\text{activation by dendritic cells}} - \underbrace{\mu T}_{\text{natural death}} + \underbrace{\frac{e_3 T I_L}{g_3 + I_L}}_{\text{clonal expansion}}, \tag{5.81}$$

$$\frac{dI_L}{dt} = \underbrace{\frac{e_4 T (X_1 + X_2)}{g_4 + X1 + X2}}_{\text{production by stimulated T cells}} - \underbrace{\omega I_L}_{\text{clearance}}, \tag{5.82}$$

$$\frac{dA}{dt} = \underbrace{\gamma(a_0 - A)}_{\text{homeostasis}} - \underbrace{\gamma a_0 u(t)}_{\text{deprivation therapy}}, \tag{5.83}$$

$$\frac{dD}{dt} = \underbrace{-cD}_{\text{natural death}}. \tag{5.84}$$

The variables used in the model and their meanings are listed in Table 5.6. Parameter interpretations and estimates are given in Table 5.7.

As in the previous section, the androgen-dependent functions for AD cell

TABLE 5.7:　Parameters (Para.) in the Portz-Kuang model.

Para.	Meaning	Value	Ref.
α_1	AD cell proliferation rate	0.025/day	[4]
β_1	AD cell death rate	0.008/day	[4]
k_1	AD cell proliferation rate dependence on androgen	2 ng/ml	[23]
k_2	effect of low androgen level on AD cell death rate	8	[5]
k_3	AD cell death rate dependence on androgen	0.5 ng/ml	[23]
r_2	AI cell net growth rate	0.006/day	[4]
m_1	maximum mutation rate	0.00005/day	[23]
a_0	normal androgen concentration	30 ng/ml	[23]
γ	androgen clearance and production rate	0.08/day	[23]
ω	cytokine clearance rate	10/day	[52]
μ	T cell death rate	0.03/day	[29]
c	dendritic cell death rate	0.14/day	[35]
e_1	maximum rate T cells kill cancer cells	0 – 1/day	[29]
g_1	cancer cell saturation level for T cell kill rate	10×10^9 cells	[29]
e_2	maximum T cell activation rate	20×10^6 cells/day	[29]
g_2	DC saturation level for T cell activation	400×10^6 cells	[57]
e_3	maximum rate of clonal expansion	0.1245/day	[29]
g_3	IL-2 saturation level for T cell clonal expansion	1000 ng/ml	[29]
e_4	maximum rate T cells produce IL-2	5×10^{-6} ng/ml/cell/day	[29]
g_4	cancer cell saturation level for T cell stimulation	10×10^9 cells	[29]
D_1	DC vaccine dosage	300×10^6 cells	[57]
c_1	AD cell PSA level correlation	10^{-9} ng/ml/cell	[23]
c_2	AI cell PSA level correlation	10^{-9} ng/ml/cell	[23]

growth and mutation are defined as follows:

$$r_1(A) = \alpha_1 \frac{A}{A + k_1} - \beta_1 \left(k_2 + (1 - k_2)\frac{A}{A + k_3} \right), \qquad (5.85)$$

$$m(A) = m_1(1 - \frac{A}{a_0}). \qquad (5.86)$$

The parameters in the expression for $r_1(A)$ are chosen such that the net growth rate of AD cells is $\alpha_1 - \beta_1$ when $A = a_0$ or $-\beta_1 k_2$ when $A = 0$. The value of parameter k_2 is chosen such that $\beta_1 k_2$ matches the rate of decline of serum PSA concentration during continuous ADT [23]. The net growth rate of AI cells, r_2, is a constant in this version of the model. Ideta et al. proposed two alternatives for r_2 which assumed that androgen had a negative effect on the proliferation rate of AI cells [23]. Mutation from AD to AI occurs at a rate m_1 when $A = 0$, and no mutation occurs when $A = a_0$. Larger values of m_1 result in a shorter time to androgen-independent relapse; thus, relapse time can be used to estimate the value of m_1 [23].

Intermittent androgen deprivation therapy is modeled by equation (5.83) where $u(t) = 0$ indicates an on-treatment period, and $u(t) = 1$ indicates an off-treatment episode. During off-treatment periods, the androgen level tends toward the set-point androgen level, a_0. Androgen decays at a rate γ during on-treatment periods. The treatment function, $u(t)$, is controlled by monitoring the serum PSA level as we saw earlier:

$$y(t) = c_1 X_1 + c_2 X_2, \tag{5.87}$$

$$u(t) = \begin{cases} 0 \to 1 & \text{when } y(t) > L_1 \text{ and } dy/dt > 0, \\ 1 \to 0 & \text{when } y(t) < L_0 \text{ and } dy/dt < 0, \end{cases} \tag{5.88}$$

where $y(t)$ is the serum PSA concentration. Androgen deprivation is switched on when the serum PSA concentration exceeds some level, L_1, and switched off when the serum PSA concentration drops below some level, L_0, with $L_0 < L_1$.

Since T cells are activated and stimulated through interactions between proteins (antigens and cytokines) and receptors [31], Michaelis-Menten kinetics are used for all immune response terms in the model. This is the approach taken by Kirschner and Panetta and is reasonable given that high levels of antigens and cytokines are likely to have a saturation effect on the T cells. IL-2 is included in the model to provide the clonal expansion dynamics of helper T cells. When stimulated by the antigens presented on tumor cells, the helper T cells produce IL-2. The IL-2 then stimulates the clonal expansion of T cells in a positive feedback loop [31]. The model assumes a constant ratio of cytotoxic and helper T cells, which greatly simplifies the model and should not have a significant impact on the long-term behavior of the system. The cytotoxic T cells interact with and kill the tumor cells based on antigen stimulation. The rate of interaction is assumed to be the same for both AD and AI cells. There is no biological reason to assume otherwise.

The antigen-loaded dendritic cells are modeled by equation (5.84), which assumes that the DCs undergo apoptosis at a constant rate and are not being replenished by any mechanisms other than further vaccinations. Vaccinations are administered every 30 days in model simulations. Each vaccination contains D_1 antigen-loaded DCs. The DCs are assumed to activate naïve T cells based on Michaelis-Menten kinetics, as shown in (5.81). The model assumes that there are always naïve T cells available for activation.

The Michaelis-Menten terms could be replaced by simpler mass action terms to make the non-zero steady states easier to find analytically. However, the system is repeatedly being perturbed by the administration of DC vaccines, so steady-state analysis has limited use. For this and several other reasons, the model was analyzed primarily through numerical simulations in Portz and Kuang [44] with Michaelis-Menten terms.

When a dendritic cell vaccine is combined with continuous androgen deprivation, the model shows that the cancer can be eliminated with a relatively strong (but still within a reasonable parameter range) antitumor immune response. Clinical studies have shown that DC vaccines are able to stabilize disease progression in some patients with hormone refractory prostate cancer. Since these patients have androgen-independent cancer, we can safely assume that DC vaccines are capable of stopping the net growth of AI cells. When administered to a patient actively undergoing hormone therapy, one would expect the DC vaccine to prevent an androgen-independent relapse while allowing continuous androgen deprivation therapy to eliminate the AD cell population. Thus the results of the above model are reasonable in the case of DC vaccination combined with continuous androgen deprivation.

An interesting result of the model is that the DC vaccine is able to prevent relapse with a slightly weaker antitumor immune response when intermittent androgen deprivation is used instead of continuous androgen deprivation. While the difference was only small ($e_1 = 0.68973$ compared to $e_1 = 0.69197$), the result was still surprising considering the effect that intermittent therapy has on relapse time without immunotherapy. This small difference can likely be attributed to the ability of the immune response to offset the higher mutation rate when intermittent therapy is used and also to the consistently larger cancer cell population which is necessary to stimulate the clonal expansion of T cells. With the low toxicity of DC vaccines [57] and the quality-of-life benefits of IAD, the combination of these two treatments could be very advantageous over continuous ADT for the treatment of advanced prostate cancer.

The hybrid nature of the model—combing continuous treatment dynamics with discrete treatment—makes standard analysis difficult. This situation forced Portz and Kuang [44] to use specific parameter values to find the desired values of e_1, preventing the possibility of having an algebraic expression for those values of e_1. Simplifying the model may be possible by using some simple continuous functions for the $D(t)$ and $u(t)$, although it would likely make clinical application of the model infeasible.

Accuracy in the estimation of parameter values for the immune response is a significant limitation of the model. Most of the parameter estimates were based on those used by Kirschner and Panetta [29]. While many of these parameter estimates were based on data from biological studies, the estimates are for generic antitumor immune responses. The proper parameter values for an immune response against prostate cancer may be significantly different.

Another serious limitation of the model is the lack of regulatory T cells, which reduce sensitivity to self-antigens and prevent self-destructive lymphocyte responses [41]. Since targeting cancer cells requires targeting self-antigens, regulatory T cells may have a significant suppressive effect on anti-tumor immunity [72]. Several studies have examined ways to target regulatory T cells to reduce their suppressive effect on immunotherapy [72]. Including regulatory T cells in the model is a possible direction for future work, and the therapies which target the regulatory T cells could then be examined.

5.11 Other prostate models

While we have primarily restricted our focus to prostate cancer, we would be remiss if we did not mention the model of Barton and Anderson [3] and its extension by Potter et al. [45] on androgen regulation of prostate growth. It is a complicated, compartmentalized pharmacokinetics model that comprehensively describes androgen dynamics and may serve as a valuable resource in developing future models more focused on cancer dynamics.

Jain et al. [26] formulated a comprehensive mathematical model of prostate cancer progression in response to androgen ablation therapy based on the model framework described in this chapter, which is adapted from the work presented in [14]. Their model includes patient-dependent parameters and captures a variety of clinically observed outcomes for typical patient data under various intermittent schedules. They fit their simulations to data reported in the literature, and then project the future course of the disease for the next 5 to 10-year period under either intermittent or continuous therapy. Their model predicts that intermittent scheduling will yield more benefit if hormone-sensitive cells have a competitive advantage since it may delay the acquisition of genetic or epigenetic alterations empowering androgen resistance. Subsequently, Jain and Friedman [27] presented some simplifications and variations to the model and carried out systematic computational and mathematical analysis of their simplified models. Most noteworthy is their novel mathematical definitions of treatment viability, treatment failure and treatment failure time. These rigorous definitions enable the authors to effectively compare and contrast prostate cancer response to continuous versus intermittent androgen ablation therapy via mathematical models. Friedman and Jain [16] also formulated and studied a free boundary partial differential equation model of metastasized prostatic cancer where they were able to establish the existence and uniqueness of solution for the model. In addition, they also established the global existence of solutions for the radially symmetric case.

Dimonte, an advanced stage prostate cancer patient and an experienced mathematical modeler, developed a sophisticated cell kinetics model for prostate cancer progression from diagnosis to final clinical outcome subject to various treatment strategies [11]. To describe hormone ablation therapy, Dimonte assumed the existence of three cancer cell populations: (1) cells local to the prostate and sensitive to hormones, (2) regional and hormone sensitive cells, and (3) systemic and hormone resistant cells. The model comprises three coupled first-order differential equations, each describing dynamics of one of these three populations. Time is scaled to the doubling time of the prostate specific antigen concentration. The rate at which local cells transition to the systemic compartment is associated with the patient's Gleason score. The model also assumes three threshold cell population sizes that represent (1) initiation of tumor spread out of the local compartment, (2) saturation capacity of the local tumor, and (3) the cell count likely to cause PC-specific death. These parameters can be calibrated using published PC clinical data and survival tables. Dimonte then applies the model to two individuals (himself and another) with complete PC diagnostic data and calculates the time to PC-specific death. In a subsequent model, Dimonte et al. [12] introduced a simplification and employed it to clarify the observed variability among biochemical recurrence nomograms for prostate cancer and suggested that nomograms should be stratified by PSA doubling time to improve predictive power.

In the same issue of the journal published the work of Dimonte [11], Hirata et al. [22] introduced a mathematical model for serum PSA dynamics in patients receiving intermittent androgen suppression (IAS) for prostate cancer. The validity of their model is supported by patient data obtained from a clinical trial of IAS. More specifically, Hirata et al. [21] formulated an intermittent androgen ablation model that comprises two submodels: an on-treatment linear model; and an off-treatment linear model. Their model considered an AD cell population (x_1), a reversible AI cell population (x_2), and an irreversible AI cell population (x_3), modeled by the following linear system

$$\frac{d}{dt} \begin{pmatrix} x_1(t) \\ x_2(t) \\ x_3(t) \end{pmatrix} = \begin{pmatrix} w_{1,1}^1 & 0 & 0 \\ w_{2,1}^1 & w_{2,2}^1 & 0 \\ w_{3,1}^1 & w_{3,2}^1 & 0 \end{pmatrix} \begin{pmatrix} x_1(t) \\ x_2(t) \\ x_3(t) \end{pmatrix} \tag{5.89}$$

for the on-treatment periods, and

$$\frac{d}{dt} \begin{pmatrix} x_1(t) \\ x_2(t) \\ x_3(t) \end{pmatrix} = \begin{pmatrix} w_{1,1}^0 & w_{1,2}^0 & 0 \\ 0 & w_{2,2}^0 & 0 \\ 0 & 0 & w_{3,3}^0 \end{pmatrix} \begin{pmatrix} x_1(t) \\ x_2(t) \\ x_3(t) \end{pmatrix} \tag{5.90}$$

for the off-treatment periods. Hirata et al. constrained the parameters so that all those off the diagonal are non-negative, while $w_{3,3}^0 > 0$, and the cell class can change its volume by at most 20% per day. They model androgen dependence by allowing parameters to vary for each on and off treatment period as well as among treatment cycles. Their model fits highly nonlinear individual patient data very well. See [21, 22] for further details.

In contrast, a more mechanistic model of intermittent androgen ablation treatment was formulated by Portz et al. [43]. This model uses androgen and

PSA data from individual patients to calibrate the various model parameters, which are clinically measurable or can be estimated from data in literature. Their model fits multiple cycles of individual patient PSA data very well, with parameter values obtained from the data well within observed physiological ranges. We will cover their model and its variations in some detail in the next chapter.

5.12   Exercises

Exercise 5.1: Systematically study the global dynamics of the ODE version of Jackson's model, consisting of equations (5.55)–(5.56), with three different treatment scenarios as described by the following androgen profiles.

$$a(t) = \begin{cases} a_0 & ; t < T, \text{for no treatment}, \\ a_0 e^{-bt} + a_s & ; t \geq T, \text{for ADT}, \\ a_0 e^{-bt} & ; t \geq T, \text{for TAB therapy}. \end{cases} \tag{5.91}$$

Exercise 5.2: Reproduce the Figure 3 in [24].

Exercise 5.3: Systematically study the global dynamics of the Ideta et al. model under hypothesis (1) (i.e., $p_2(a) \equiv 1$). Show that intermittent ADT reduces time to relapse as illustrated by the Figure 5.5. In other words, continuous ADT is slightly superior in terms of delaying treatment resistance.

Exercise 5.4: Systematically study the global dynamics of the Ideta et al. model under hypothesis (2) (i.e., $p_2(a) \equiv 1$). Show that intermittent ADT increases time to relapse as illustrated by the Figure 5.5. In other words, both intermittent and continuous ADT will result in treatment resistance, with intermittent ADT being superior in delaying treatment resistance.

Exercise 5.5: Systematically study the global dynamics of the Ideta et al.'s model under hypothesis (3) (i.e., $p_2(a) \equiv 1$). Show that any treatment cycling at all permanently prevents relapse as illustrated by the Figure 5.5.

Exercise 5.6: In a recent paper by Jain and Friedman [27], the authors formulated a model for prostate carcinoma growth and treatment with androgen deprivation therapy. The principal species modeled are the following: number of androgen-dependent cancer cells in millions (N), number of castration-resistant cancer cells in millions (M), and serum PSA concentration in ng/ml (P). When treatment is off $(u(t) = 0)$, N and M grow at rates $\alpha_N(1 - \epsilon_M)N$ and $\alpha_M M + \alpha_N \epsilon_M N$, respectively, and have growth rates $-\beta_N N$ and $\beta_M M$, respectively, when the treatment is on. Jain and Friedman assume that androgen-dependent cells mutate irreversibly to an androgen-independent phenotype with probability ϵ_M. Both androgen-dependent and -independent cells are assumed to produce PSA. These assumptions yield the following model:

$$\frac{dN}{dt} = \alpha_N(1 - \epsilon_M)N(1 - u(t)) - \beta_N N u(t), \tag{5.92}$$

$$\frac{dM}{dt} = (\alpha_M M + \alpha_N \epsilon_M N)(1 - u(t)) + \beta_M M u(t), \tag{5.93}$$

$$\frac{dP}{dt} = \theta_N N + \theta_M M - \lambda_P P. \tag{5.94}$$

Solve the above system explicitly in cases when the treatment is locked on and locked off.

Exercise 5.7: Reproduce Figure 4 in the paper by Jain and Friedman [27].

Exercise 5.8: Simulate the model of Portz and Kuang [44] with parameters given in Table 5.7. You may use or adapt the following MATLAB® program.

```
--------------------------------------------------------------
% MATLAB program to simulate a prostate cancer model with intermittent
% ADT and dendritic cell vaccine immunotherapy
tstart = 0; tfinal = 1200;
x0=[15;0.1;0;0;30;0]; % initial conditions (billions of cells)
% treatment parameters
u = 1;       % androgen-ablation on/off state
L0 = 3;      % androgen-ablation off PSA value
L1 = 13;     % androgen-ablation on PSA value
%D1 = 0.3; % number of DCs (billions) in single infusion
D1 = 0;
tstep = 1200;
%tstep = 30; % days between immunotherapy injections
% tumor size to PSA level correlation parameters
c1 = 1; c2 = c1;
tout = tstart; xout = x0.';
teout = []; xeout = []; ieout = [];
options = odeset('Events', @(t,x)events(t,x,c1,c2,L0,L1,u));
first = true;
tprev = tstart;
while tstart < tfinal
    [t,x,te,xe,ie] = ode45(@(t,x)f(t,x,u),...
        [tstart min([tprev+tstep tfinal])],x0,options);
    nt = length(t);
    tout = [tout; t(2:nt)];
    xout = [xout; x(2:nt,:)];
    if isempty(ie)
        % administer DC vaccine
        D = x(nt,6)+D1;
        tprev = t(nt);
    else
        % toggle androgen ablation therapy
```

```
        u = 1 - u;
        D = x(nt,6);
    end
    % adjust initial conditions
    x0 = x(nt,1:5);
    x0(6) = D;
    options = odeset('Events', @(t,x)events(t,x,c1,c2,L0,L1,u),...
        'InitialStep',t(nt)-t(nt-1),'MaxStep',t(nt)-t(1));
    tstart = t(nt);
end
y = c1.*xout(:,1) + c2.*xout(:,2);

figure(1);
hold off
plot(tout,y,'k','LineWidth',1.5);
xlabel('time (days)'); ylabel('PSA (ng/mL)');
set(gca,'xlim',[0 tfinal],'ylim',[0 40]);
box on

figure(2);
hold off
plot(tout,xout(:,5),'k-','LineWidth',1.5);
xlabel('time (days)');
ylabel('Androgen (nmol/L)');
set(gca,'xlim',[0 tfinal],'ylim',[0 40]);
box on

figure(3);
hold off
subplot(2,1,1);
plot(tout,xout(:,1),'k','LineWidth',1.5);
xlabel('time (days)'); ylabel('AD cells');
subplot(2,1,2);
plot(tout,xout(:,2),'k-','LineWidth',1.5);
xlabel('time (days)'); ylabel('AI cells');
% ----------------------------------------------------------------
function dxdt = f(t,x,u)
X1 = x(1); X2 = x(2); T = x(3); IL = x(4); A = x(5); D = x(6);
alpha1 = 0.025; beta1 = 0.008;
k1 = 2; k2 = 8; k3 = 0.5; alpha2 = 0.025; beta2 = 0.019;
m1 = 0.00005;   % mutation rate
a0 = 30;        % normal androgen level
gamma = 0.08;   % androgen clearance rate
c = 0.14;       % DC death rate
e1 = 0.6897325;
g1 = 10; e2 = e1; g2 = g1;
e3 = 0.02;      % T-cell sensitivity to dendritic cells
g3 = 0.4;
e4 = 0.1245;    % T-cell sensitivity to IL-2
```

```
g4 = 1e3;
e5 = 5e3;          % IL-2 production from T cell stimulation
g5 = 10;
mu = 0.03;         % T-cell death rate
omega = 10;        % IL-2 clearance rate

r1 = alpha1*A/(A+k1) - beta1*(k2+(1-k2)*A/(A+k3));
r2 = alpha2 - beta2;
m = m1*(1-A/a0);
dX1 = r1*X1 - m*X1 - e1*X1*T/(g1+X1);
dX2 = r2*X2 + m*X1 - e2*X2*T/(g2+X2);
dT = e3*D/(g3+D) - mu*T + e4*T*IL/(g4+IL);
dIL = e5*T*(X1+X2)/(g5+X1+X2) - omega*IL;
dA = gamma*(a0-A)-gamma*a0*u;
dD = -c*D;
dxdt = [dX1; dX2; dT; dIL; dA; dD];
% ------------------------------------------------------------------
function [value,isterminal,direction] = events(t,x,c1,c2,L0,L1,u)
% hysteresis relay
isterminal = 1;
if u == 1
    value = c1*x(1) + c2*x(2) - L0;
    direction = -1;
else
    value = c1*x(1) + c2*x(2) - L1;
    direction = 1;
end
% ------------------------------------------------------------
```

Exercise 5.9: Consider a special case of the model in Portz and Kuang [44] in which there is just one type of tumor cells X_1. This can be achieved by assuming $X_2(0) = 0$ and $m(A) = 0$. Assume also that both the dendritic cell density and the androgen level are constant so that their model is reduced to a system of just three equations. In addition, assume that the term describing T-cells killing cancer cells can be modeled by a mass action term and same for the term describing the IL-2 production due to T-cell stimulation. Study the boundedness of solutions, existence of steady states and the stability of the steady states of this reduced system.

Exercise 5.10: In addition to the assumptions presented in the previous exercise, apply a quasi-steady state approximation on the IL-2 equation. This shall reduce the model to a system of two equations. Study the boundedness of solutions, existence of steady states, the local and global stability of the steady states of this dramatically reduced system.

5.13 Projects

Here, we briefly suggest some modifications or extensions of existing dynamical models that could be pursued as meaningful projects by the interested students.

5.13.1 The epithelial-vascular interface and serum PSA

A surprising and important prediction of Vollmer and Humphrey's model is that at steady state the serum PSA does not depend upon β, the rate of transport from the interstitium to the serum, which is a marker for the extent of the epithelial-vascular interface. Moreover, we have found that when PSA is increasing according to an empirical exponential model, β has a minimal effect on PSA over most biologically reasonable parameter space. It is important to determine if this prediction is robust under different model constructions. For example, one can model transport as a function of the vascular surface area, A, the PSA concentration difference in the tissue and serum compartments, $(f - y)$, and a permeability coefficient P. This gives the new model,

$$\frac{\mathrm{d}f}{\mathrm{d}t} = \alpha - PA(f - y)\frac{1}{V_p}, \tag{5.95}$$

$$\frac{\mathrm{d}y}{\mathrm{d}t} = PA(f - y)\frac{1}{V_s} - ky. \tag{5.96}$$

Some simple calculations show that the steady state for this model is

$$f_s = \frac{\alpha V_p}{PA} + \frac{\alpha V_p}{kV_s}, \tag{5.97}$$

$$y_s = \frac{\alpha V_p}{kV_s}. \tag{5.98}$$

Thus, while the tissue PSA steady state is more complex and explicitly depends on the transport parameters P and A, serum PSA is unchanged from the original model, remaining independent of the vasculature. An important advantage of this formulation is that the parameters P and A may be estimated directly from empirical biological data.

We now return to the original model, but suggest that the PSA in the tissue compartment should also degrade at some rate μ. This yields another modified model:

$$\frac{\mathrm{d}f}{\mathrm{d}t} = \alpha - \beta f - \mu f, \tag{5.99}$$

$$\frac{\mathrm{d}y}{\mathrm{d}t} = \beta f - ky. \tag{5.100}$$

The steady state is now

$$f_s = \frac{\alpha}{\beta + \mu},$$ (5.101)

$$y_s = \frac{\alpha\beta}{k(\beta + \mu)} \frac{V_p}{V_s}.$$ (5.102)

Thus, PSA degradation in the tissue compartment causes steady state serum PSA to depend upon β, and therefore the epithelial-vascular interface. We suggest that the interested student more thoroughly examine and characterize the dynamics of these modified models, and as a final modification, combine the two. This would allow a more detailed understanding of how tissue compartment PSA dynamics and transport parameters interact to determine serum PSA, hopefully giving deeper insight into the importance of the tissue-vascular interface in determining serum PSA.

5.13.2   A clinical algorithm based on a dynamical model

Our analysis of the models of both Swanson et al. [61] and Vollmer and Humphrey [67], among other things, suggests that serum PSA is not a reliable predictor of tumor volume in individual cases. Indeed, serum PSA in the 4 to 10 ng/ml range is not a reliable predictor of the presence of clinically meaningful cancer. Therefore, in order to determine if an individual patient's PSA indicates the existence of cancer, we propose the following example algorithm:

1. Assume cancer does not exist.

2. Use several serial PSA measurements to determine PSA velocity.

3. Measure prostate volume by transrectal ultrasound and estimate epithelial fraction.

4. Calculate the likely range of serum PSA under equation (5.25) derived by Vollmer and Humphrey [67] (but with $V_c = 0$).

In other words, we are deriving a PSA interval implied by a null hypothesis—*viz.* that cancer is not present—based on several individually measured parameter values and the normal value range for the other parameters. If the PSA is outside the range predicted by the null hypothesis, then we may accept the alternative hypothesis that cancer is present. An algorithm along these lines, tested on a clinical data-set, could be of clinical use.

5.13.3   An extension of Vollmer and Humphrey's model

As it stands, γ, the relative PSA velocity, reflects the growth of both benign and cancerous tissues. Both c_1 and c_2 are affected by γ. One could productively extend Vollmer and Humphrey's model by adding an explicit model of tumor growth. A simple exponential growth model could be used as a first approximation; e.g.,

$$V_c = V_0 e^{\gamma t}. \tag{5.103}$$

Total prostate volume,

$$V_p = V_b + V_c, \tag{5.104}$$

would cease to be constant. Further analysis of this or a similar model could give a better understanding of the relative contribution to PSA change by cancerous tissue compared to benign tissue. Growth in the benign tissue compartment could also be considered explicitly.

5.13.4 Androgens positively regulating AI cell proliferation

In the model of Ideta et al. [23], we can assume that androgens increase AI proliferation, and even in the absence of androgens there is some small but positive proliferation rate. Therefore, AI cells will have a proliferative advantage when androgen deprivation is in effect, but disease may flare differently when androgens are re-introduced as compared to the predictions of other hypotheses. This hypothesis can be formalized and used to examine, both numerically and mathematically, how intermittent androgen ablation affects AI recurrence.

5.13.5 Combining androgen ablation with other therapies

Another interesting extension of Ideta et al.'s model [23] would include an examination of AI recurrence following androgen ablation therapy combined with surgical resection of a portion of the tumor. In a numerical ODE setting, surgical treatment can easily be simulated by setting x_1 and x_2 to some fraction of their previous value; a lower fraction implies more complete surgery.

It is not intuitively obvious whether cancer would be better controlled by surgical resection of a portion of the tumor at the beginning of therapy or at some later time, nor how different schedules of androgen ablation may interact with surgery. Numerical experiments may give insights into this problem.

Cytotoxic chemotherapy of the tumor in combination with intermittent androgen ablation therapy may also be studied. As a first approximation, one can model chemotherapy very simply by adding an additional, first-order death term to the governing ODE for x_1 and x_2. Some simple schedules of chemotherapy that could be studied include:

1. A course of chemotherapy at the onset of treatment.

2. A course of chemotherapy upon AI recurrence.

3. Chemotherapy only concurrent with androgen ablation therapy.

4. Chemotherapy and androgen ablation therapy in alternating schedules.

References

[1] Banerjee PP, Banerjee S, Lai JM, Strandberg JD, Zirkin BR, Brown TR: Age-dependent and lobe-specific spontaneous hyperplasia in the brown Norway rat prostate. *Biol Reprod* 1998, 59:1163–1170.

[2] Banerjee S, Banerjee PP, Brown TR: Castration-induced apoptotic cell death in the Brown Norway rat prostate decreases as a function of age. *Endocrinology* 2000, 141:821–832.

[3] Barton HA, Andersen ME: A model for pharmacokinetics and physiological feedback among hormones of the testicular-pituitary axis in adult male rats: A framework for evaluating effects of endocrine active compounds. *Toxicol Sci* 1998, 45:174–187.

[4] Berges RR, Vukanovic J, Epstein JI, CarMichel M, Cisek L, Johnson DE, Veltri RW, Walsh PC, Isaacs JT: Implication of cell kinetic changes during the progression of human prostatic cancer. *Clin Cancer Res* 1995, 1:473–480.

[5] Bruchovsky N, Klotz L, Crook J, Goldenberg SL: Locally advanced prostate cancer-biochemical results from a prospective phase II study of intermittent androgen suppression for men with evidence of prostate-specific antigen recurrence after radiotherapy. *Cancer* 2007, 109:858–867.

[6] Cayatte C, Pons C, Guigonis JM, Pizzol J, Elies L, Kennel P, Rouqui D, Bars R, Rossi B, Samson M: Protein profiling of rat ventral prostate following chronic finasteride administration: Identification and localization of a novel putative androgen-regulated protein. *Mol Cell Proteomics* 2006, 5:2031–2043.

[7] Chan LW, Stamey TA: Calculating prostate cancer volume preoperatively: The D'Amico equation and some other observations. *J Urol* 1998, 159:1998–2003.

[8] Chen CD, Welsbie DS, Tran C, Baek SH, Chen R, Vessella R, Rosenfeld MG, Sawyers CL: Molecular determinants of resistance to antiandrogen therapy. *Nat Med* 2004, 10:33–39.

[9] D'Amico AV, Chang H, Holupka E, Renshaw A, Desjarden A, Chen M, Loughlin KR, Richie JP: Calculated prostate cancer volume: The optimal predictor of actual cancer volume and pathologic stage. *Urology* 1997, 49:385–391.

[10] Davies KJ: The broad spectrum of responses to oxidants in proliferating cells: A new paradigm for oxidative stress. *IUBMB Life* 1999, 48:41–47.

[11] Dimonte G: A cell kinetics model for prostate cancer and its application to clinical data and individual patients. *J Theor Biol* 2010, 264:420–442.

[12] Dimonte G, Bergstralh EJ, Bolander ME, Karnes RJ, Tindall DJ: Use of tumor dynamics to clarify the observed variability among biochemical recurrence nomograms for prostate cancer. *Prostate* 2012, 72:280–290.

[13] Dobzhansky T: Nothing in biology makes sense except in the light of evolution. *Am Biol Teacher* 1973, 35:125–129.

[14] Eikenberry SE, Nagy JD, Kuang Y: The evolutionary impact of androgen levels on prostate cancer in a multi-scale mathematical model. *Biol Direct* 2010, 5:24.

[15] Ellis WJ, Vessella RL, Buhler KR, Baldou F, True LD, Bigler SA, Curtis D, Lange PH: Characterization of a novel androgen-sensitive, prostate-specific antigen-producing prostatic carcinoma xenograft: luCaP 23. *Clin Cancer Res* 1996, 2:1039–1048.

[16] Friedman A, Jain HV: A partial differential equation model of metastasized prostatic cancer. *Math Biosc Eng* 2013, 10:591–608.

[17] Fukatsu A, Ono Y, Ito M, Yoshino Y, Hattori R, Gotoh M, Ohshima S: Relationship between serum prostate-specific antigen and calculated epithelial volume. *Urology* 2003, 61:370–374.

[18] Gregory CW, Johnson RT Jr, Mohler JL, French FS, Wilson EM: Androgen receptor stabilization in recurrent prostate cancer is associated with hypersensitivity to low androgen. *Cancer Res* 2001, 61:2892–2898.

[19] Haab F, Meulemans A, Boccon-Gibod L, Dauge MC, Delmas V, Boccon-Gibod L: Clearance of serum PSA after open surgery for benign prostatic hypertrophy, radical cystectomy, and radical prostatectomy. *Prostate* 1995, 26:334–338.

[20] Heinlein CA, Chang C: Androgen receptor in prostate cancer. *Endocr Rev* 2004, 25:276–308.

[21] Hirata Y, Akakura K, Higano CS, Bruchovsky N, Aihara K: Quantitative mathematical modeling of PSA dynamics of prostate cancer patients treated with intermittent androgen suppression. *J Mol Cell Biol* 2012, 4:127–132.

[22] Hirata Y, Bruchovsky N, Aihara K: Development of a mathematical model that predicts the outcome of hormone therapy for prostate cancer. *J Theor Biol* 2010, 264:517–527.

[23] Ideta A, Tanaka G, Takeuchi T, Aihara K: A Mathematical Model of Intermittent Androgen Suppression for Prostate Cancer. *J Nonlinear Sci* 2008, 18:593–614.

[24] Jackson TL: A mathematical model of prostate tumor growth and androgen-independent relapse. *Disc Cont Dyn Sys B* 2004, 4:187–201.

[25] Jackson TL: A mathematical investigation of the multiple pathways to recurrent prostate cancer: Comparison with experimental data. *Neoplasia* 2004, 6:697–704.

[26] Jain HV, Clinton SK, Bhinder A, Friedman A: Mathematical modeling of prostate cancer progression in response to androgen ablation therapy. *Proc Natl Acad Sci USA* 2011, 108:19701–19706.

[27] Jain HV, Fridman A: Modeling prostate cancer response to continuous versus intermittent androgen ablation therapy. *Disc Cont Dyn Sys B* 2013, 18:945–967.

[28] Jemal A, Siegel R, Ward E, Hao Y, Xu J, Murray T, Thun MJ: Cancer statistics, 2008. *CA Cancer J Clin* 2008, 58:71–96.

[29] Kirschner D, Panetta JC: Modeling immunotherapy of the tumor-immune interaction. *J Math Biol* 1998, 37:235–252.

[30] Koivisto P, Kolmer M, Visakorpi T, Kallioniemi OP: Androgen receptor gene and hormonal therapy failure of prostate cancer. *Am J Pathol* 1998, 152:1–9.

[31] Lanzavecchia A, Sallusto F: Regulation of T cell immunity by dendritic cells. *Cell* 2001, 106:263–266.

[32] Lepor H, Wang B, Shapiro E: Relationship between prostatic epithelial volume and serum prostate-specific antigen levels. *Urology* 1994, 44:199–205.

[33] Lin K, Lipsitz R, Miller T, Janakiraman S; U.S. Preventive Services Task Force: Benefits and harms of prostate-specific antigen screening for prostate cancer: An evidence update for the U.S. Preventive Services Task Force. *Ann Intern Med* 2008, 149:192–199.

[34] Lotan Y, Roehrborn CG: Clearance rates of total prostate specific antigen (PSA) after radical prostatectomy in African-Americans and Caucasians. *Prostate Cancer Prostatic Dis* 2002, 5:111–114.

[35] Lotze MT, Thomson AW: *Dendritic Cells: Biology and Clinical Applications.* London: Academic Press, 2001.

[36] Marks LS, Treiger B, Dorey FJ, Fu YS, deKernion JB. Morphometry of the prostate: I. Distribution of tissue components in hyperplastic glands. *Urology* 1994 44:486–492.

[37] Martel C, Trudel C, Couet J, Labrie C, Blanger A, Labrie F: Blockade of androstenedione-induced stimulation of androgen-sensitive parameters in the rat prostate by combination of Flutamide and 4-MA. *Mol Cell Endocrinol* 1993, 91:43–49.

[38] McLaren DB, McKenzie M, Duncan G, Pickles T: Watchful waiting or watchful progression?: Prostate specific antigen doubling times and clinical behavior in patients with early untreated prostate carcinoma. *Cancer* 1998, 82:342–348.

[39] Noldus J, Stamey TA: Limitations of serum prostate specific antigen in predicting peripheral and transition zone cancer volumes as measured by correlation coefficients. *J Urol* 1996, 155:232–237.

[40] Normington K, Russell DW: Tissue distribution and kinetic characteristics of rat steroid 5 alpha-reductase isozymes: Evidence for distinct physiological functions. *J Biol Chem* 1992, 267:19548–19554.

[41] O'Garra A, Vieira P: Regulatory T cells and mechanisms of immune system control. *Na Med* 2004, 10:801–805.

[42] Pang ST, Dillner K, Wu X, Pousette A, Norstedt G, Flores-Morales A: Gene expression profiling of androgen deficiency predicts a pathway of prostate apoptosis that involves genes related to oxidative stress. *Endocrinology* 2002, 143:4897–4906.

[43] Portz T, Kuang Y, Nagy JD: A clinical data validated mathematical model of prostate cancer growth under intermittent androgen suppression therapy. *AIP Advances*, 2012, 2:011002; doi: 10.1063/1.3697848

[44] Portz T, Kuang Y: A mathematical model for the immunotherapy of advanced prostate cancer, *BIOMAT 2012*, pp. 70-85. Edited by Rubem P Mondaini. World Scientific, 2013.

[45] Potter LK, Zager MG, Barton HA: Mathematical model for the androgenic regulation of the prostate in intact and castrated adult male rats. *Am J Physiol Endocrinol Metab* 2006, 291:E952–E964.

[46] Prehn RT: On the prevention and therapy of prostate cancer by androgen administration. *Cancer Res* 1999, 59:4161–4164.

[47] Prostate Cancer Trialists' Collaborative Group: Maximum androgen blockade in advanced prostate cancer: an overview of the randomised trials. *Lancet* 2000, 355:1491–1498.

[48] Raynaud JP: Prostate cancer risk in testosterone-treated men. *J Steroid Biochem Mol Biol* 2006, 102:261–266.

[49] Ripple MO, Henry WF, Rago RP, Wilding G: Prooxidant-antioxidant shift induced by androgen treatment of human prostate carcinoma cells. *J Natl Cancer Inst* 1997, 89:40–48.

[50] Rittmaster RS, Manning AP, Wright AS, Thomas LN, Whitefield S, Norman RW, Lazier CB, Rowden G: Evidence for atrophy and apoptosis in the ventral prostate of rats given the 5 alpha-reductase inhibitor finasteride. *Endocrinology* 1995, 136:741–748.

[51] Rocchi P, Muracciole X, Fina F, Mulholland DJ, Karsenty G, Palmari J, Ouafik L, Bladou F, Martin PM: Molecular analysis integrating different pathways associated with androgen-independent progression in LuCaP 23.1 xenograft. *Oncogene* 2004, 23:9111–9119.

[52] Rosenberg SA, Lotze MT: Cancer immunotherapy using interleukin-2 and interleukin-2-activated lymphocytes. *Annu Rev Immunol* 1986, 4:681–709.

[53] Ross PL, Scardino PT, Kattan MW: A catalog of prostate cancer nomograms. *J Urol* 2001, 165:1562–1568.

[54] Scheffer M, Hosper SH, Meijer ML, Moss B, Jeppesen E: Alternative equilibria in shallow lakes. *Trends Ecol Evol* 1993, 8:275–279.

[55] Severi G, Morris HA, MacInnis RJ, English DR, Tilley W, Hopper JL, Boyle P, Giles GG: Circulating steroid hormones and the risk of prostate cancer. *Cancer Epidemiol Biomarkers Prev* 2006, 15:86–91.

[56] Shabsigh A, Ghafar MA, de la Taille A, Burchardt M, Kaplan SA, Anastasiadis AG, Buttyan R: Biomarker analysis demonstrates a hypoxic environment in the castrated rat ventral prostate gland. *J Cell Biochem* 2001, 81:437–444.

[57] Small EJ, Fratesi P, Reese DM, Strang G, Laus R, Peshwa MV, Valone FH: Immunotherapy of hormone-refractory prostate cancer with antigen-loaded dendritic cells. *J Clin Oncol* 2000, 18:3894–3903.

[58] Sofikerim M, Eskicorapci S, Oruç O, Ozen H: Hormonal predictors of prostate cancer. *Urol Int* 2007, 79:13–18.

[59] Stacewicz-Sapuntzakis M, Borthakur G, Burns JL, Bowen PE: Correlations of dietary patterns with prostate health. *Mol Nutr Food Res* 2008, 52:114–130.

[60] Story MT: Regulation of prostate growth by fibroblast growth factors. *World J Urol* 1995, 13:297–305.

[61] Swanson KR, True LD, Lin DW, Buhler KR, Vessella R, Murray JD: A quantitative model for the dynamics of serum prostate-specific antigen as a marker for cancerous growth: An explanation for a medical anomaly. *Am J Pathol* 2001, 158:2195–2199.

[62] Tam NN, Gao Y, Leung YK, Ho SM: Androgenic regulation of oxidative stress in the rat prostate: Involvement of NAD(P)H oxidases and antioxidant defense machinery during prostatic involution and regrowth. *Am J Pathol* 2003, 163:2513–2522.

[63] Tam NN, Leav I, Ho SM: Sex hormones induce direct epithelial and inflammation-mediated oxidative/nitrosative stress that favors prostatic carcinogenesis in the noble rat. *Am J Pathol* 2007, 171:1334–1341.

[64] Thompson IM, Goodman PJ, Tangen CM, Lucia MS, Miller GJ, Ford LG, Lieber MM, Cespedes RD, Atkins JN, Lippman SM, Carlin SM, Ryan A, Szczepanek CM, Crowley JJ, Coltman CA Jr: The influence of finasteride on the development of prostate cancer. *N Engl J Med* 2003, 349:215–24.

[65] Vesely S, Knutson T, Damber JE, Dicuio M, Dahlstrand C: Relationship between age, prostate volume, prostate-specific antigen, symptom score and uroflowmetry in men with lower urinary tract symptoms. *Scand J Urol Nephrol* 2003, 37:322–328.

[66] Vollmer RT, Egawa S, Kuwao S, Baba S: The dynamics of prostate specific antigen during watchful waiting of prostate carcinoma: A study of 94 Japanese men. *Cancer* 2002, 94:1692–1698.

[67] Vollmer RT, Humphrey PA: Tumor volume in prostate cancer and serum prostate-specific antigen: Analysis from a kinetic viewpoint. *Am J Clin Pathol* 2003 119:80–89.

[68] Wang Z, Tufts R, Haleem R, Cai X: Genes regulated by androgen in the rat ventral prostate. *Proc Natl Acad Sci USA* 1997, 94:12999–13004.

[69] Wilson EM, French FS: Binding properties of androgen receptors: Evidence for identical receptors in rat testis, epididymis, and prostate. *J Biol Chem* 1976, 251:5620–5629.

[70] Wright AS, Douglas RC, Thomas LN, Lazier CB, Rittmaster RS: Androgen-induced regrowth in the castrated rat ventral prostate: Role of 5-α-reductase. *Endocrinology* 1999, 140:4509–4515.

[71] Wright AS, Thomas LN, Douglas RC, Lazier CB, Rittmaster RS: Relative potency of testosterone and dihydrotestosterone in preventing atrophy and apoptosis in the prostate of the castrated rat. *J Clin Invest* 1996, 98:2558–2563.

[72] Zou W: Regulatory T cells, tumour immunity and immunotherapy. *Nature Rev Immunol* 2006, 6:295–307.

Chapter 6

Resource Competition and Cell Quota in Cancer Models

6.1 Introduction

In its early growth phase, a malignant tumor often exhibits rapid growth due primarily to a high proliferation rate. Although this may not be true for all cancers (see [11]), rapid proliferation must be fueled by abundant resources. As the tumor expands, it is widely thought that competition for resources tends to increase, because more cells compete for fewer resources as tumor vasculature becomes increasingly deranged. As a result, tumor growth rate tends to decrease as the tumor ages and key resources become limiting (see Chapter 2). This limitation also provides the impetus for a long lasting and intensive evolutionary process that enables cancer cells to eventually resist almost any known treatments.

In order to reasonably model limiting resource driven evolutionary cancer dynamics, we must model the resource dynamics explicitly and accurately. In view of the fact that evolutionary cell population growth is intrinsically a multi-scale process in both time and structure, a natural question is how to realistically model these multi-scale phenomena and validate the resulting model. To deal with the evolutionary nature of the cancer cell growth, one may start by modeling just two types of competing cancer cells. To reasonably and accurately model cell population dynamics, one may explicitly keep track of the most limiting nutrient driving the cell growth. A natural way to validate models of complex biological processes is to compare the model outputs to good clinical data which enables modelers to exclude both simplistic and overly complex models.

In this chapter, we present several case studies where the rigorously tested cell quota based Droop model and the ideas from the growing field of ecological stoichiometry are successfully applied to model cancer growth and treatments. Ecological stoichiometry is the study of the balance of energy and multiple chemical resources (elements) in ecological interactions [41]. Readers are referred to [9] for a concise introduction to the concept and applications of the theory of ecological stoichiometry. In the following, we present and compare some cell quota based and clinical date validated treatment models of prostate cancer and chronic myeloid leukemia with some standard population level models in the literature. Some noteworthy advantages of cell quota based population models include: (1) the models use well-defined and measurable biological parameters [33]; (2) the models generate good data fit with biologically realistic parameter values [14]; (3) the model parameters are actually functions of the dynamic cell quota and hence generate rare and valuable evolutionary insights from the parameter dynamics [31]; (4) the models can be expected to simultaneously fit multiple highly nonlinear or oscillatory data sets depicting several variables observed in the same experimental settings [32].

6.2   A cell-quota based population growth model

Cell population growth is determined by its birth and death processes. Cells grow through division, but can die in many ways. Hence, in general, cell death mechanisms are more numerous and difficult to study than cell division mechanisms in a lab or field setting. In a very short time period, growth dynamics can be approximated by a linear differential equation with slope representing the net growth rate. Longer term, this growth rate shall be regarded as time dependent since it is often density dependent. The so-called Droop equation [6, 7] provides a time and experiment tested simple mathematical expression for biomass growth rate. It provides a natural starting point for formulating mechanistic cell population growth models. Indeed, it provides a simple mechanism for us to understand the classical logistic growth equation as pointed out by Kuang et al. [21] and which will be fully explored in the next section.

In 1968, Droop [6] published some ground-breaking experimental and theoretical results describing the kinetics of vitamin B_{12} limited growth in the photosynthetic alga *Monochrysis lutheri*. Contrary to earlier belief, Droop found that the specific growth rate did *not* depend directly on the medium substrate concentration, but was rather a function of the intracellular vitamin B_{12} concentration, the so-called cell quota , Q. Before continuing, we define Q as C/x, where C is the concentration of substrate across all cells (nM), and x is the cell mass concentration measured in cells L^{-1}. However in many applications, Q can be a scalar representing the percentage of the nutrient weight in a unit weight of cell mass. See Table 6.1 for all variables and parameters.

Androgens represent a limiting "nutrient" for prostate epithelial cell growth. This motivates our discussion of the Droop (or cell quota) model. But first, we must give history its due. Justus von Liebig, who in his 1840 treatise [25], studied the "matters which supply the nutriment of plants," recognized the vital role of nitrogen and various minerals in plant growth. And his *Law of the Minimum* endures, as discussed in "The Natural Laws of Husbandry," [26]

Every field contains...a *minimum* of one or several other nutritive substances. It is by the *minimum* that the crops are governed...Where lime or magnesia, for instance, is the minimum constituent, the produce of corn and straw, turnips, potatoes, or clover, will not be increased by a supply of even a hundred times the actual store of potash, phosphoric acid, silicic acid, etc., in the ground.

It was also recognized by other early investigators such as Lotka in 1925 [29] that one essential nutritional currency could limit *growth rate*, similar to crop *yield*.

The Monod model for nutrient-limited growth preceded Droop's, and takes the specific growth rate, μ, to be a saturating function of medium concentration, s,

$$\mu = \mu_m \frac{s}{K_s + s}. \tag{6.1}$$

In a closed growth environment, the total nutrient in the environment is constant C_0. In such an environment, the Monod model also stipulates that nutrient consumption is converted to cell growth at a constant rate. The conservation law of nutrient yields

$$\frac{1}{Y}x(t) + s(t) = C_0. \tag{6.2}$$

where x is cell mass, and Y is the yield constant. The conservation of nutrient in a closed growth environment can also be expressed in terms of the cell quota $Q(t)$ in the following form

$$Q(t)x(t) + s(t) = C_0. \tag{6.3}$$

Together with equation (6.2), we see that $Q(t) = 1/Y$.

Differentiate the equation (6.2) yields

$$\frac{dx}{dt} = -Y\frac{ds}{dt} \Rightarrow Y = -\frac{dx}{ds}. \tag{6.4}$$

A note on the units: it is equally valid to think of the cell quota as either an amount of substrate per cell (nmol cell^{-1}, as in Table 6.1) or as an intracellular concentration (convert using volume per cell). Now, Droop empirically found the relationship between cell quota and specific growth rate to take the form:

$$\mu(Q) = \mu_m \left(1 - \frac{q}{Q}\right). \tag{6.5}$$

This is the *Droop model* or *Droop equation*. The subsistence quota, q, is the minimum quota necessary to sustain life. Figure 6.1 depicts the function $\mu(Q)$.

TABLE 6.1:   Variables and parameters for the Droop model.

Symbol	Meaning	Units
Q	Cell quota, equals C/x	nmol cell^{-1}
Y	Yield constant, x/C	cells nmol^{-1}
x	Cell mass	cells l^{-1}
C	Concentration of substrate in cells	nM (nmol l^{-1})
s	Medium substrate concentration	nM (nmol l^{-1})
K_s	Michaelis constant for uptake	nM (nmol l^{-1}
μ	Specific growth rate	day^{-1}
μ_m	Max specific growth rate	day^{-1}
u	Specific overall rate of uptake	nmol cell^{-1} day^{-1}
u_m	Maximum specific rate of uptake	nmol cell^{-1} day^{-1}
q	Subsistence quota	nmol cell^{-1}

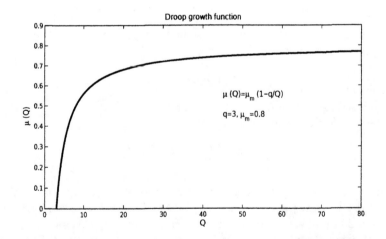

FIGURE 6.1: Plot of Droop function $\mu_m\left(1 - \frac{q}{Q}\right)$ with $\mu_m = 0.8$ and $q = 3$.

Note that

$$\frac{dx}{dt} = \mu x. \tag{6.6}$$

We still need to determine an expression for the cell quota, $Q = Q(t)$. In general,

$$\frac{dQ}{dt} = u - \Phi. \tag{6.7}$$

Where u is the specific rate of substrate uptake, and Φ is the rate of substrate depletion inside the cells. Assume the cell quota is at a steady-state. Since Q is unchanging, we have that the specific growth rate is constant, and the relation in equation (6.4), which states that medium substrate depletion is converted to (net) growth at a constant rate, holds, giving:

$$Y = -\frac{dx}{ds} \Rightarrow Q = -\frac{ds}{dx}. \tag{6.8}$$

The conservation of substrate implies that

$$\frac{ds}{dt} = -ux \tag{6.9}$$

which is equivalent to

$$u = -\frac{ds}{dt}\frac{1}{x} = Q\frac{dx}{dt}\frac{1}{x} = \mu Q. \tag{6.10}$$

This follows from the chain rule,

$$\frac{ds}{dt} = \frac{ds}{dx}\frac{dx}{dt} = -Q\frac{dx}{dt} \tag{6.11}$$

and from equation (6.6). Finally, from the steady-state assumption for Q, we have

$$\frac{dQ}{dt} = 0 = u - \Phi = \mu Q - \Phi \Rightarrow \Phi = \mu Q = \mu_m(Q - q). \tag{6.12}$$

As to the form of the uptake term, u, Droop found a simple saturating function to describe the data well:

$$u = u_m \frac{s}{K_s + s}. \tag{6.13}$$

This completes the derivation of the basic cell quota model, which we restate in full:

$$\frac{dx}{dt} = \mu x, \tag{6.14}$$

$$\frac{dQ}{dt} = u - \mu Q = u - \mu_m(Q - q), \tag{6.15}$$

$$\frac{ds}{dt} = -ux, \tag{6.16}$$

$$\mu = \mu_m\left(1 - \frac{q}{Q}\right),$$

$$u = u_m\frac{s}{K_s + s}.$$

The above model shall be supplemented with appropriate nonnegative initial conditions for all the variables and the additional requirement that $Q(0) \geq q$. Note that as long as $Q(0) \geq q$, then $Q(t) \geq q$.

There are several straightforward modifications to the basic framework that we may consider. The most obvious follows from the reality that the cell quota must have some upper limit, giving the common modification:

$$u = u_m\frac{s}{K_s + s}\frac{q_m - Q}{q_m - q}, \tag{6.17}$$

where q_m is the maximum cell quota. In this case, the rate of uptake decreases linearly with cell quota. Alternatives include the addition of a Michaelis-Menton (Hill) term.

Growth rate subject to two potentially limiting nutrients may be modeled by the following cell quota based equation

$$\mu = \mu_m \min\left(1 - \frac{q_A}{Q_A}, 1 - \frac{q_B}{Q_B}\right). \tag{6.18}$$

6.3   From Droop cell-quota model to logistic equation

The main purpose of this section is to derive the logistic model via Droop equation. We consider a single species growing in a closed environment where there is a single most limiting nutrient. For convenience, we assume below that this species is an species of algae and the limiting nutrient is phosphorous P. Observe that the total amount of phosphorus P_T in this closed growth environment remains constant.

If we let P_x and P_f be the phosphorus in the algal cells and the free phosphorus respectively, then $P_T = P_x + P_f$. Let $x = x(t)$ be the algal cell density and $Q = Q(t)$ be the algae's cell phosphorus quota. Then $P_x = Qx$. Hence

$$P_T = P_f + Qx. \tag{6.19}$$

In the following, we let q be the algae's minimal cell quota of P, μ_m be the species' maximal growth rate, D be its natural death rate. By (6.5), we have the following equation for the species growth:

$$\frac{dx}{dt} = \mu_m\left(1 - \frac{q}{Q}\right)x - Dx. \qquad (6.20)$$

The free phosphorus changes according to

$$\frac{dP_f}{dt} = -\alpha P_f x + DxQ. \qquad (6.21)$$

The first term describes the loss of phosphorous due to the uptake by x cells which may be simply approximated by a mass action process. The second term assumes that the dead algal cells immediately release their phosphorous back to the growth environment.

We still need an equation governing the dynamics of Q, the species' cell quota for P. However, this equation can be derived from the nutrient conservation equation (6.19). We have

$$0 = P_T' = P_f' + (Qx)' = -\alpha P_f x + DxQ + \mu_m(Q - q)x - DxQ + xQ'. \qquad (6.22)$$

This results in the following simple equation

$$\frac{dQ}{dt} = \alpha P_f - \mu_m(Q - q) \equiv \alpha(P_T - xQ) - \mu_m(Q - q). \qquad (6.23)$$

We assume that $Q(0) \geq q$ and $P_T > x(0)Q(0)$. Mathematically, this ensures that $Q(t) \geq q$ for all $t > 0$.

Since the cell metabolic process operates at a much faster pace than the growth of total biomass of a species, the quasi-steady-state argument allows us to approximate $Q(t)$ by the solution of

$$\alpha P_f - \mu_m(Q - q) = 0, \qquad (6.24)$$

which takes the form of

$$Q = \frac{\alpha P_f + q\mu_m}{\mu_m}. \qquad (6.25)$$

This together with (6.19) yields

$$P_f = \frac{\mu_m}{\mu_m + \alpha x}\left(P_T - qx\right). \qquad (6.26)$$

Substituting (6.26) into (6.25) yields

$$Q = q + \frac{\alpha}{\mu_m + \alpha x}\left(P_T - qx\right). \qquad (6.27)$$

Substituting the above into (6.20) yields

$$\frac{dx}{dt} = \mu_m x \frac{P_T - qx}{P_T + \mu_m q\alpha^{-1}} - Dx. \tag{6.28}$$

The above equation can be rewritten as

$$\frac{dx}{dt} = \mu_m x \left(1 - \frac{qx + \mu_m q\alpha^{-1}}{P_T + \mu_m q\alpha^{-1}}\right) - Dx. \tag{6.29}$$

We can further rewrite the above equation as

$$\frac{dx}{dt} = (\mu_m - D)x \left[1 - \frac{x + \mu_m \alpha^{-1}}{[(\mu_m - D)/\mu_m][\mu_m \alpha^{-1} + P_T/q]}\right]. \tag{6.30}$$

Or equivalently,

$$\frac{dx}{dt} = \frac{(\mu_m - D)P_T q^{-1} - D\mu_m \alpha^{-1}}{P_T q^{-1} + \mu_m \alpha^{-1}} x \left[1 - \frac{x}{(\mu_m - D)P_T(q\mu_m)^{-1} - D\alpha^{-1}}\right]. \tag{6.31}$$

It clearly takes the form of the classical logistic model

$$\frac{dx}{dt} = rx \left[1 - \frac{x}{K}\right] \tag{6.32}$$

with

$$r = \frac{(\mu_m - D)P_T q^{-1} - D\mu_m \alpha^{-1}}{P_T q^{-1} + \mu_m \alpha^{-1}} \tag{6.33}$$

and

$$K = (\mu_m - D)P_T(q\mu_m)^{-1} - D\alpha^{-1} = \left(\frac{P_T}{q\mu_m} + \alpha^{-1}\right)r. \tag{6.34}$$

From Eq. (6.34), we see that r and K are not independent. Indeed, they are linearly dependent. It is easy to observe that both r and K are increasing functions of α and decreasing function of D. This makes good biological sense. It should be pointed out here that we did not assume the population suffers from a crowding effect explicitly. However, this crowding effect is implicitly provided by the fact that the total nutrient in the system (here P) is fixed, and individuals have to compete for this resource. Observe that the expression of K is different from the intuitive carrying capacity of the form P_T/q. The expression of K says that while in theory the environment may sustain a maximum of P_T/q cells, the actual upper limit for the algae biomass can attain is $K = (\mu_m - D)P_T/(q\mu_m) - D\alpha^{-1}$, which is less than P_T/q. The reason that the intuitive carrying capacity P_T/q cannot be reached in practice is that the death process in a population keeps the population below its potential maximum. However, the equation (6.34) says clearly that a population with a relatively low death rate will likely amass more biomass than a population with a relatively high death rate.

Droop equation based population models are capable of generating rich and often intriguing dynamics that realistically match those observed in field and lab experiments ([27, 28], [41]).

6.4 Cell-quota models for prostate cancer hormone treatment

In the following, we present two cell-quota based models for prostate cancer hormone treatment due to Portz, Kuang and Nagy [33]. Their models aim to produce solutions match the data of clinical trials. With a data validated model, we can hope to gain a greater understanding of the processes at work in prostate cancer and androgen suppression therapy. Moreover, a model capable of predicting the course of an individual case of prostate cancer would be useful in developing a treatment schedule in a clinical setting. Their models are based on the works of Jackson et al.[18] and Ideta et al.[17]. These earlier models include an androgen-dependent (AD) cell population and an androgen independent (AI) cell population with mutation from the AD population to the AI population at a rate based on the androgen concentration. The material of this section is adapted from the content of Portz, Kuang and Nagy [33].

6.4.1 Preliminary model

In an initial model formulated by Portz, Kuang and Nagy [33], the growth rate of the AD cell population is given by Droop's cell quota model. The cell quota model introduces a new variable, $Q(t)$, which is the cell quota for androgen. The AD and AI cell populations are modeled by

$$\frac{dX_1}{dt} = \mu_m \left(1 - \frac{q}{Q}\right) X_1 - \delta X_1 - m_1(Q)X_1 + m_2(Q)X_2, \qquad (6.35)$$

$$\frac{dX_2}{dt} = rX_2 - m_2(Q)X_2 + m_1(Q)X_1. \qquad (6.36)$$

The proliferation rate of the AD cell population is zero when $Q(t)$ is at the minimum cell quota q. As $Q(t)$ increases, the growth rate approaches its maximum, μ_m. The apoptosis rate of the AD cell population and the net growth rate of the AI population excluding mutation are assumed to be constant.

This model includes mutation between both cell populations. This change is made under the hypothesis that androgen dependence can be regained by AI cells in an androgen-rich environment with some sort of switching behavior. The switching rates are given by hill equations,

$$m_1(Q) = k_1 \frac{K_1^n}{Q^n + K_1^n}, \qquad (6.37)$$

$$m_2(Q) = k_2 \frac{Q^n}{Q^n + K_2^n}. \qquad (6.38)$$

The AD to AI mutation rate, $m_1(Q)$, is low for normal and high androgen levels and high for low androgen levels. In contrast, the AI to AD mutation rate, $m_2(Q)$, is high for normal and high androgen levels and low for low androgen levels.

The cell quota for androgen within the AD cells is modeled by

$$\frac{dQ}{dt} = v_m \frac{q_m - Q}{q_m - q} \frac{A}{A + v_h} - \mu_m(Q - q) - bQ. \tag{6.39}$$

Androgen within the cells is used for growth up to the minimum cell quota at a rate μ_m and is also assumed to degrade at a constant rate b.

The serum PSA concentration, $P(t)$, is modeled as a linear function of the two cancer cell populations:

$$P(t) = c_1 X_1(t) + c_2 X_2(t). \tag{6.40}$$

6.4.2 Final model

In the final model, Portz, Kuang and Nagy use the cell quota model for both the AD and AI cell populations [33]:

$$\frac{dX_1}{dt} = \mu_m \left(1 - \frac{q_1}{Q_1}\right) X_1 - d_1 X_1 - \lambda_1(Q_1)X_1 + \lambda_2(Q_2)X_2, \tag{6.41}$$

$$\frac{dX_2}{dt} = \mu_m \left(1 - \frac{q_2}{Q_2}\right) X_2 - d_2 X_2 - \lambda_2(Q_2)X_2 + \lambda_1(Q_1)X_1. \tag{6.42}$$

Both cell populations have the same maximum proliferation rate μ_m. To give the AI cells greater capacity for proliferation in low androgen environments compared to AD cells, they select a lower minimum cell quota for the AI cells, $q_2 < q_1$.

The relevance of the cell quota model follows from the nature of androgen's action and how its signal is transduced. The AR is intracellular, so only intracellular androgen can be sensed, and proliferation depends on AR:androgen binding. Hence androgen is clearly a resource. For the AI cells, androgen receptors are typically either activated in an androgen-independent way, are amplified, or are otherwise activated by mutated regulators [22]. In any of these cases, the androgen receptor is still active and some AR:androgen response remains. However, far less androgen is required to achieve the same level of proliferation, which the Droop formalism captures very nicely with the constraint $q_2 < q_1$.

The cell quotas for androgen are modeled by the same equation as the preliminary model,

$$\frac{dQ_i}{dt} = v_m \frac{q_m - Q_i}{q_m - q_i} \frac{A}{A + v_h} - \mu_m(Q_i - q_i) - bQ_i. \tag{6.43}$$

However, there are now two cell quota variables, $Q_1(t)$ for the AD population and $Q_2(t)$ for the AI population.

The switching rates between the AD and AI cell populations take the same form as in the preliminary model,

$$\lambda_1(Q) = c_1 \frac{K_1^n}{Q^n + K_1^n}, \qquad (6.44)$$

$$\lambda_2(Q) = c_2 \frac{Q^n}{Q^n + K_2^n}. \qquad (6.45)$$

With this final model, a new model for the serum PSA concentration is introduced:

$$\frac{dP}{dt} = \sigma_0(X_1 + X_2) + \sigma_1 X_1 \frac{Q_1^m}{Q_1^m + \rho_1^m} + \sigma_2 X_2 \frac{Q_2^m}{Q_2^m + \rho_2^m} - \delta P. \qquad (6.46)$$

It is assumed that PSA is produced by both AD and AI cells at a baseline rate σ_0 plus an additional androgen-dependent rate. Experimental evidence supports the assumption that PSA production is dependent on androgen levels [15]. Hill functions are used for the androgen-dependent rate so that it increases with the cell quota toward a maximum rate σ_i. The androgen-dependent rate is split into two terms, one for each cell population, because the two populations respond differently to androgen. It is also assumed that PSA is cleared from the blood at a constant rate δ.

6.4.3 Simulation

In a clinical study [1], seven men with stage C and stage D prostate cancer were treated with intermittent androgen suppression therapy. Androgen withdrawal was maintained for 6 or more months and then interrupted for 2 to 11 months. Serum androgen and PSA concentrations were measured on a monthly basis. When serum PSA concentrations exceeded a threshold of about 20 ng/mL, androgen withdrawal was resumed. This treatment cycle was continued over periods of 21 to 47 months. Portz et al. used the androgen and PSA time series data from the seven cases to simulate and fit the models [33]. Using clinical data from an intermittent androgen suppression trial provides a better assessment of the dynamics of prostate cancer as responses to both the initiation and withdrawal of treatment can be observed as opposed to continuous androgen suppression where only a single on-treatment period is observed.

The serum androgen data from the clinical cases is used directly as $A(t)$ for fitting rather than modeling the androgen concentration. However, the coarse androgen data must be interpolated before it can be used as an input to the numerical simulations. Using linear interpolation results in the peaks of the PSA concentrations being significantly delayed. These delays are caused by the slow linear decline in androgen concentration between off-treatment and on-treatment periods in the simulations when the actual androgen concentrations drop very quickly. To obtain more accurate results, an exponential fit

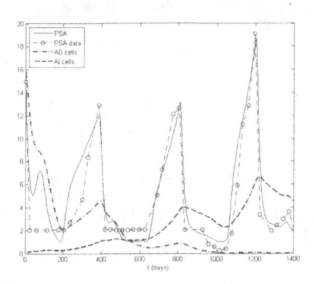

FIGURE 6.2: Final model results, case 1. In general, the PSA data is matched very accurately.

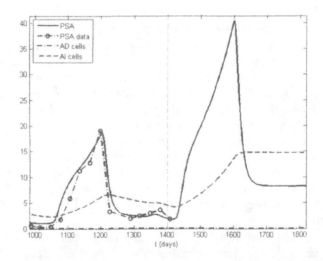

FIGURE 6.3: Prediction of the final model, case 1. The dashed vertical line separates the clinical fit and the prediction. The PSA concentration and AI population increase significantly with another off-treatment period. However, the subsequent on-treatment period remains effective in stopping further growth.

TABLE 6.2: Comparison of model fits, including mean squared error (MSE) and Schwarz Bayesian Criterion (SBC). Lower values indicate better fits for both MSE and SBC.

Case	MSE			SBC		
	Ideta	Prelim.	Final	Ideta	Prelim.	Final
1	21.21	30.61	2.960	145	145	68.0
2	25.65	96.05	2.790	85.9	112	41.5
3	216.6	238.2	46.42	185	188	139
4	12.26	13.60	3.924	93.5	96.4	61.6
5	358.1	497.3	41.81	101	105	70.7
6	44.59	41.91	0.3985	96.9	95.7	2.57
7	6.655	6.588	3.710	67.7	67.5	53.7
Overall	81.40	106.3	13.46	806	853	491

is used between the last off-treatment data points and the first on-treatment data points,

$$A(t) = A(t_f) + (A(t_i) - A(t_f)) e^{-\gamma(t-t_i)}, \qquad (6.47)$$

where t_i is the time of off-treatment data point, t_f is the time of the on-treatment data point, and γ is the serum androgen clearance rate. The remaining segments of the androgen data are interpolated using piecewise cubic hermite splines to give smoother responses in the simulations.

The parameters of the models are initially fit by hand to provide good qualitative fits with the clinical PSA data. Once a reasonably close fit has been obtained, a simplex search method is used to minimize the mean square error between the clinical PSA data and the simulated PSA concentrations. This parameter fitting is performed for each of the seven cases on all three models.

The simulation result for the final model for subject 1 described in [1] is shown in Figure 6.2. This model is better at fitting the clinical PSA data than either the preliminary model or the model of Ideta et al. [33].

Table 6.2 shows the mean squared error between the simulated PSA levels and the clinical PSA levels for each case and model as reported in [33]. The table also presents a comparison based on the Schwarz Bayesian Criterion, which includes an adjustment for the number of free parameters. The results clearly confirm that the final model produces the best fits out of the three.

6.4.4 Predictions

The result of running the final model for another treatment cycle beyond the clinical data for subject 1 is shown in Figure 6.3 and the results for the other subjects can be found in [33]. It is known that the patients in cases 1, 2, 3, and 5 had stage C cancer, while the patients in cases 4, 6, and 7 had stage D (metastatic) cancer [1]. The final model can be used to predict uncontrolled growth in the AI population for the stage D cases even though the PSA concentrations do respond to the final on-treatment period in cases 6 and 7. The model also predicts a poor response to another treatment cycle for the patient in case 3, who had already undergone two long treatment cycles.

In [13], Everett, Packer and Kuang extended the above final model in two ways and tested their models predictive accuracy, using only a subset of the data to find parameter values. The results are compared with the piecewise linear model of Hirata et al. [22] which does not use testosterone as an input (see Section 5.11). Based on a small set of data from seven patients contained in [1], their results showed that the piecewise linear model actually produced slightly more accurate results while the two more biologically plausible predictive methods are comparable. However, this comparison did not penalize the excessive number of parameters (since each on and off treatment requires a new set of parameter values) used in the piecewise linear model of Hirata et al. [22]. Nevertheless, it suggests that a simple piecewise linear model may still be useful for a predictive use despite its simplicity and lack of close biological connections.

6.5   Other cell-quota models for prostate cancer hormone treatment

In Section 6.4, we described two cell-quota models for prostate cancer hormone treatment due to Portz et al. [33]. They modeled androgen-limited growth in prostate cancer using a Droop model, with androgen-dependent and independent cell lines characterized by different subsistence quotas (q), and modeled intermittent androgen suppression therapy. Excellent agreement between the model and clinical data was achieved with this approach. We develop below several similar models for competition between AI and AD cells.

6.5.1   Basic model

We consider a model very similar to that of Portz et al. [33]. We have two strains of cell, $X_1(t)$ and $X_2(t)$, with differing sensitivities to androgen. We let these variables represent the absolute numbers of cells, rather than cells l^{-1} as above (which follows Droop's original derivation). The cell-quota model above can be reformulated using absolute cell count and substrate amount, and it remains essentially identical, although care must be taken to appropriately convert substrate amount to concentration, based on medium volume. For parametrization, we take cell quota to be in units of concentration (nM), by converting from nmol cell^{-1} to nmol l^{-1} (nM).

The cell quota for this model is intracellular androgen concentration (presumably complexed to its receptor, with units nM), still represented by Q. The basic cell quota model for androgen-mediated growth is:

$$\frac{dX_1}{dt} = \mu_1 X_1 - d_1 X_1 - m_1 X_1 + m_2 X_2 \tag{6.48}$$

$$\frac{dX_2}{dt} = \mu_2 X_2 - d_2 X_1 + m_1 X_1 - m_2 X_2 \tag{6.49}$$

$$\mu_1 = \mu_m \left(1 - \frac{q_1}{Q_1}\right)$$

$$\mu_2 = \mu_m \left(1 - \frac{q_2}{Q_2}\right).$$

The mutation rates between strains are given by m_1 and m_2, and, unlike in Ideta's or Portz et al.'s models, are constants, and background cell death occurs at rates d_1 and d_2. We have the cell quotas governed by:

$$\frac{dQ_i}{dt} = u_{m_i} \frac{A}{K_{A_i} + A} \frac{K_{Q_i}}{K_{Q_i} + Q_i} - \mu_i Q_i - bQ_i, \ i = 1, 2. \tag{6.50}$$

Androgen uptake is saturable according to the K_{A_i} parameters, and cell quota is limited by the second Hill function in the uptake term. Finally, androgens positively regulate the expression of PSA by prostate epithelium, so rather than take the serum PSA level as a simple linear conversion from cell mass as done by Ideta et al. [17], we model PSA dynamic quantity with the same governing equation as done by Portz et al. [33] which is identical to (6.39):

$$\frac{dP}{dt} = \sigma_0(X_1 + X_2) + \sigma_1 X_1 \frac{Q_1^m}{Q_1^m + \rho_1^m} + \sigma_2 X_2 \frac{Q_2^m}{Q_2^m + \rho_2^m} - \delta P \tag{6.51}$$

where σ_0 is the baseline level of PSA production by all cells, and androgen induces each strain to increase production according to their respective Hill functions.

6.5.2 Long-term competition in the basic model

Portz et al. [33] modeled increased sensitivity to androgens by AI cells, $X_2(t)$, by making the subsistence quota for such cells smaller than that for AD cells, namely $q_2 > q_1$. While apparently reasonable, such a choice results in the cell quota for AI cells being *smaller* than for AD cells (a growth advantage does still exist for such AI cells). We argue that, under the basic model, modifying the uptake parameters K_{A_2} and K_{Q_2} may better represent increased sensitivity to androgen via such AR upregulation or other common mechanisms. And indeed, making either $K_{A_2} < K_{A_1}$ or $K_{Q_2} > K_{Q_1}$ results in a greater Q_2 and a selective advantage for AI cells that stems from their larger cell quota.

Before examining intermittent androgen deprivation, we briefly re-address the evolutionary question posed in Section 5.4: how do serum androgen levels affect the early evolution of prostate cancers? The model of Eikenberry et al. [8] suggested that low systemic androgen may induce selective pressure for highly androgen-sensitive strains, which may have higher malignant potential or give rise to cancers poorly responsive to treatment. We apply the basic model just presented to this problem by letting $K_{A_2} = K_{A_1}/10$. We also make growth space-limited as follows:

$$\mu_i = \mu_m \left(1 - \frac{q_i}{Q_i}\right)\left(1 - \frac{X_1 + X_2}{K}\right) \tag{6.52}$$

where K is the carrying capacity of the normal prostate. This change also necessarily affects the cell quota loss term, $-\mu_i Q_i$. We simulate the model starting with AD cells at their approximate equilibrium point with no AI cells present, and we set the serum androgen level, A, to a constant.

Figure 6.4 shows the evolution of the two strains under different values for A. As in Section 5.4, this model, independently derived from quite different principles, suggests that low systemic androgen selects for cell strains with reduced androgen dependence. Again, more androgen-sensitive strains are also ultimately selected for in normal and high-androgen environments, but in such environments the selective pressure is weaker. The time from the start of the simulation until AI cells become a majority is plotted in Figure 6.5, and this time increases nearly linearly with serum androgen.

Reasonable parameter values for the model, based on the chemical kinetics model of Section 5.4.2, Ideta et al. [17], and Portz et al. [33], are reported in the caption of Figure 6.4.

6.5.3　Intermittent androgen deprivation

Akakura et al. [1] used the following algorithm in a clinical study of intermittent androgen deprivation: androgen withdrawal was initiated pharmacologically. Following 6 months of PSA in the normal range, treatment was stopped; it was resumed once serum PSA exceeded 20 ng/ml. Portz et al. [33] achieved an excellent match between his model and the clinical data in this study. Figure 6.6 shows the results of intermittent androgen suppression with A set at either 0.5 nM or 0.1 nM during therapy and at 15 nM off therapy. Treatment is given for a fixed period of 180, 90, or 30 days, and resumed when PSA > 20 ng/ml.

From Figure 6.6, it can be seen that all suppression periods generate essentially identical results. When androgen deprivation is severe, continuous deprivation may be superior to intermittent therapy. However, this result depends entirely on the androgen sensitivity of the AI line. If the AI line is capable of responding to even very low androgen levels, i.e. K_{A_2} is made sufficiently small, this advantage disappears.

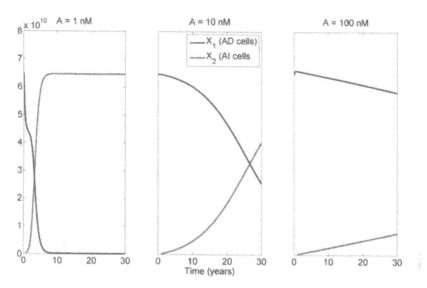

FIGURE 6.4: Competition between AD and AI cells in a prostate under the basic cell quota model with space-limited (logistic) growth. AI cells have greater uptake of androgen, with $K_{A_1} = 3$ and $K_{A_2} = 0.3$ nM. From left to right, A is fixed at 1 nM, 10 nM, and 100 nM. Parameter values are $\mu_m = 0.035$, $d_1 = d_2 = 0.01$, $q_1 = q_2 = 0.3$, $m_1 = m_2 = 10^{-5}$, $b = 0.09$, $K_{Q_1} = K_{Q_2} = 5$, $K_{A_1} = 3$, $K_{A_2} = 0.3$, $u_m = 0.275$, $\sigma_0 = 10^{-10}$, $\sigma_1 = \sigma_2 = 5 \times 10^{-10}$, $m = 2$, $\rho_1 = \rho_2 = 1.2$, $\delta = 0.2$.

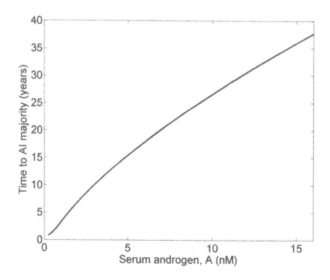

FIGURE 6.5: Under the same model as in Figure 6.4, the time until AI cells become a majority as a function of serum androgen, A.

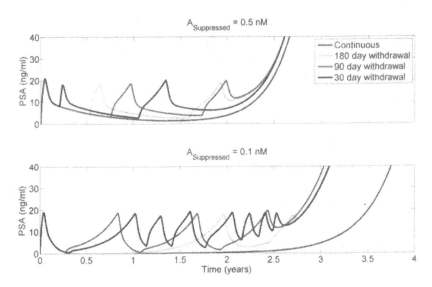

FIGURE 6.6: Intermittent androgen deprivation therapy under the Droop model presented in Section 6.5.1. Parameter values are as in Figure 6.4, except $\mu_m = 0.025$, $m_1 = m_2 = 10^{-6}$, and $K_{A_1} = 0.1$. Note that the advantage to continuous depravation in the lower panel disappears if K_{A_2} is made sufficiently small.

The Droop model predicts then, that even very brief periods of androgen deprivation are likely to suppress tumor growth as well as other schedules. Since clinically maximal androgen blockade has not proven superior to standard therapy, it is likely that brief intermittent therapy is nearly as good as continuous therapy.

6.5.4 Cell quota with chemical kinetics

We now propose a somewhat more complex model that takes the intracellular chemical kinetics of androgens into account. For simplicity, we consider only testosterone (T) and disregard DHT for the moment. The cell quota, Q, is now taken to be the intracellular T:AR complex concentration (nM). From first principles, we have that T is a steroid hormone that freely crosses cells membranes, and we have a system for intracellular free testosterone, T_i, free AR, R_i, and T:AR complex Q_i, all in nM units:

$$\frac{dT_i}{dt} = PA_i(T_S - T_i)\gamma - k_a T_i R_i + k_d Q_i - \beta_T T \quad (6.53)$$

$$\frac{dR_i}{dt} = -k_a T_i R_i + k_d Q_i \quad (6.54)$$

$$\frac{dQ_i}{dt} = k_a T_i R_i - k_d Q_i. \quad (6.55)$$

Note that we disregard production and loss of both free and complexed AR. In the chemical kinetics model of Section 5.4.2 we did consider these behaviors, but the model was such that the total AR concentration within the cell remains constant. Our simplification here is therefore a "shortcut" to the same essential behavior.

The parameter A_i is the surface area of vasculature associated with strain i and can be calculated as:

$$A_i = \alpha X_i \quad (6.56)$$

where α is a constant with units m^2 $cell^{-1}$. The permeability coefficient of testosterone is P (m hr^{-1}), and γ is a constant giving the volume in L per cell. The other kinetic parameters are as in Section 5.4.2.

This model allows us to more realistically model upregulation of the AR as a mode of androgen independence. We leave such investigation to the student. We also leave it as an exercise to extend the model to consider conversion of T to DHT by 5α-reductase, and the appropriate modification of the cell quota.

6.6 Stoichiometry and competition in cancer

Tumors compete for resources with their hosts. As we have discussed at various points in this and other chapters, healthy cells and multiple strains of cancer cell must compete for space, oxygen, glucose, growth factors, and, as we discuss here, elemental nutrients. A striking demonstration of a tumor's effect on the host physiology is cancer *cachexia*, a complex wasting syndrome characterized by weight loss, anorexia, and neurohormonal abnormalities. A combination of tumor and host factors cause progressive breakdown of skeletal muscle, fat, and bone that is *not* reversible by nutritional supplementation; resting energy expenditure is often increased and there is increased glucose production and catabolism [43].

Ecological stoichiometry is an increasingly important field which studies the balance of elements in ecological systems. In particular, organisms vary greatly in their relative nitrogen (N), carbon (C), and phosphorus (P) content. Phosphorus is a key element in ecological systems, and multiple lines of evidence point to it being a key resource that healthy and cancerous cells may compete for. Phosphorus is an essential component of DNA and RNA, and the majority (about 80%) of RNA is incorporated into ribosomes, vast molecular factories that translate mRNAs into proteins. The proteins produced by ribosomes are essential for rapid cell growth and proliferation, and multiple studies indicate that ribosome synthesis is essential to rapid cancer cell proliferation [23]. Hence, the importance of phosphorus. Note also that nitrogen is an essential component of both the nitrogenous bases that make up RNA and DNA, and of the amino acids that compose proteins. Glucose, a six-carbon chain, provides energy. Thus, the three elements studied in ecological ecology are essential factors in the tumor-host ecology as well.

The *growth rate hypothesis* posits that organisms with a rapid growth rate require a high P content in biomass, due to the requirement for high-P content ribosomes [10]. Thus, rapid growth as an evolutionary strategy has the cost of increased dependence on environmental or dietary phosphorus. Translating this to tumor biology, the hypothesis predicts that rapidly growing tumors should have a high P content, but there is an evolutionary cost in P-limited environments.

6.6.1 KNE model

Kuang, Nagy and Elser have recently applied ideas from ecological stoichiometry to tumor growth, and have proposed a mathematical model for phosphate-limited growth in human cancer [23]. The model considers a cancer of some organ which, for concreteness, we take it to be the lung. Let x represent the mass of healthy cells in the organ, y_i represents cancerous cells of strain i, and z is the total tumor microvessel mass. We assume the organ has an overall mass of K_H, and the tumor's maximum size is K_T.

Phosphate-limited growth. Now, considering phosphorus mass, P, we assume the total phosphorus in the organ remains constant. We assume n represents the average amount (grams) of phosphorus per kg of healthy tissue; likewise m_i is phosphorus mass per kg of cancerous tissue of strain i. That is, these are units of concentration. Thus, we have:

$$P - \left(nx + nz + \sum_i m_i y_i\right) \equiv P_e \qquad (6.57)$$

where P_e is mass of extracellular P within the organ. The extracellular volume is given by $f \times K_H$, where f is the fraction of organ mass that is extracellular (about 1/3), and K_H is the organ mass. Therefore, extracellular concentration is simply:

$$[P_e] = \frac{P_e}{f K_H} \qquad (6.58)$$

Now, we assume that if the extracellular P concentration, $[P_e]$, falls below the mean for healthy cells, n, then the maximal proliferation rate, a, is impaired. That is, cells proliferate at rate a when sufficient P is present ($[P_e] \geq n$), and at rate

$$a\frac{[P_e]}{n} = a\frac{P_e}{fK_Hn} \tag{6.59}$$

when $[P_e] < n$. Malignant cells similarly depend on $[P_e]$ and m_i and have maximum growth rate b_i.

Vasculature-limited growth. Tumor growth requires a sufficient vasculature. We define L as follows:

$$L = \frac{g\left(z - \alpha\sum_i y_i\right)}{\sum_i y_i} \tag{6.60}$$

Here, α is the mass of cancer cells that can just barely be supported by a unit of blood vessels, and g measures the sensitivity of tumor cells to hypoxia. If $L \geq 1$, then proliferation is unimpaired, but if $L < 1$, then the growth rate is modified by the factor L. We also assume that cancer cells generate some signal that recruits immature endothelial cells to form blood vessels at rate c. We can incorporate a delay into the model by assuming it takes τ time units for vessels to form in response to this signal.

Space-limited growth. We have growth limited by organ size. Healthy cells grow to a carrying capacity of K_H, with all cells competing for this space. The tumor mass grows to a carrying capacity of K_T. Healthy cells do not affect the maximum tumor size, from the assumption that cancer cells better compete for space. We also assume that healthy, malignant, and endothelial cells undergo death at per-capita rates d_x, d_i, and d_z, respectively.

Model. The above considerations give the model for two strains of cancer cells, $i = 1, 2$:

$$
\begin{aligned}
\frac{dx}{dt} &= x\left(a\min\left(1, \tfrac{[P_e]}{n}\right) - d_x - (a - d_x)\tfrac{x+u}{K_H}\right) \\
\frac{dy_i}{dt} &= y_i\left(b_i\min\left(1, \tfrac{\beta_i[P_e]}{m_i}\right)\min(1, L) - d_i - (b_i - d_i)\tfrac{u}{K_T}\right) \\
\frac{dz}{dt} &= c\min\left(1, \tfrac{[P_e]}{n}\right)(y_1(t-\tau) + y_2(t-\tau)) - d_z z \\
u &= y_1 + y_2 + z.
\end{aligned} \tag{6.61}
$$

Note that we have added the coefficient β_i to represent the relative efficiency of P uptake by cancer cells. Finally, the model may be modified to consider a changing phosphate load in the organ. We assume a constant influx of P at rate r, representing dietary intake. When cells die, they liberate P; a small amount of the liberated P will not be recycled, but will be lost to the circulation. We let γ represent the fraction lost, and have:

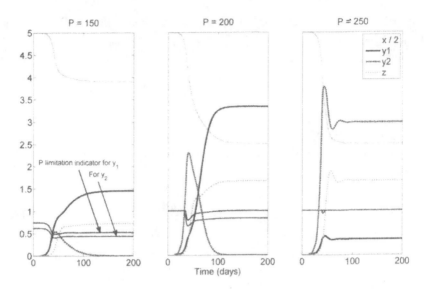

FIGURE 6.7: KNE model for phosphate-limited tumor growth under different values of total organ phosphorus, P. Parameter values are $a = 0.3$; $n = 10$, $b_1 = 0.6$, $m_1 = 20$, $b_2 = 0.72$, $m_2 = 24$, $d_x = d_1 = d_2 = 0.1$, $d_z = 0.2$, $c = 0.1$, $\alpha = 0.05$, $g = 100$, $K_T = 5$, and $K_H = 10$.

$$\frac{dP}{dt} = r - \gamma \left(n(d_x + d_z z) + (a - d_x)nx\frac{x + y_1 + y_2 + z}{K_H} \right.$$
$$\left. + \sum_{i=1}^{2} m_i d_i y_y + \frac{y_1 + y_2 + z}{K_T} \sum_{i=1}^{2} m_i(b_i - d_i) \right). \qquad (6.62)$$

6.6.2 Predictions

Kuang et al. [23] estimated the phosphorus content of the lung to be roughly 150 g. We also assume that growth rate for different cell lines scales linearly with cellular phosphate concentration (i.e., a 10% increase in P increases the specific growth rate by 10%). From numerical simulation of the model with fixed P (results are essentially the same using equation (6.62) for P dynamics), it is clear that when total organ phosphorus is below some threshold, rapidly growing cancer strains may dominate early in time, but slower growing lines dominate with large time, and the faster growers are driven to extinction. If organ phosphorus is sufficiently plentiful that space becomes limiting before phosphorus, then rapid growers dominate. However, in this case there is coexistence of the slow and rapid growers. These results are summarized graphically in Figures 6.7 and 6.8.

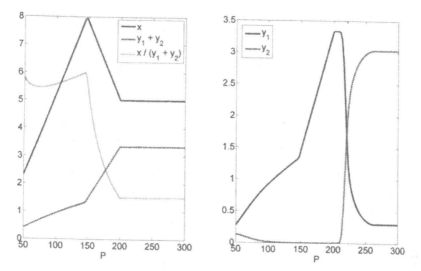

FIGURE 6.8: The left panel gives long-term total tumor mass $(y_1 + y_2)$, healthy cell mass (x), and the ratio of healthy to cancerous tissue $(x/(y_1+y_2))$, for different values of total organ phosphorus under the KNE model. The right panel gives the long-term population size of slowly (y_1) and rapidly growing (y_2) tumor cell strains. See Figure 6.7 for parameter values.

This gives the counterintuitive prediction that slow-growing cell lines can dominate the tumor over time and continually threaten faster-growing lines with extinction. Kuang et al. suggested that this may provide the evolutionary impetus for aggressive growers to spread metastatically to distant sites, replete with phosphorus.

On the surface, phosphate-limited tumor growth suggests limiting phosphate by dietary restriction or with pharmacologic phosphate binders as a clinical treatment strategy. However, limiting P also harms the healthy tissue. From Figure 6.8, it can be seen that the ratio of healthy to tumor tissue is maximized when P is about 150 g (at least for this parameter set). Decreasing P reduces both tumor and healthy cell mass, but is relatively more deleterious to healthy cells. Increasing P fuels tumor growth to the detriment of healthy tissue, and there is a region where tumor mass is quite sensitive to increases in P.

6.7   Mathematical analysis of a simplified KNE model

For realistic parameter values and initial conditions, the ultimate outcome of the KNE model (6.61) is that solutions tend to a positive steady state where phosphorus limits both healthy and tumor cell growth. Unfortunately, it is difficult to find an analytic expression of this positive steady state and even more daunting to determine its stability properties. However, near this steady state, tumor growth shall be minimally limited by its blood vessel infrastructure since it has enough time for the infrastructure to be put in place. This presents a realistic simplification of the KNE model. Therefore, in this section we will assume that

(A1): The construction of blood vessels is not limited by phosphorus supply.

With (A1), the KNE model (6.61) becomes the following simplified model:

$$
\begin{aligned}
\frac{dx}{dt} &= x\left(a\min\left(1, \frac{P_e}{fnK_h}\right) - d_x - (a - d_x)\frac{x + y + z}{K_h}\right), \\
\frac{dy}{dt} &= y\left(b\min\left(1, \beta\frac{P_e}{fmK_h}\right)\min(1, L) - d_y - (b - d_y)\frac{y + z}{K_t}\right), \\
\frac{dz}{dt} &= cy(t - \tau) - d_z z, \\
L &= \frac{g(z - \alpha y)}{y}.
\end{aligned}
\tag{6.63}
$$

Additional support for the validity of assumption (A1) comes from the fact that the difference between ultimate sizes of tumors described by the KNE model (6.61) and the simplified KNE model (6.63) are negligible.

In the following, we will study the stability of the positive steady state E^* of model (6.63). Our analysis is simplified by the following observation from simulation results: at this steady state E^*, $\dfrac{P_e}{fnK_h} < 1$ and $L > 1$ (see Figure 6.9). Hence we assume further that

(A2): For model (6.63), $\dfrac{P_e}{fnK_h} < 1$ and $L > 1$ at E^*.

Clearly (A2) implies that $\dfrac{\beta P_e}{fmK_h} < 1$. With this additional assumption,

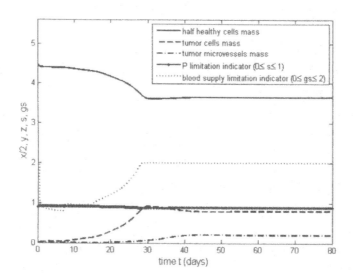

FIGURE 6.9: A solution for model (4.1) with $a = 3, m = 20, n = 10, K_h = 10, K_t = 3, f = 0.6667, P = 150, \alpha = 0.05, b = 6, \tau = 7, c = 0.05, d_x = d_y = 1, d_z = 0.2, g = 100$ and $(x(0), y(0), z(0)) = (9, 0.01, 0.001)$. Here we assume no treatment blocking phosphorus uptake by tumor cells ($\beta = 1$) and the construction of blood vessel is NOT phosphorus limited. Notice that $\frac{P_e}{fnK_h} < 1$ and $L > 1$ at E^* for model (6.63).

model (6.63) is further reduced to

$$
\begin{aligned}
\frac{dx}{dt} &= x\left(a\frac{P_e}{fnK_h} - d_x - (a - d_x)\frac{x + y + z}{K_h}\right), \\
\frac{dy}{dt} &= y\left(b\beta\frac{P_e}{fmK_h} - d_y - (b - d_y)\frac{y + z}{K_t}\right), \\
\frac{dz}{dt} &= cy(t - \tau) - d_z z.
\end{aligned}
\tag{6.64}
$$

This reduced model has a unique positive steady state $E^* = (x^*, y^*, z^*)$:

$$
\begin{aligned}
x^* &= \frac{K_h}{a - d_x}\left[\frac{an}{\beta bm}\left(d_y + (b - d_y)\frac{y^* + z^*}{K_t}\right) - d_x\right] - y^* - z^*, \\
y^* &= d_z K_t N/D, \\
z^* &= cK_t N/D.
\end{aligned}
\tag{6.65}
$$

where

$$
N = aPb\beta - d_x Pb\beta - d_y K_h ma - afmK_h d_y + b\beta nd_x K_h + fd_x mK_h d_y
$$

and

$$
\begin{aligned}
D = {} & -fd_x d_z mK_h b - ad_z fmK_h d_y - fd_x cmK_h b - acfmK_h d_y - \\
& ad_z mK_h d_y + ad_z mK_h b + acfmK_h b + ad_z mK_t b\beta + fd_x d_z mK_h d_y - \\
& d_x d_z mK_t b\beta + ad_z fmK_h b - K_t b\beta d_z na + K_t b\beta d_z nd_x + \\
& fd_x cmK_h d_y + acmK_h b - acmK_h d_y.
\end{aligned}
\tag{6.66}
$$

Notice that $y^* = 0$ if and only if

$$
N = aPb\beta - d_x Pb\beta - d_y K_h ma - afmK_h d_y + b\beta nd_x K_h + fd_x mK_h d_y = 0.
$$

This yields a threshold value for β, which we denote by β^*,

$$
\beta^* = \frac{d_y K_h ma + afmK_h d_y - fd_x mK_h d_y}{(a - d_x)Pb + bnd_x K_h} = \frac{m}{b}K_h d_y \frac{a(f + 1) - fd_x}{(a - d_x)P + nd_x K_h}.
\tag{6.67}
$$

In order to study stability aspects of the steady state E^* of model (6.64), appropriate methods from the theory of delay differential equations are needed. The recent textbook on this subject by Smith [39] is a timely and easy to follow reference. However, to obtain some quick result, we can apply the following lemma, the proof of which follows directly from that of Theorem 6.5.2 (page 227) in Kuang, 1993 [20].

LEMMA 6.1

Assume that the parameters in the following system are positive, and $x^* > 0, y^* > 0, z^* > 0$:

$$\frac{dx}{dt} = -x(A_1(x - x^*) + A_2(y - y^*) + A_3(z - z^*)),$$

$$\frac{dy}{dt} = -y(B_1(x - x^*) + B_2(y - y^*) + B_3(z - z^*)), \qquad (6.68)$$

$$\frac{dz}{dt} = -(-c(y(t - \tau) - y^*) + d_z(z - z^*)).$$

If there are positive constants c_1, c_2 such that
(1): $d_z > c/c_2$,
(2): $B_2/c_2 > B_3 + B_1/c_1$,
(3): $A_1/c_1 > A_1 + A_2/c_2$
then the steady state $E^* = (x^*, y^*, z^*)$ is globally asymptotically stable.

Near the steady state E^*, model (6.63) can be rewritten in the form of system (6.68) with

$$A_1 = \frac{1}{K_h}(\frac{a}{f} + a - d_x) = A_3, \qquad A_2 = \frac{1}{K_h}(\frac{am}{fn} + a - d_x) \qquad (6.69)$$

and

$$B_1 = \frac{b\beta n}{fK_h m}, \qquad B_2 = \frac{b\beta}{fK_h} + \frac{b - d_y}{K_t}, \qquad B_3 = \frac{b\beta n}{fK_h m} + \frac{b - d_y}{K_t}. \qquad (6.70)$$

For $a = 3, m = 20, n = 10, K_h = 10, f = 0.6667, P = 150, b = 6, d_x = 1, d_y = 1, \beta = 1$, we can chose $c_1 = 0.5$ and $c_2 = 0.2$ to satisfy conditions 1) through 3) in Lemma 6.1. In other words, we have shown that for this set of parameters, the positive steady state E^* of model (6.64) is locally asymptotically stable. However, simulation suggests that it is actually globally asymptotically stable. So, this mathematical question remains open.

As we increase P in model (6.64), the condition $\dfrac{P_e}{fnK_h} < 1$ may be violated, and the positive steady state may become the positive solution of

$$x^* + y^* + z^* = K_h,$$

$$B_1 x^* + B_2 y^* + B_3 z^* = \frac{b\beta P}{fmK_h} - d_y, \qquad (6.71)$$

$$cy^* - d_z z^* = 0,$$

where $B_i, i = 1, 2, 3$ are given by equation (6.70). In this case, we have

$$y^* = \frac{(b\beta P)/(fmK_h) - d_y - B_1 K_h}{B_2 - B_1 + (B_3 - B_1)c/d_z}. \qquad (6.72)$$

Using Lemma 6.1, we can also show that this steady state is locally asymptotically stable.

A sufficiently large increase in P will lead to a scenario in which $\dfrac{\beta P_e}{fmK_h} > 1$. For example, when $a = 3, m = 20, n = 10, K_h = 10, f = 0.6667, P = 150, b = 6, d_x = 1, d_y = 1, \beta = 1$, we need $P > 257.34$. In such a case, the positive steady state is simply $E^* = (K_h - K_t, d_z K_t/(c + d_z), cK_t/(c + d_z))$, which again by Lemma 6.1, is locally asymptotically stable.

We summarize the above statements into the following theorem.

THEOREM 6.1

Assume that in model (6.64) there is a unique positive steady state $E^* = (x^*, y^*, z^*)$. Assume further that there are positive constants c_1, c_2 such that
(1): $d_z > c/c_2$,
(2): $B_2/c_2 > B_3 + B_1/c_1$,
(3): $A_1/c_1 > A_1 + A_2/c_2$
where A_1, A_2, B_1, B_2, B_3 are given by equations (6.69) and (6.70). Then the steady state $E^* = (x^*, y^*, z^*)$ is locally asymptotically stable.

In an interesting special case when $ma = nb, d_x = d_y, \beta = 1$, we have

$$
\begin{aligned}
y^* &= \frac{aP - fnK_h d_x}{fn(a - d_x)(\rho + 1 + \sigma) + a(n\rho + m + n\sigma)} \\
&= \frac{bP - fmK_h d_y}{fm(a - d_y)(\rho + 1 + \sigma) + b(n\rho + m + n\sigma)},
\end{aligned}
\tag{6.73}
$$

where

$$
\sigma = \frac{c}{d_z}, \qquad \rho = \left(\frac{b - d_y}{K_t} - \frac{a - d_x}{K_h}\right) \frac{K_h(1 + \sigma)}{a - d_x}.
$$

This expression of tumor steady state size shows that P plays a prominent role in determining its value. We observe that the tumor dies out if one can increase the tumor's death rate or the tumor's P requirement m, or lower the tumor's proliferation rate to certain threshold levels.

6.8   Exercises

Exercise 6.1: Explicitly derive a version of the Droop model that considers cell mass in terms of absolute cell count (units cells) rather than cells l^{-1}. Confirm that the cell quota Q has the same units as before.

Exercise 6.2: The basic cell quota model of biomass growth in a batch culture (a closed system in which cells are grown in a fixed volume of nutrient culture medium under specific environmental conditions) takes the form of

$$\frac{dx}{dt} = \mu_m \left(1 - \frac{q}{Q}\right) x, \qquad (6.74)$$

$$\frac{dQ}{dt} = u_m \frac{s}{K_s + s} - \mu_m (Q - q), \qquad (6.75)$$

$$\frac{ds}{dt} = -u_m \frac{s}{K_s + s} x, \qquad (6.76)$$

with positive initial values. All parameters are assumed to be positive constants. Show that

i): Explain the model formulation. (Hint: $s(0) = s(t) + x(t)Q(t)$)

ii): If $Q(0) \geq q$, then $Q(t) \geq q$, $\forall t \geq 0$;

iii): The solution tends to a nonnegative steady state that is dependent on its initial condition.

Exercise 6.3: Use MATLAB® to simulate model system (6.74)-(6.76) with your own positive parameter values and appropriate positive initial value. Print out several representative simulation figures.

Exercise 6.4: Chronic myeloid leukemia (CML) is a cancer of the white blood cells. CML can be molecularly diagnosed by detecting the presence of the Philadelphia (Ph) chromosome and the fusion oncogene BCR-ABL. This oncogene is the result of translocation of the BCR, or breakpoint cluster, gene located on chromosome 22 and the ABL, or Ableson leukemia virus, gene located on chromosome 9. The growth rate of the BCR-ABL dependent and independent cell populations are modeled using the Droop's cell quota model, where $Q(t)$ represents the cell quota for BCR-ABL. The BCR-ABL dependent, BCR-ABL independent, and normal populations are modeled respectively by the following [14]:

$$\frac{dx_1}{dt} = r_1\left(1 - \frac{q_1}{Q}\right)x_1 - d_0 x_1 - m_{12}(Q)x_1 + m_{21}(Q)x_2, \qquad (6.77)$$

$$\frac{dx_2}{dt} = r_2\left(1 - \frac{q_2}{Q}\right)x_2 - d_0 x_2 + m_{12}(Q)x_1 - m_{21}(Q)x_2, \qquad (6.78)$$

$$\frac{dx_3}{dt} = \left(\frac{r_3}{1 + p_3(x_1 + x_2 + x_3)}\right)x_3 - d_0 x_3, \qquad (6.79)$$

$$\frac{dQ}{dt} = v_m \frac{q_{m1} - Q}{q_{m1} - q_1} \frac{B}{B + v_h} - \mu_m(Q - q_1) - bQ. \qquad (6.80)$$

We assume that the proliferation rates, $r_i(1 - \frac{q_i}{Q}), i = 1, 2$, of both BCR-ABL dependent and independent populations are BCR-ABL cell quota dependent while the proliferation rate, $\frac{r_3}{1 + p_3(x_1 + x_2 + x_3)}$, for the normal population is density dependent. We assume $q_1 > q_2$,

$$m_{12}(Q) = k_1 \frac{K_1^n}{Q^n + K_1^n}, \quad m_{21}(Q) = k_2 \frac{Q^n}{Q^n + K_2^n}, \quad B = \frac{p}{(d_1 + u(t))}. \qquad (6.81)$$

All parameters are positive constants. Show that the solutions of (6.77)-(6.80) stay in the region $\{(x_1, x_2, x_3, Q) : x_1 \geq 0, x_2 \geq 0, 0 \leq x_3 \leq \max\{\frac{1}{d_0 p_3}(r_3 - d_0), x_3(0)\}, q_1 \frac{\mu_m}{\mu_m + b} \leq Q \leq q_{m1}\}$ provided that $x_1(0) \geq 0, x_2(0) \geq 0, x_3(0) \geq 0, q_{m1} \geq Q(0) \geq q_1$.

Exercise 6.5: Show that the system (6.77)-(6.80) has two possible boundary equilibria: $E_0 = (0, 0, 0, Q_1)$, $E_1 = (0, 0, \frac{r_3 - d_0}{p_3 d_0}, Q_1)$. Can it have any interior equilibrium? Find the local stability conditions for the boundary equilibria. Show that the system (6.77)–(6.80) has no positive periodic solutions.

Exercise 6.6: Consider the following case of phosphorus-limited growth in some species. Let $P_t = $ total environmental P, which remains constant. Let P_f be the free P, and we have the total intracellular phosphorus as $P_x = Qx$. Hence

$$P_t = P_f + Qx \tag{6.82}$$

Assume that P uptake increases linearly with free phosphorus P_f and the difference of the maximum cell quota and the current cell quota of the phosphorus, giving:

$$\frac{dQ}{dt} = \alpha P_f \frac{Q_M - Q}{Q_M - q} - \mu_m (Q - q) \tag{6.83}$$

and use the following equation for species growth,

$$\frac{dx}{dt} = \mu_m \left(1 - \frac{q}{Q}\right) x - dx \tag{6.84}$$

which is simply the Droop equation for growth supplemented by a constant death rate.

1. Using a quasi-steady-state argument for Q, show that when $Q_M \gg 1$, this model can be approximated by the classical logistic growth model of the form

$$\frac{dx}{dt} = rx \left(1 - \frac{x}{K}\right) \tag{6.85}$$

where

$$r = \frac{q(\mu_m - d)P_t^{-1} - d\mu_m \alpha^{-1}}{P_t q^{-1} + \mu_m \alpha^{-1}}$$

$$K = (\mu_m - d)P_t(q\mu_m)^{-1} - d\mu_m \alpha^{-1}.$$

2. Show that both r and K are increasing functions of α and decreasing function of d.

3. Give a biological interpretation of these results.

   (a) Consider the simplified case where $d = 0$. What do r and K reduce to, and give a biological interpretation. What is the significance of the ratio P_t/q?

   (b) How do you expect a species with high turnover (i.e., high specific growth rate and death rate) would compete with one with low turnover?

   (c) How does increasing μ_m affect r and K if phosphorus uptake (α) is not increased?

   (d) How does increasing α affect r and K?

Exercise 6.7: Derive the equation (6.67) for the expression of β^* and the equation (6.72) for the expression of y^*.

Exercise 6.8: Reproduce Figure 6.9.

Exercise 6.9: Apply the Theorem 6.5.2 (page 227) in Kuang, 1993 [20] to prove Lemma 6.1.

Exercise 6.10: The following simple delay differential equation model was introduced in Everett et al. [12] to describes ovarian tumor growth and tumor induced angiogenesis, subject to on and off anti-angiogenesis treatment. Let y represent the vascularized tumor volume and Q represent the intracellular concentration of necessary nutrients provided by angiogenesis, or the cell quota of some limiting nutrient from angiogenesis. The model takes the following form:

$$y' = \underbrace{\mu_m \left(1 - \frac{q}{Q}\right) y}_{\text{growth}} - \underbrace{dy}_{\text{death}}, \qquad (6.86a)$$

$$Q' = \underbrace{\alpha \frac{y(t-\tau)}{y(t)}}_{\text{nutrient uptake}} - \underbrace{\mu_m (Q-q)}_{\text{dilution}}. \qquad (6.86b)$$

The above model assumes that it takes τ units of time for the vascular endothelial cells to respond to the angiogenic signal and mature to fully functional vessels and that the nutrient uptake rate is proportional to the nutrient concentration in the interstitial fluid, which in turn is proportional to the blood vessel density τ time units in the past. The delay arises because the tumor is assumed to grow into regions that are unvascularized, and it takes τ units of time for them to vascularize. Parameter α represents both uptake rate of the nutrients in the interstitial fluid and resulting nutrient concentration per tumor unit.

Show that the solutions of the system (6.86) with the initial conditions $Q_M > Q(t) > q$ and $y(t) > 0$ for $t \in [-\tau, 0]$ will remain in that region for all $t > 0$, where $Q_M = \max\left\{Q(s), q + \frac{\alpha}{\mu_m} e^{d\tau}, s \in [0, \tau]\right\}$. If $\mu_m \le d$, then $\lim_{t\to\infty} y(t) = 0$.

Exercise 6.11: The uptake of nutrients is usually on a faster time scale than the population growth dynamics. If we apply a quasi-steady state argument on the cell quota equation in system (6.86) by allowing $Q' = 0$, we have

$$Q'(t) = 0 = \frac{\alpha y(t-\tau)}{y(t)} - \mu_m(Q^*(t) - q)$$

and so

$$Q^*(t) = \frac{\alpha y(t-\tau)}{\mu_m y(t)} + q.$$

1. Show that

$$y'(t) = f(y(t), y(t-\tau)) \equiv \left(\frac{\mu_m \alpha y(t-\tau)}{\alpha y(t-\tau) + q\mu_m y(t)} - d\right) y(t). \qquad (6.87)$$

2. Show that if $d > \frac{\mu_m \alpha}{\alpha + q\mu_m}$, then the solutions of the limiting system (6.87) tend to $y = 0$.

Exercise 6.12: Motivated by the off-treatment tumor growth data in [19], Everett et al. [12] looked for the existence of a dominating exponential solution. Let $y(t) = y_0 e^{\lambda t}$. Then

$$e^{\lambda \tau} = \frac{\alpha(\mu_m - \lambda - d)}{\mu_m q(\lambda + d)}. \tag{6.88}$$

1. Show that if $\alpha(\mu_m - d) > \mu_m qd$, then there exists a unique real eigenvalue, $0 < \lambda_1 < \mu_m - d$, that satisfies (6.88).

2. Show that $(y_1(t), Q^*)$ is a solution to the limiting system (6.87) where

$$(y_1(t), Q^*) = \left(y_0 e^{\lambda_1 t}, \frac{\alpha e^{-\lambda_1 \tau} + \mu_m q}{\mu_m} \right). \tag{6.89}$$

Exercise 6.13: Show that if $\alpha(\mu_m - d) > \mu_m qd$ and $\alpha < \mu_m q e^{\lambda_1 \tau}$, then $\lambda_1 \geq \sup\{\text{Re}(\lambda) : \lambda \text{ is any solution of (6.88)}\}$.

This result provides a sufficient condition that ensures λ_1 as the dominant eigenvalue, i.e. when the solution can be approximated by (6.89) for some $y_0 > 0$. Also, note that since $y_1(t) = y_0 e^{\lambda_1 t}$ is a solution to the system, we know that $y(t)$ is not bounded above.

Exercise 6.14: Show that if $\frac{\mu_m \alpha(\alpha e^{-2\lambda_1 \tau} + q\mu_m)}{(\alpha e^{-\lambda_1 \tau} + q\mu_m)^2} < d < \frac{\mu_m \alpha}{\alpha + q\mu_m}$, then the solution $y_1 = y_0 e^{\lambda_1 t}$ of equation (6.88) is unstable.

Exercise 6.15: During an anti-VEGF treatment, the blood vessel growth will be impaired due to the inhibition of VEGF, but existing vasculature is likely not to be affected by the treatment. In that case nutrient delivery remains constant due to the static vasculature. One may assume that blood vessel sprouts that began forming within τ time units before the onset of treatment will not be fully formed and functional. Let t_0 represent the time of treatment onset and $\bar{y} = y(t_0 - \tau)$. Then nutrient delivery is dependent upon $\bar{y}$, and the delay differential equation model (6.86) becomes an ordinary differential equation model:

$$y' = \underbrace{\mu_m \left(1 - \frac{q}{Q} \right) y}_{\text{growth}} - \underbrace{dy}_{\text{death}} \tag{6.90a}$$

$$Q' = \underbrace{\alpha p \frac{\bar{y}}{y(t)}}_{\text{nutrient uptake}} - \underbrace{\mu_m (Q - q)}_{\text{dilution}} \tag{6.90b}$$

1. Show that the solutions of the system (6.90) are bounded away from zero.

2. Show that the solutions to the system (6.90) are bounded from above.

3. The only positive equilibrium point E^*, when exists, is globally asymptotically stable with respect to positive initial value (y_0, Q_0) such that $Q_M > Q(0) = Q_0 > q$ and $y_0 = y(0) > 0$ where $Q_M = \max \left\{ Q(s), q + \frac{\alpha}{\mu_m} e^{d\tau}, s \in [0, \tau] \right\}$.

Exercise 6.16: Fit model system (6.86) to the on and off treatment preclinical data sets from Mesiano et al. [19] with $\tau = 10, \mu_m = 0.41, d = 0.28, q = 0.0064, \alpha = 0.050, p = 0.17, Q_0 = 0.014$. You can use some computer program such as Plot Digitizer to approximate the on and off treatment data from Mesiano et al. [19].

6.9   Projects

The KNE model described in this chapter shall be viewed as only an initial attempt to understand the growth dynamics of a single vascularized solid tumor growing within the confines of an organ, such as a primary lung or breast tumor. In this section, we point out one of its main limitations and some opportunities for alternative model formulations and some natural extensions. In addition, we briefly mention a few other biomedical examples of nutrient-limited growth or resource competition that may be of interest to the readers.

6.9.1   Beyond the KNE model

6.9.1.1   Phosphate homeostasis

The most significant limitation of the KNE model is that it makes phosphate regulated at the level of total organ phosphate, rather than extracellular serum phosphate. In reality, serum phosphate concentration is held within a narrow range by endocrine regulation, about 3–4.5 mg/l. Because bone mineral is made of calcium-phosphate complexes in the form of hydroxyapatite, $Ca_5(PO_4)_3(OH)$, phosphate homeostasis is intimately linked to calcium homeostasis, and skeletal bone serves as a dynamic reservoir and buffer for both phosphate and calcium. Indeed, about 85% of the body's phosphate is stored in the bone, and total body phosphate is at a dynamic steady state governed mainly by dietary intake (about 16 mg/kg/day), liberation and storage in the bone (about 3 mg/kg/day), and renal excretion (about 16 mg/kg/day) [3].

6.9.1.2   Intracellular phosphate: A Droop approach?

Extracellular phosphate levels are hormonally regulated to remain within a narrow range, and the intracellular concentration of ribosomal phosphate is clearly the relevant metric for determining cellular growth rate, suggesting that a Droop-like model may be well-suited to this problem. Indeed, our Droop model for androgen-limited growth discussed in Section 6.5.4 could be adapted.

6.9.1.3   Tumor lysis syndrome

Tumor lysis syndrome (TLS) is a complication of chemotherapy where large numbers of dying cancer cells release their intracellular contents into the systemic circulation, leading to high levels of phosphate, potassium, uric acid, and consequence hypocalcemia. These derangements can lead to life-threatening kidney failure and cardiac arrhythmias [16]. We can speculate that the phosphate liberated by chemotherapy, even if it does not lead to overt

TLS, could fuel fast-growing cancer cell lines, causing a post-therapy tumor growth burst. Moreover, this abundant phosphate could give fast growers at least a transient selective advantage, possibly increasing the aggressiveness of a recurring tumor. Such an idea could be tested mathematically with a model that considers the positive contribution of cell death to serum phosphate concentration.

6.9.2 Iodine and thyroid cancer

In response to pituitary thyroid stimulating hormone (TSH), the thyroid gland produces thyroid hormone (TH), which modulates metabolic activity by nearly every cell of the body. Iodine is an essential component of the thyroid hormones, and iodine dietary intake clearly affects the incidence of thyroid cancer. Chronic iodine deficiency greatly increases the risk of benign thyroid goiters and nodules, which in turn are susceptible to malignant transformation. Thus, iodine deficiency is clearly associated with an increased cancer risk. The effect of excess iodine is less clear, but it too may increase cancer risk [30].

Iodine deficiency continues to affect a significant proportion of the globe, although iodization of salt has ameliorated the problem in much of the world. Introducing iodine to chronically iodine-deficient populations can also result in reactive hyperthyroidism, as thyroids adapted to low iodine levels are suddenly flush with the element.

6.9.3 Iron and microbes

Iron is an essential nutrient required by all pathogenic microorganisms as well as the hosts they infect. Unlike essentially all other nutrients, iron is not freely available to pathogens, and its transport and storage is tightly regulated by the host. Thus, iron can be viewed as *the* critical limiting element in pathogen growth [34].

Iron acts on several levels in infection. It is an essential cofactor in a number of metabolic pathways required for energy production and cellular replication. Iron is also essential to innate host immune responses, being required for energy production and the generation of nitric oxide and other reactive oxygen species (ROS) that play an essential role in destroying intracellular pathogens [5]. In a variety of microbial infections both iron overload and deficiency can increase morbidity and mortality [5]. Thus, competition between the host and pathogen for iron is essential in determining the course of disease.

Salmonella, for example, infect macrophages which destroy most bacteria within hours by iron dependent mechanisms (e.g. the respiratory burst). Surviving bacteria enter a cytostatic state of bacterial persistence, and iron load may be crucial to the switch to bacteria persistence. Epidemiologically, iron overload and deficiency increase both salmonella infection and disease virulence.

Intracellular pathogens compete with and/or exploit host iron trafficking mechanisms in a number of ways. For an exhaustive review of iron trafficking and metabolism in bacterial pathogens, see [34].

6.9.3.1   Salmonella infection

Salmonella primarily infects mononuclear macrophages, and intracellular survival in these cells is essential to in vivo virulence. When responding to a salmonella infection, macrophages execute two killing programs. The first is the respiratory burst, which primarily contributes to early killing. Inducible nitric oxide synthase (iNOS) is one of three key enzymes generating nitric oxide (NO). NO generation by iNOS is the second major pathway, and contributes to both early and late stage killing [44]. Both pathways are dependent upon iron, and NO may act synergistically with oxygen radicals generated by the respiratory burst.

This response leads to a pattern of infection where 99% of bacteria are killed within the first few hours of infection, while the survivors enter into a cytostatic state of bacterial persistence [44]. There is a delicate balance between pro-oxidant and antioxidant effects of NO in salmonella; this balance is largely mediated by iron, and iron load may be crucial to the switch to bacteria persistence. This is particularly relevant as 2–5% of infections can lead to asymptomatic chronic carrier states.

Epidemiologically, iron overload and deficiency both increase salmonella infection and disease virulence. Treatment with the intracellular iron chelator deferoxamine resulted in a 2–3 log increase in bacterial load, while extracellular iron chelation did not affect the disease [5].

An interesting area that could be investigated with modeling is understanding how iron load affects progression to a bacteriostatic state, and how dietary intervention or treatment with chelators could influence the chronic infection state.

6.9.3.2   Malaria

Malaria induces a cytokine mediated host response that induces iNOS and NO production. Malaria infection causes anaemia, but can increase iron concentration in the liver and spleen due to recycling of heme-bound iron stored in both infected and uninfected erythrocytes eliminated by macrophages. This inhomogeneous iron load may affect the course of disease, and salmonella co-infection with malaria may be an interesting and important area to study.

References

[1] Akakura K, Bruchovsky N, Goldenberg SL, Rennie PS, Buckley AR, Sullivan LD: Effects of intermittent androgen suppression on androgen-dependent tumors. Apoptosis and serum prostate-specific antigen. *Cancer* 1993, 71:2782–2790.

[2] Berges RR, Vukanovic J, Epstein JI, CarMichel M, Cisek L, Johnson DE, Veltri RW, Walsh PC, Isaacs JT: Implication of cell kinetic changes during the progression of human prostatic cancer. *Clin Cancer Res* 1995, 1:473–480.

[3] Bergwitz C, Jüppner H: Regulation of phosphate homeostasis by PTH, vitamin D, and FGF23. *Annu Rev Med* 2010, 61:91–104.

[4] Bruchovsky N, Klotz L, Crook J, Goldenberg SL: Locally advanced prostate cancer-biochemical results from a prospective phase II study of intermittent androgen suppression for men with evidence of prostate-specific antigen recurrence after radiotherapy. *Cancer* 2007, 109:858–867.

[5] Collins HL: The role of iron in infections with intracellular bacteria. *Immunol Lett* 2003, 85:193–195.

[6] Droop MR: Vitamin B12 and marine ecology, IV: The kinetics of uptake, growth and inhibition in *Monochrysis lutheri*. J Mar Biol Assoc, UK 1968, 48:689–733.

[7] Droop MR: Some thoughts on nutrient limitation in algae. *J Phycol* 1973, 9:264–272.

[8] Eikenberry SE, Nagy JD, Kuang Y: The evolutionary impact of androgen levels on prostate cancer in a multi-scale mathematical model. *Biol Direct* 2010, 5:24.

[9] Elser JJ, Kuang Y: Ecological stoichiometry. *Encyclopedia of Theoretical Ecology*, Hastings and Gross eds. University of California Press, 2012, 718–722.

[10] Elser JJ, Nagy JD, Kuang Y: Biological stoichiometry: An ecological perspective on tumor dynamics. *BioScience* 2003, 53:1112–1120.

[11] Elser, JJ, Kyle MM, Smith JS, Nagy JD: Biological stoichiometry in human cancer. *PLoS ONE* 2007, 2:e1028.doi:10.1371/journal.pone.0001028.

[12] Everett RA, Nagy JD, Kuang Y: Dynamics of a data based ovarian cancer growth and treatment model with time delay. *J. Dyn and Diff Equat*, 2015, DOI: 10.1007/s10884-015-9498-y.

[13] Everett RA, Packer A, Kuang Y: Can Mathematical models predict the outcomes of prostate cancer patients undergoing intermittent androgen deprivation therapy? *Biophys Rev and Lett* 2014, 9:173–191.

[14] Everett RA, Zhao Y, Flores KB, Kuang Y: Data and implication based comparison of two chronic myeloid leukemia models. *Math Biosc Eng* 2013, 10:1501–1518.

[15] Feldman BJ, Feldman D: The development of androgen-independent prostate cancer. *Nat Rev Cancer* 2001, 1:34–45.

[16] Gemici C: Tumour lysis syndrome in solid tumours. *Clin Oncol* (R Coll Radiol) 2006, 18:773–80.

[17] Ideta A, Tanaka G, Takeuchi T, Aihara K: A mathematical model of intermittent androgen suppression for prostate cancer. *J Nonlinear Sci* 2008, 18:593–614.

[18] Jackson TL: A mathematical model of prostate tumor growth and androgen-independent relapse. *Disc Cont Dyn Sys B* 2004, 4:187–201.

[19] Mesiano S, Napoleone Ferrara N, and Robert B. Jaffe RB: Role of vascular endothelial growth factor in ovarian cancer: Inhibition of ascites formation by immunoneutralization. *Am J Pathol* 1998, 153:1249–1256.

[20] Kuang Y: *Delay differential equations with applications in population dynamics.* New York: Academic Press, 1993.

[21] Kuang Y, Huisman J, Elser JJ: Stoichiometric plant-herbivore models and their interpretation, *Math Biosc Eng* 2004, 1:215–222.

[22] Lamont KR, Tindall DJ: Minireview: Alternative activation pathways for the androgen receptor in prostate cancer. *Mol Endocrinol* 2011, 25:897–907.

[23] Kuang Y, Nagy JD, Elser JJ: Biological stoichiometry of tumor dynamics: mathematical models and analysis. *Disc Cont Dyn Sys B* 2004, 4:221–240.

[24] Leadbeater BSC: The "Droop equation": Michael Droop and the legacy of the "cell-quota model" of phytoplankton growth. *Protist* 2006, 157:345–358.

[25] Liebig J: *Organic Chemistry in Its Application to Agriculture and Physiology.* London: Taylor Walton, 1840.

[26] Liebig J: *The Natural Laws of Husbandry.* London: Walton Maberly, 1863.

[27] Loladze I, Kuang Y, Elser JJ: Stoichiometry in producer-grazer systems: Linking energy flow with element cycling. *Bull Math Biol* 2000, 62:1137–1162.

[28] Loladze I, Kuang Y, Elser JJ, Fagan W: Coexistence of two predators on one prey mediated by stoichiometry. *Theor Popul Biol* 2004, 65:1–15.

[29] Lotka AJ: *Elements of Physical Biology.* Baltimore: Williams & Wilkins, 1925.

[30] Maso DL, Bosetti C, La Vecchia C, Franceschi S: Risk factors for thyroid cancer: an epidemiological review focused on nutritional factors. *Cancer Causes Control* 2009, 20:75–86.

[31] Morken JD, Packer A, Everett RA, Nagy JD, Kuang Y: Mechanisms of resistance to intermittent androgen deprivation therapy identified in prostate cancer patients by novel computational method. *Cancer Res* 2014, 74:2673–2683.

[32] Packer A, Li Y, Andersen T, Hu Q, Kuang Y, Sommerfeld M: Growth and neutral lipid synthesis in green microalgae: A mathematical model. *Biore Technol* 2011, 102:111–117.

[33] Portz T, Kuang Y, Nagy JD: A clinical data validated mathematical model of prostate cancer growth under intermittent androgen suppression therapy. *AIP Advances* 2012, 2:011002; doi: 10.1063/1.3697848

[34] Ratledge C, Dover LG: Iron metabolism in pathogenic bacteria. *Annu Rev Microbiol* 2000, 54:881–941.

[35] Raynaud JP: Prostate cancer risk in testosterone-treated men. *J Steroid Biochem Mol Biol* 2006, 102:261–266.

[36] Rocchi P, Muracciole X, Fina F, Mulholland DJ, Karsenty G, Palmari J, Ouafik L, Bladou F, Martin PM: Molecular analysis integrating different pathways associated with androgen-independent progression in LuCaP 23.1 xenograft. *Oncogene* 2004, 23:9111–9119.

[37] Schaible UE, Kaufmann SH: Iron and microbial infection. *Nat Rev Microbiol* 2004, 2:946–953.

[38] Severi G, Morris HA, MacInnis RJ, English DR, Tilley W, Hopper JL, Boyle P, Giles GG: Circulating steroid hormones and the risk of prostate cancer. *Cancer Epidemiol Biomarkers Prev* 2006, 15:86–91.

[39] Smith HL: *An Introduction to Delay Differential Equations with Applications to the Life Sciences*. Texts in Applied Mathematics. New York: Springer, 2011.

[40] Sofikerim M, Eskicorapci S, Oruç O, Ozen H: Hormonal predictors of prostate cancer. *Urol Int* 2007, 79:13–18.

[41] Sterner RW, Elser JJ: *Ecological Stoichiometry: The Biology of Elements from Molecules to the Biosphere*. Princeton: Princeton University Press, 2002.

[42] Thompson IM, Goodman PJ, Tangen CM, Lucia MS, Miller GJ, Ford LG, Lieber MM, Cespedes RD, Atkins JN, Lippman SM, Carlin SM, Ryan A, Szczepanek CM, Crowley JJ, Coltman CA Jr: The influence of finasteride on the development of prostate cancer. *N Engl J Med* 2003, 349:215–224.

[43] Tisdale MJ: Mechanisms of cancer cachexia. *Physiol Rev* 2009, 89:381–410.

[44] Vazquez-Torres A, Jones-Carson J, Mastroeni P, Ischiropoulos H, Fang FC: Antimicrobial actions of the NADPH phagocyte oxidase and inducible nitric oxide synthase in experimental salmonellosis. I. Effects on microbial killing by activated peritoneal aacrophages in vitro. *J Exp Med* 2000, 192:227–236.

Chapter 7

Natural History of Clinical Cancer

7.1 Introduction

The **natural history** of a disease refers to its uninterrupted course in the absence of intervention, from initiation through its end stage—resolution, chronic illness or death. Any model for natural history of clinical cancer, either conceptual or formal, must address at least two central dynamics: (1) the growth kinetics of the primary tumor, and (2) the dynamics of metastatic spread to both regional (e.g., lymph nodes) and distant sites. Ultimately, our motivation to study cancer natural history is not purely intellectual interest. After all, cancer is very rarely left to follow its unperturbed natural history, so at first glance this might be considered an irrelevant problem. However, rational *a priori* design of treatment protocols and optimized screening programs requires models that accurately reproduce disease dynamics in the absence of intervention. It is with such clinical goals in mind that we proceed.

Breast cancer, more than any other malignancy, has undergone radical shifts in the prevailing conceptual model for its natural history. Furthermore, these models have motivated shifting treatment paradigms. From the end of the 19th century to the middle of the 20th, breast cancer was viewed as a local disease that spread outward from its epicenter. This view informed the decision to treat the disease with radical mastectomy. This paradigm was supplanted by the notion that breast cancer is primarily a systemic disease that metastasizes very early in its history, prior to clinical detection. Under the latter view, local therapy was minimized and systemic chemotherapy was used to target occult metastases. However, the systemic theory cannot successfully explain all clinical observations either, and many authors now view breast cancer as something between a purely local and purely systemic phenomenon. More recently, interest has turned from the local/systemic debate to understanding the role of the host environment in cancer progression. Of particular interest is the hypothesis that metastasis dormancy, which is influenced by interaction with the primary tumor and host environment, plays an important role in the dynamics of disease recurrence following local treatment.

In recent years, formal dynamical models have begun to contribute to our understanding of the natural history of breast cancer and the optimal treatment strategy. The Gompertz model has a relatively long history of motivating chemotherapy strategies. Other models have been used to study treatment strategies to control the primary tumor and development of metastases. More recent models suggest that tumor dormancy plays an important role in delaying metastatic recurrence following initial therapy. One prominent recent example is a set of models designed to predict the effect of different mammography screening regimes on breast cancer mortality, which have led to new screening guidelines [54, 77].

In this chapter we focus on the evolution of both conceptual and formal models of cancer's long-term natural history and inferences about proper treatment strategies that may be deduced from these models about proper treatment strategies. Breast cancer will serve as the primary illustration. We also include relevant models for metastatic spread in prostate and melanoma cancer, as the metastatic cascade is believed to be similar in all neoplasms. We further address the issues of tumor dormancy and primary tumor-metastasis interaction, and the evolutionary dynamics that drive metastatic spread. We finally address the role of modeling in understanding the effect of mammography screening on breast cancer mortality and optimization of screening schedules.

7.2 Conceptual models for the natural history of breast cancer: Halsted vs. Fisher

Models for the natural history of breast cancer have focused primarily on the relationship between the primary tumor and metastatic spread. Classically, breast cancer has been viewed as either (1) a primarily local lesion that spreads continuously and predictably away from the primary tumor (Halsted model), or (2) a systemic disease with metastases already present at the time of diagnosis (systemic/Fisher model) [67]. More recently, tumor ecology's role—including its micro-, global, and host hormonal environments— in modulating tumor growth has come to the fore. Early treatment paradigms focused on the Halsted concept, and therefore called for disfiguring radical mastectomy to obtain maximal local control of the primary tumor. Later, as the Fisher concept gained prominence, standard of care shifted away from radical treatment of the primary lesion in favor of distal disease control using systemic treatments. The most recent concepts now call for growth suppression by alteration of the host environment. Here we review the history of these paradigms; the reader may consult similar reviews in [46, 67] and especially [5].

7.2.1 Surgery and the Halsted model

From antiquity to the late 19th century, surgical removal of a breast cancer lesion was largely regarded as a futile endeavor. While removal of small tumors was sometimes performed, the disease inevitably and rapidly recurred. Surgery was often considered more detrimental than beneficial [5]. Such (justified) pessimism persisted through the late 19th century. It was not until the advent of anaesthetics and antisepsis that radical operations with any hope of cure could be attempted.

Significant progress was not made until the very end of the 19th century, when the great Johns Hopkins surgeon William S. Halsted reported success in preventing local cancer recurrence by the "complete radical mastectomy" [41]—i.e., mastectomy with complete removal of the underlying pectoralis major muscle and complete dissection of the fascia and glands of the axilla (armpit). This "cleaning of the axilla" (generally the first site for lymphatic metastases) was regarded as especially important if hope for a cure was to be entertained. Before this, nihilism concerning the course of disease was the rule. In his 1894 paper [41], Halsted reported that

> Every one knows how dreadful the results were before the cleaning out of the axilla became recognized as an essential part of the operation. Most of us have heard our teachers in surgery admit that they have never cured a case of cancer of the breast. The younger Gross did not save one case in his first hundred. D. Hayes Agnew stated in a lecture, a very short time before his death, that he operated on breast cancers solely for the moral effect on the patients, that he believed the operation shortened rather than prolonged life. H. B. Sands once said to me that he could not boast of having cured more than a single case, and in this case a microscopical examination of the tumor had not been made.

The success of the Halsted operation (Fig. 7.1) in controlling local disease provided validation for the paradigm of breast cancer as a local disease that spread progressively away from the primary lesion. It was understood that cancer cells invaded first through local lymphatic channels to regional lymph nodes, from which they invaded secondary lymphatics, before finally spreading as distant metastatic disease. As described by Baum et al. [5], various lymphatic levels (sentinal, secondary, tertiary, etc.) were viewed as defenses against metastatic spread "like the curtain walls around a medieval citadel." Metastatic disease occurred when local layers of defense were at last overwhelmed.

The Halsted model and its corollary of radical local cancer resection dominated treatment paradigms until the latter half of the 20th century. Yet, despite preventing local recurrence, such radical operations failed to yield long-term survival for the majority of patients because tumor recurrence in distant metastases was almost universal.

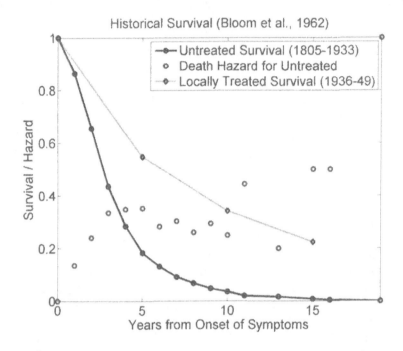

FIGURE 7.1: Bloom et al. [7] reported the survival data for 250 women diagnosed with breast cancer between 1805 and 1933 who refused treatment. Median survival time from onset of symptoms was 2.7 years, and one woman lived 18 years and 3 months. For comparison, 5, 10, and 15 year survival for women treated locally by radical mastectomy with or without axillary radiation between 1936 and 1949 is included. Note, however, that many women in the Bloom data set were only identified when their symptoms became serious, years after the initial onset. Therefore, this data set is biased and does not establish that untreated breast cancer is uniformly fatal, as has been asserted elsewhere.

7.2.2 Systemic chemotherapy and the Fisher model

The Fisher concept arose in the 1970s in response to Halsted's.[1] Fisher took the view that breast cancer metastasizes very early in its natural history, with metastasis occurring generally throughout the systemic vasculature. Local lymphatics were thought to be no defense against distant spread. This notion recommended systemic chemotherapy to eliminate occult metastases expected to be present at time of diagnosis. Surgical treatment of the primary tumor was deemphasized since local recurrence was considered unlikely to affect distant recurrence and overall survival. Moreover, the classical Fisher model asserts a kind of "metastatic predestination"—i.e., there exist two distinct classes of tumors: one that metastasizes essentially immediately, and another that never does.

Several factors influenced development of the Fisher concept. One was the development of effective systemic chemotheapeutics following the Second World War. But most importantly, progressively more radical operations inspired by the Halsted approach routinely failed to control the disease. Under the Fisher model, this failure was not surprising since the Halsted model is faulty. Two early clinical trials, NSABP B-04 and NSABP B-06, found that increased local control did not translate into improved survival or lower metastatic burden [67, 28]. These trials were very important in establishing the Fisher paradigm. More recently, meta-analysis of 194 trials by the Early Breast Cancer Trialists Collaborative Group (EBCTCG) demonstrated that anthracycline-based chemotherapy regimens reduce mortality.

Even though these results largely support the Fisher model, more recent data shows that local tumor control improves long-term (i.e., 15-year) survival. As pointed out in [11], this benefit is likely to accrue only in cases where a locally recurring tumor serves as a source for metastatic spread. If so, a significant lag-time between metastatic seeding and the emergence of clinically detectable metastases is expected. Therefore, local recurrence is likely to have little effect on short-term survival, and its importance should only become apparent when examining long-term (e.g. 10- or 15-year) survival. A separate meta-analysis by the EBCTCG [11], incorporating 78 trials with 42,000 patients, indeed demonstrated that improved local control has a small but real effect on long-term survival—less local recurrence at five years was associated with a proportional decrease in mortality at 15 years.

Furthermore, it is now well established that mammography screening can reduce breast cancer mortality, implying that at least a subset of tumors have long-term metastatic potential that can be averted if the primary tumor is

---

[1] In a 1992 "personal perspective" paper [28], Bernard Fisher discusses the transition from the Halsted to the systemic paradigm. It is notable how completely he rejects any Halstedian notion that the local tumor matters at all, and dismisses surgery more radical than a lumpectomy as "an effort to compromise by using principles of both paradigms simultaneously, which is antithetical to Kuhn's principle that a scientific community can be governed by only a single paradigm."

removed early enough [67]. This contradicts Fisher's notion of "metastatic predestination."

Finally, from first principles it would seem that the Halsted/Fisher dichotomy is a false one. Of course metastases arise and spread from a local lesion (Halsted view), but the dynamics may be such that distant metastasis has occurred long before clinical detection (Fisher view). The real question becomes, how do both local and systemic therapies perturb an ongoing metastatic process and affect existing metastases?

We can conclude that the evidence supports neither "pure" version of the Halsted nor Fisher models. Nevertheless, both are useful *conceptual models* that offer explanations for various therapeutic successes and failures. We might conclude, then, that the truth is somewhere between these two extremes, and indeed, modern standard of care refers to both models simultaneously.

7.2.3   Integration of Halsted and Fisher concepts: Surgery with adjuvant chemotherapy

Early research into tumor kinetics and response to cytotoxic chemotherapy by Howard Skipper and colleagues (e.g. [62]), among others, led to one of the most influential concepts in the clinical management of malignancy. Inspiration for this concept came from studies that established the following observations: (1) rapidly proliferating cells are preferentially sensitive to cytotoxic drugs; (2) the fractional cell-kill in such populations increases logarithmically with linear increases in drug dose; (3) the doubling time and growth fraction—therefore, the fraction of cells vulnerable to drugs—decreases with tumor size. The second observation is closely related to the *log-kill* (or log-linear) model of chemotherapy, which states that a given dose of a cytotoxic drug kills a fixed fraction of reproducing cells exposed to the drug (see Section 2.5). Together, these observations suggest that, since small metastatic foci are less crowded and therefore grow faster than the primary tumor, rapidly growing metastatic disease could be eliminated by cytotoxic therapy, even if the primary tumor was resistant. Support for this idea came from studies showing that metastatic tumors could be "cured" in animal models if therapy were initiated early enough or if the primary tumor were surgically resected prior to systemic therapy [62], which argues in favor of a combined Halsted-Fisher paradigm.

These notions were extended and modified into the *Norton-Simon hypothesis* (see Section 2.5), which proposes that the cytotoxic effect of chemotherapy is proportional to the intrinsic (unperturbed) tumor growth rate as described by the Gompertz model [64]. Both log-kill model and Norton-Simon hypotheses helped motivate the "hit'em hard and hit'em fast" approach to adjuvant chemotherapy that has come to dominate clinical thinking. These hypotheses also serve as the basis for the hypothesis of kinetic resistance to chemotherapy, contra the notion of acquired resistance driven by the preferential survival of resistant mutant clones. The former suggests that recurrent tumors arise largely from cancer cell populations untouched by the chemotherapy agent, e.g., because the cells were quiescent or in a tissue compartment into which the agent could not enter. The latter posits that recurrence occurs when mutant clones that resist the cytotoxic action of the agent arise within the tumor. Here we have yet another interesting, if too simple-minded, dichotomy that promises to advance cancer biology and treatment.

7.3 A simple model for breast cancer growth kinetics

Mathematical models provide a rigorous framework for formalizing conceptual models like those described above. In this chapter we introduce models that can serve as guide to understanding (1) the different possible dynamics of breast cancer growth, (2) the treatment implications of various hypotheses for metastatic growth and spread.

Here we focus on kinetics of primary tumor growth without worrying about metastatic disease. Although relatively simple and limited mainly to early tumor growth, these models have informed treatment strategies, although not without controversy. Also, they can be used to evaluate mammography screening protocols, which we discuss later.

We begin our study of the natural history with the Gompertz model, perhaps the most commonly used model of primary tumor growth. The discussion in Chapter 2 makes several points of immediate importance here:

1. In general, the Gompertz model has been fit to aggregated and therefore "smoothed out" tumor growth curves. Individual growth curves, even in animal models, are likely much more erratic, with many inflections. Such behavior cannot be captured in an unmodified Gompertz model.

2. The Gompertz model has been applied to clinically detected cancers or xenografts, which do not necessarily (and likely do not) have the same growth dynamics as early pre-clinical cancers.

3. It is unsurprising that a population growth model with several degrees of freedom can be made to fit aggregated growth data with a sigmoidal form; therefore, any such fit may be uninformative.

4. Finally, the widespread acceptance of the Gompertz model as an almost universal description of tumor growth is based on a limited amount of early, primarily animal, data.

The growth dynamics of pre-clinical cancers remains poorly understood. It was long believed that cancer inevitably progressed from a single altered cell to an invasive malignancy. However, newer evidence suggests that many, perhaps the vast majority, of nascent cancers never grow beyond a very small size. Such tumors may remain static for the lifetime of the host, spontaneously regress, be destroyed by the immune system, or progress to a clinical cancer.

In addition, it is widely believed, at least for cancers of epithelial origin (i.e., carcinomas), that tumors pass through at least two distinct growth phases: avascular followed by vascular. Nascent avascular tumors are thought to grow asymptotically to some small size determined by the ability of diffusion to transport oxygen and nutrients, primarily glucose, from the tumor's surface to the cancer cells. (See Chapter 3 for a discussion of avascular tumor growth and Greenspan's seminal model.) This hypothesis suggests that avascular tumors may remain in this static state for many years before undergoing the so-called "angiogenic switch," when tumor cells begin promoting neoangiogenesis constitutively. This flush of new blood vessels initiates a second phase of growth characterized by tumor angiogenesis and tissue invasion.

The unmodified Gompertz model simply cannot yield such stop-and-go behavior, nor can any model with simple sigmoidal solutions (e.g., the models of von Bertalanffy or Verhulst; see chapter 2). Nevertheless, it is hard to overstate the importance of the Gompertz model in our developing understanding of cancer natural history—most work in this direction is either based upon or is, at least in part, a response to the Gompertz model.

7.3.1 Speer model: Irregular Gompertzian growth

In 1984, John Speer and colleagues [74] developed a primary breast cancer model in which tumors grow by simple Gompertzian kinetics punctuated with occasional sudden changes in parameter values. Such a model is intuitively appealing, seemingly capturing the "multiple genetic hits" leading to malignant transformation. Indeed, Speer et al. were motivated by earlier Gompertzian models that implied implausibly short durations of pre-clinical disease originating from a single cell.

Recall from Chapter 2, equation (2.13), that the solution to the Gompertz differential equation model takes the form,

$$N(t) = N_0 \exp\left[\frac{G_0}{\alpha}\left(1 - \exp(-\alpha t)\right)\right],$$

and that the limiting tumor size

$$N_\infty = N_0 \exp\left(\frac{G_0}{\alpha}\right),$$

where $N_0 = N(0)$ and G_0 and α are constants. Speer et al. simulate saltatory changes in growth rate by introducing a stochastic parameter, R, taken from the unit interval that modifies the parameter α. (G_0 stays fixed.) Every five (simulated) days the simulation changes parameter R with probability p. If

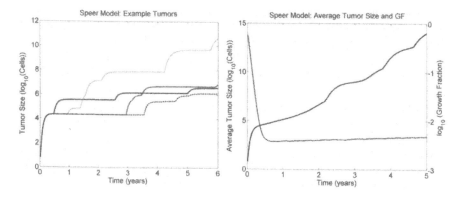

(a) Four tumor growth profiles under the Speer model.

(b) Average tumor size and growth fraction on a log-scale. Results are the average of 100,000 individual growth curves.

FIGURE 7.2: Tumor growth curves under the Speer model.

R is scheduled to change at a given 5th time step, then a new R is chosen with probability uniformly distributed over the unit interval. A new alpha is set at

$$\alpha = \frac{\hat{\alpha}}{1 + aR},\tag{7.1}$$

where $\hat{\alpha}$ and a are constants. The simulated tumor then grows according to the Gompertz solution (2.13) with this new parametrization and initial N_0 equal to the tumor size at the end of the previous 5 "days." After another 5 "days," the simulation checks to see if R changes (always with probability p) and the process repeats indefinitely. Figure 7.2(a) shows an example of several typical simulations (compare to Chart 2 in [74]).

7.3.2 Calibration and predictions of the Speer model

Speer et al. calibrated the model parameters using several clinical data-sets: Bloom et al.'s [7] data on survival of 250 untreated breast cancer cases (Fig. 7.1), serial mammography data from Heuser et al. [47], and data from Fisher et al. [29] giving survival following surgical treatment of the primary tumor. Tumor size at detection was between 1×10^9 and 5×10^9 cells, and lethal tumor size was assumed to be 10^{12} cells. Parameter optimizations for the Bloom and Heuser data imply that tumors grow 7.8 ± 3.9 years before clinical detection, with post-detection patient survival of 3.3 ± 2.5 years. The Fisher data predicts that the number of metastatic sites, S, is related to the number of positive lymph nodes N at diagnosis by the equation $S = 0.24 + 0.35N$.

Retsky et al. [71] later used the Speer model to study several data-sets of tumors treated by chemotherapy. In this work, they concluded that chemotherapy reduces the tumor burden by 1 or 2 orders of magnitude, and while it increases median survival time, long-term survival is unaffected. Based on the fact that individual tumors experience many growth plateaus in the model, both Retsky [71] and Speer [74] suggested that long-term maintenance chemotherapy may be a viable alternative to short-term intense chemotherapy. The Speer model has also been used to study tumor dormancy, as discussed later in this chapter.

One should be careful not to rely on these modeling predictions. For example, a large meta-analysis of clinical trials clearly demonstrates that (1) multi-agent chemotherapy significantly improves long-term survival, and (2) compared to short chemotherapy regimes (mean 5.0 months), prolonged episodes (mean 10.7 months) offer little to no benefit [23].

7.3.3 Limitations of the Speer approach

While the Speer model can produce survival curves that match clinical data, its dynamics are somewhat problematic. Although individual growth spurts are Gompertzian, tumor growth is ultimately unbounded. Furthermore, aggregated simulated growth curves do not reproduce the expected sigmoidal pattern. Instead, they show an initial Gompertzian growth phase that rapidly equilibrates, followed by irregular but roughly exponential growth (Figure 7.2(b)). This pattern is contradicted by observations from experimental systems, e.g. [15].

The Speer model also predicts a different dynamic for the tumor growth fraction (GF) (fraction of cells actively proliferating at the given time) compared to sigmoidal growth models like the simple Gompertz model. Figure 7.2(b) shows the approximate GF, which is estimated by the formula:

$$F = \frac{\tau N'}{N \ln 2}, \tag{7.2}$$

where F is the GF, τ is the time it takes for a cell to pass through the cell cycle, and N' is the time derivative of the tumor size. We leave the derivation of this formula as an exercise. (Hint: first consider simple exponential growth. In that case, $N' = \alpha N$ and we have $F = \tau\alpha/\ln 2$. If all cells are proliferating, then τ is the same as the cell cycle time and $F = 1$.) The GF is an important prediction because it estimates the tumor's chemosensitivity, since actively proliferating cells are much more susceptible to most cytotoxic agents.

The Speer model predicts that, on average, the GF for all tumors is roughly fixed, regardless of their size, except very early in their growth. On the other hand, the Gompertz model predicts a continuously decreasing GF, which has more support from experimental data. Therefore, we conclude that, despite making some interesting predictions, the Speer model is an unsatisfactory description of a tumor's natural growth history. However, we will return to the Speer model when we examine the role of multiple genetic hits in the evolution of cancer later in this chapter.

There is also a spirited debate between those championing the Speer approach and those in the Norton-Simon camp [70] that arises because both Speer and Gompertz models can be made to fit data. The ability of two very different models to describe at least some clinical survival data equally well suggests that curve-fitting is a poor way to judge the merits of a model. We are likely better served by examining the hypotheses that different models generate. Furthermore, we may need to move from such simple phenomenological models to more complex, mechanistic models for tumor growth.

7.4   Metastatic spread and distant recurrence

Following the Halsted-Fisher pattern, we now move on to some formal models of metastatic spread from a primary tumor. We also take this opportunity to discuss several models that have addressed the evolutionary pressures that give rise to the metastatic phenotype in the first place. We close this section with a discussion of some open questions related to metastatic spread and the potential for models to address these.

7.4.1   The Yorke et al. model

In 1993, Yorke et al. [82] proposed a model for metastasis development in prostate cancer which represents a good baseline approach to the problem. Although this model was motivated by attempts to understand metastasis in prostate cancer, it addresses one of the central questions in breast cancer treatment—does local tumor control play a role in preventing distant metastasis? The key assumptions of this model are the following:

1. The primary tumor grows by Gompertzian kinetics from initial size N_0 to the size at which it is detected, N_D. Time T_D represents the sojourn time from tumor initiation to detection.

2. Mutants with metastatic capability arise within the primary tumor according to the Goldie-Coldman model [28]. Metastatic foci are seeded at a rate directly proportional to the number of metastatically capable cells.

3. Metastases grow by simple Gompertzian kinetics and themselves become detectable after time T_M.

4. Following clinical detection at time T_D, the primary tumor is treated. If treatment is successful, we set $n(T_D) = 0$. If not, the tumor regrows

and continues to seed metastases.

In this model, $N(t)$ denotes the total number of primary tumor cells, $\mu(t)$ is the (mean) number of metastatically capable mutant cells within the primary tumor, and $m(t)$ represents the number of metastatic foci at time t. The following form of the Gompertz model is used to model the growth of the primary tumor:

$$N(t) = N_0 \exp\left(k\left(1 - e^{-\alpha t}\right)\right), \tag{7.3}$$

where

$$k = \frac{G_0}{\alpha} = \ln\left(\frac{N_\infty}{N_0}\right).$$

As before, $N(0) = N_0$, the asymptotic tumor size is N_∞, and α is the growth rate. Since the tumor is detected at size N_D,

$$T_D = -\frac{1}{\alpha} \ln\left(\frac{-\ln(N_D/N_\infty)}{k}\right), \tag{7.4}$$

where, again, T_D is the time from tumor initiation to detection. As the tumor grows, mutants with metastatic capability arise at a small stochastic rate. This process is modeled by a modified Goldie-Coldman model, which originally was used to describe dynamics of chemotherapy-resistant mutants [28, 29]. In particular, the number of metastatically capable cells within the primary tumor, $\mu(t)$, is determined by the following equation:

$$\frac{d\mu}{dN} = \frac{\mu}{N} + p\left(1 - \frac{\mu}{N}\right). \tag{7.5}$$

where p is the probability that a dividing cell mutates to the metastatic phenotype. Assuming an initial condition $\mu(N_0) = \mu_0$, we have the following solution for μ in terms of $N(t)$:

$$\mu(N) = N\left(1 - N_0^p N^{-p}\left[1 - \frac{\mu_0}{N_0}\right]\right). \tag{7.6}$$

If $\mu_0 = 0$ and $N_0 = 1$, solution (7.6) reduces to

$$\mu(N) = N\left(1 - N^{-p}\right) \tag{7.7}$$

Furthermore, for small p, we have that $N^{-p} = \exp(-p\ln N) \simeq 1 - p\ln N$, so

$$\mu(N) \simeq pN\ln N. \tag{7.8}$$

The above form of the model is useful, but it may be easier to understand the development of metastatic capability in terms of its change with respect to time. The time dependent change in mutant cells is:

$$\frac{d\mu}{dt} = p\frac{dN}{dt}\left(1 - \frac{\mu}{N}\right) + \left(\frac{\mu}{N}\right)\frac{dN}{dt}. \tag{7.9}$$

The first term represents mutation from wild-type cells at a rate proportional to the tumor growth rate, and the second expresses proliferation of existing mutants. Since dN/dt represents the overall tumor growth rate, while $(1 - \mu/N)$ and (μ/N) are the tumor fractions that are wild-type and mutant, respectively, mutant cells are assumed to grow with the same kinetics as the wild-type cells. Rearranging equation (7.9) yields equation (7.5). A direct derivation of equation (7.5) is given in Section 9.5, where we discus the Goldie-Coldman model in depth. This formulation is useful in that it makes the rate of metastatic spread a function of primary tumor mass and growth rate in a natural and mechanistic, rather than empirical, manner.

Now that we have the number of metastatically capable cells in the primary tumor, we must determine the actual number of metastatic foci seeded at time t which we represent it by $m(t)$. Yorke et al. modeled this by (implicitly) assuming that as metastatically capable cells arise they are instantaneously converted into metastases according to some efficiency parameter η. Metastatic seeding ceases when treatment of the primary tumor begins (time T_D). Formally,

$$\frac{dm}{dt} = \begin{cases} \eta \dfrac{d\mu}{dt} & ; t < T_D, \\ 0 & ; t \geq T_D, \end{cases} \tag{7.10}$$

which implies

$$m(t) = \begin{cases} \eta\mu(t) \, (\simeq \eta p N \ln N) & ; t < T_D, \\ m(T_D) & ; t > T_D. \end{cases} \tag{7.11}$$

So this model predicts that the expected number of metastases is (approximately) proportional to ηp, which serves as a measure of the intrinsic tendency of a tumor to metastasize. While this formulation's simplicity makes it useful, biologically we expect the rate of metastasis seeding to be a function of the absolute number of metastatic capable cells.

The variable $m(t)$ represents an expectation or average number of metastatic foci across an infinite number of essentially identical patients. However, comparing the model results to patient data requires calculation of the *probability* that metastases are clinically detectable at a given time *in a given patient* following diagnosis of the primary tumor. Let $X(t)$ be a random variable expressing the number of metastases in our patient at time t. We assume that seeding events occur independently at some time-dependent rate. From these assumptions and the model formulation, we naturally assume that $\{X(t), t \geq 0\}$ is a nonhomogeneous Poisson stochastic process with rate $m(t)$. Therefore, the probability that no metastases are present at time t is simply

$$P(X(t) = 0) = \exp(-m(t)). \tag{7.12}$$

Once established, each metastasis grows by Gompertzian kinetics until it becomes detectable at time T_M with size N_M. The equation for T_M is similar to that for T_D (equation (7.4)). Therefore, the probability that no metastases are detectable at time t is simply the probability that no metastases are present at time $t - T_M$ (with, of course, $t \geq T_M$ since no metastasis could be detectable before then). We denote this probability by S_p (Survival, no metastasis from the primary tumor), so

$$S_p(t) = \exp(-m(t - T_M)). \tag{7.13}$$

The general approach taken here is common; that is, metastasis seeding is widely modeled as a (nonhomogeneous) Poisson process where the rate of seeding (intensity function) often depends upon tumor size. This model has the advantage of expressing this dependency in a mechanistic way.

Up to this point we have treated only pre-clinical tumor growth and metastasis. When the primary tumor is detected at time T_D, we assume that local treatment is initiated. To describe recurrent growth and metastasis from the residual primary tumor, we simply introduce a new set of variables, $\hat{N}(t)$, $\hat{\mu}(t)$, and $\hat{m}(t)$ with definitions analogous to those above. The initial size of the residual primary tumor is $\hat{N}_0$, with some number of these, $\hat{\mu}_0$, metastatically capable. We assume that the fraction of metastatic cells is unchanged by local treatment, so

$$\frac{\hat{\mu}_0}{\hat{N}_0} = \frac{\mu(N_D)}{N_D}. \tag{7.14}$$

Applying the above reasoning to the post-treatment situation, the probability that no detectable metastases have been spawned by the recurring tumor (rather than the primary tumor) by time t is equal to the probability that no metastases are present at time $t - \hat{T}_M$, where $\hat{T}_M$ is the time needed for a metastasis seeded by the locally recurrent tumor to become clinically detectable (where, again, $t \geq \hat{T}_M$). We denote this probability by S_r, so

$$S_r(t) = \exp(-\hat{m}(t - \hat{T}_M)). \tag{7.15}$$

Assuming that the probabilities of being free from metastases seeded by the primary and locally recurrent tumor are independent, the overall probability of *metastasis-free survival* (S) is simply

$$S(t) = S_p(t)S_r(t). \tag{7.16}$$

In the literature, $S(t)$ is frequently called **distant-metastasis-free survival**, or DMFS.

7.4.2   Parametrization and predictions of the Yorke model

Yorke et al. [82] followed the example of Norton [53] by assuming the Gompertz growth parameter, α, has a log-normal distribution with mean 0.0283 month^{-1} and standard deviation 0.98. However, under these parameters a significant fraction of tumors take over 50 years to become detectable, so for demonstrative purposes (i.e. in the figures) we use $\alpha \sim \text{LOGN}(\mu = -2.9, \sigma = 0.71)$ with units month^{-1}.

Yorke et al. estimated parameter values by calibrating to cumulative DMFS values from patients whose local tumors were successfully controlled. However, these parameter values achieved only poor concordance with the DMFS data from patients experiencing local tumor recurrence—typically, parameter values derived from primary tumors overestimate DMFS for recurrent tumors. From these observations, Yorke et al. concluded that locally recurrent tumors are intrinsically more metastatic than those that do not recur, and they also contribute significantly to metastatic recurrence. Therefore, they predicted that control of tumors likely to recur locally should improve long-term survival, but the prognosis remains worse in these cases compared to tumors with little propensity for recurrence. This conclusion argues against the "strong form" of Fisher's conceptual model. If this interpretation is correct, then tumors are not "predestined" to be either metastatic or not regardless of local control. This does not argue that no tumors are intrinsically more metastatic than others, but even in such cases, local control probably improves prognosis. There is a caveat—these results are derived from prostate cancer data. So these conclusions rely on the assumption that the Yorke model sufficiently describes metastasis in breast cancer.

7.4.3 Limitations of the Yorke approach

To be used, the Yorke model requires introduction of a new set of variables for every episode of treatment. Complex interventions, such as prolonged systemic chemotherapy or a fractionated course of radiotherapy, are therefore very difficult to incorporate into the basic framework. The assumption that the rate of metastatic seeding is a function of the rate at which cells acquire the metastatic phenotype, rather than the number of cells with that phenotype, should raise some eyebrows among cancer biologists and may represent another limitation of the approach.

Another weakness of the Yorke model is that it does not consider secondary metastatic spread—i.e., seeding of new metastases from existing metastases. However, an extension of Yorke et al.'s approach to include such behavior should be straightforward and yield very interesting results.

7.4.4 Iwata model

A more detailed model of metastasis, including age structure of metastatic tumors, has been introduced by Iwata et al., [49], and has received more recent attention by other authors [20, 3]. In this model, primary tumor growth is described by an ODE, while a PDE governs the size or age distribution of metastatic colonies. Both the primary tumor and metastases grow at rate $g(x)$, where x is the number of cells in the tumor. New metastases are seeded at rate $\beta(x)$ by both primary and existing metastatic tumors of size x. Let

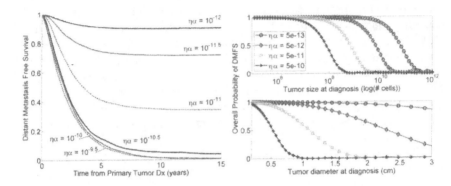

(a) Curves showing the distant metastasis free survival (DMFS) versus year from primary tumor diagnosis for different metastatic tendencies. Each curve is generated by determining DMFS probability at each time-point after diagnosis for 2,000 patients and averaging the results.

(b) Long-term (16-year) distant metastasis free survival (DMFS) for locally controlled tumors diagnosed at different diameters, and for different intrinsic tendencies to metastasize ($\eta\alpha$). Note that plotting tumor cell count on a log scale reveals a relatively rapid transition from the patient being metastasis-free to almost inevitable metastatic recurrence. Each data point represents the average value for 2,000 simulated patients.

FIGURE 7.3: Some results from the basic Yorke model with complete local control. Note especially that the DMFS curve unexpectedly appears to approach an asymptotic curve as $\eta\alpha$ increases. This is not seen when using the agent-based simulation model, as in Figure 7.4. Also, compare the prediction in (b) to that for drug resistance under the Goldie-Coldman model discussed in Chapter 9.

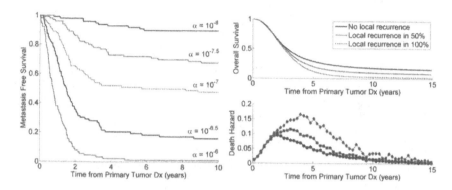

(a) Cumulative survival for 100 patients under different α, i.e. metastasis seeding rates, with η fixed at 10^{-6} and all other parameters as in Figure 7.3. Only 100 patients are simulated to help demonstrate the fundamentally coarse nature of a Kaplan-Meier plot.

(b) Overall survival (and death hazard per 100 days) when local recurrence never occurs, occurs with 50% probability, and with 100% probability. Each curve represents a very large cohort of 25,000 simulated patients. Note that short-term survival is identical, but there is long-term divergence, just as predicted by our conceptual model for breast cancer (see the discussion in Section 7.2).

FIGURE 7.4: Kaplan-Meier survival curves generated using agent-based simulation of the Yorke model. Patient survival data is often reported and displayed using a Kaplan-Meier survival curve which gives the cumulative probability of the patient survival at any time. See exercise 7.4 for a method of generating such a curve.

$x_p(t)$ represent the number of primary tumor cells at time t. Then we consider the initial value problem for the primary tumor:

$$\frac{dx_p}{dt} = g(x_p), \qquad x_p(0) = 1. \tag{7.17}$$

To describe the metastases, let $\rho(x,t)$ be density of metastatic colonies composed of x cells at time t. It is easy to see that $\rho(x(t),t)$ is a constant if colonies do not disappear on their own. Hence it obeys a von Foerster (age-structured) model, as follows:

$$\frac{\partial \rho(x,t)}{\partial t} + g(x)\frac{\partial \rho(x,t)}{\partial x} = 0, \tag{7.18}$$

with initial and boundary conditions

$$\rho(x,0) = 0, \tag{7.19a}$$

$$g(1)\rho(1,t) = \int_1^\infty \beta(x)\rho(x,t)dx + \beta(x_p(t)). \tag{7.19b}$$

Details of how one constructs and interprets such physiologically structured models can be found in chapter 4.

Initially there are no metastatic colonies of any size (condition (7.19a)). The (left) boundary condition (7.19b) is the rate of metastatic colony formation. All metastatic colonies begin with a single cell (at size 1). Integrating $\beta(x)\rho(x,t)\,dx$ over all possible sizes gives the rate of seeding by existing metastases, while the primary tumor spawns new metastases at rate $\beta(x_p)$.

Finally, we assume that the rate of metastatic seeding is proportional to the number of tumor cells in contact with blood vessels. That is,

$$\beta(x) = kx^\xi \tag{7.20}$$

where k is the "colonization coefficient," and ξ is the fractal dimension of the tumor vasculature. For a homogeneous vasculature servicing the entire tumor, $\xi = 1$, whereas a vasculature supplying only the outer tumor surface would have $\xi = 2/3$.

Iwata et al. considered both Gompertz and exponential growth functions for $g(x)$ and found explicit solutions for both cases. Furthermore, they calibrated the model using data from a case of primary hepatocellular carcinoma with multiple liver metastases tracked by three serial CT scans. Interestingly, their calibration suggested that ξ was very close to $2/3$, suggesting superficial vasculature.

This model allows the size distribution of metastatic colonies to be tracked through time. Because of its generality, it would be straightforward to explore the effect of therapeutic intervention on the size distribution of metastases. Adjuvant therapy could be easily modeled by adding a negative term to the growth function, $g(x)$.

7.4.5   Thames model

Thames et al. [76] used a model to address the question of whether it is better to treat with chemotherapy before ("neoadjuvant" therapy) or after surgical resection of the primary tumor. They concluded that for early T1 cancer it is better to treat surgically first, but more advanced tumors may be better treated by neoadjuvant chemotherapy.

7.4.6   Other models

Chen and Beck [10] recently derived a mechanistic model for metastatic spread using superstatistics, which performed well against some SEER breast cancer survival data. Several other examples of stochastic models for metastatic spread can be found in [4, 42, 83]. Withers and Lee [80] discussed the implications of different tumor growth and metastasis dissemination kinetics on the distribution of metastases and the possibility of curative therapy.

7.5   Tumor dormancy hypothesis

Most continuous growth models for metastatic spread implicitly assume that tumors grow most rapidly in the earliest phases of their history, which follows naturally from the assumption of sigmoidal growth. Support for these assumptions rests on comparisons of model predictions to patient survival curves (e.g., Speer et al. [74] and Norton [53]). However, such data miss an important aspect of malignancy—cancer recurrence—that challenges the insightfulness of simple tumor growth models. In particular, the hazard function for both local and distant metastatic recurrence of breast cancer following treatment exhibits two peaks: a large one at about 18 months followed by a smaller peak at 60 months [17]. This pattern was first observed in data from the Milan National Cancer Institute for women treated by surgery alone [14], and has since been observed in other unrelated data sets [16]. It was not seen in the Bloom data [7], which has a single peak in mortality following diagnosis [18]. However, the Bloom data includes only untreated tumors and has a slight bias. So the two-peak mortality pattern appears to be a real characteristic of treated breast cancer.

On the basis of this pattern and other data challenging the continuous growth model for primary tumors, Demicheli, Retsky, and colleagues proposed the **tumor dormancy hypothesis** in 1997 [17, 69]. More recent reviews by this group (the hypothesis is essentially unchanged) can be found in [5, 16, 68].

This hypothesis proposes that metastatic colonies experience periods of dormancy that tend to end with removal of the primary tumor, at least transiently. Such stimulation of metastatic growth following primary tumor removal has been observed in a variety of cancer types in both animal models and human data-sets.

Demicheli and Retsky et al. proposed that distant micrometastases potentially experience two distinct phases of both dormancy and growth. Following seeding, the metastatic cells are in the first dormant state, S_1, until some stimulus induces them to enter the first avascular growth phase, S_2, where they grow until reaching the maximum avascular size, estimated to be between 2×10^3 and 1.5×10^5 cells. The first dormancy phase occurs when the tumor reaches diffusion limitation. (See Chapter 3.)

The final growth phase, S_3, occurs when avascular colonies induce angiogenesis and switch to vascular growth. Demicheli and Retsky considered two possible mechanisms. The first was essentially evolutionary—metastatic cells initially lack the angiogenic phenotype and then switch to S_3 by genetic alterations that confer angiogenic potential. The second possibility came from studies by Judah Folkman [30] and others suggesting that dormant micrometastases are held in an avascular state by secretions from the primary tumor, including endostatin and angiostatin. Removal of the primary relaxes the inhibition, allowing metastatic tumors to switch to an angiogenic phenotype, reigniting their growth.

Retsky et al. [69] simulated this hypothesis and compared the results to the Milan data. The simulation's assumptions include the following:

1. The primary tumor grows according to the Speer model for irregular breast cancer growth.

2. At the onset of tumor vascularization, the primary tumor begins shedding metastatic cells.

3. Larger tumors shed more cells than smaller tumors.

4. New metastases always start in the first dormant state S_1. They transition stochastically to the first growing state S_2, where they expand according to the Gompertz model to their maximum avascular size. They then randomly transition to vascular growth, S_3, which is also Gompertzian but with a much larger limiting size. Cells may also transition directly from S_1 to S_3.

5. Upon clinical detection, the primary tumor is removed, and metastasis ceases.

6. When a distant relapse occurs—that is, when a metastatic tumor in S_3 reaches clinical detection—the simulation ends.

This model is depicted schematically in Figure 7.5. Retsky et al. did not present the model in a traditional mathematical framework. Doing so would be fairly straightforward and likely to clarify precisely how the model works. For example, it is only reported that metastases are seeded from the primary

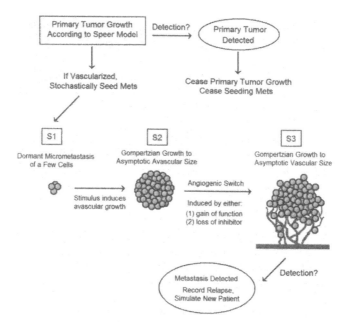

FIGURE 7.5: Schematic for the Retsky-Demicheli model for metastasis dormancy.

tumor at a stochastic rate: it is unclear what the distribution of this rate is or precisely how it depends upon the primary tumor size and node status. Presumably the stochastic process is Poisson, but how the rate is determined is unclear.

This model cannot be compared to clinical data without parametrization. Unfortunately, no details on the Gompertz growth parameters were published, other than that they were based on published values for animal (e.g., LoVo xenografts in nude mice in [15]) and human data (e.g., testicular cancer lung metastases in [13]). Details about the primary tumor growth model are also absent, as is a precise description of when in its history it begins to shed metastases. These weaknesses make it difficult to replicate or expand the results.

Retsky et al. use their model to distill the following conclusions and predictions:

1. Simple stochastic transitions between the S_1, S_2, and S_3 states can yield a two-peaked hazard function, but the first peak is unrealistically early and narrow compared to data.

2. State transitions must be transiently enhanced following surgery to satisfactory fit the data. That is, a pro-growth stimulus following surgery

spurs transition from $S_1 \to S_2$, and a pro-angiogenic signal increases $S_2 \to S_3$. Both may give an $S_1 \to S_3$ transition.

3. The first peak of relapses represents metastases whose growth was stimulated by primary tumor removal. Unperturbed micrometastases follow a slower growth pattern and give a broader second peak.

4. More advanced tumors with metastasis-positive lymph nodes may respond best to chemotherapy, as they will be characterized by $S_2 \to S_3$ transitions, implying that metastases will be vascularized and rapidly growing, and hence more responsive to chemotherapy.

5. Smaller primary or larger, node-negative tumors are likely to be characterized by early metastases and $S_1 \to S_2$ transitions following surgery. Since avascular tumors may respond poorly to chemotherapy, earlier detection may antagonize efficacy of chemotherapy. Long-term suppressive therapy is therefore a potential alternative to short-term chemotherapy.

The final two predictions are potentially important clinically. However, the notion that advanced tumors respond better to therapy because of a predominant $S_2 \to S_3$ shift is questionable. Overall, it is likely that many $S_1 \to S_2$ transitions would still occur. Thus, early stage metastases are likely to be equally present in more or less advanced tumors. Because continuous metastasis seeding by a primary tumor will always give a metastasis distribution with many early-stage metastases, early detection may antagonize chemotherapy in the short term, it still must improve the absolute benefit.

This model matches distant recurrence data fairly well and offers an interesting dynamical description of tumor dormancy, and indeed, it seems fairly certain that some departure from continuous growth kinetics must explain the extremely delayed recurrence seen in breast cancer. However, a major challenge to the conclusions and model is that, in the data, the two peaks of distant recurrence are paralleled by two peaks in *local* recurrence. How can we account for this? It may be that a similar tumor-host dynamic governs the growth of both occult metastases and residual primary tumor, and a transient pro-growth stimulus following surgery might affect both similarly. But residual local disease is unlikely to be described by the same $S_1 \to S_2 \to S_3$ sequence as metastases.

The tumor dormancy hypothesis is an intriguing one and well worth further investigation. Therefore, it would be a valuable project to develop this model in an ODE setting that can more easily be expanded upon. Such a model could be used to more formally study the effect of adjuvant therapy on metastatic recurrence and relative efficacy for different primary tumor types. Also, the model should be expanded to explicitly consider residual local disease, and the relationship between local and metastatic recurrence could be further studied, as discussed above.

7.6 The hormonal environment and cancer progression

While we present no new models here, it is important to at least mention the role of hormones and breast cancer. Breast cancer, unlike most other solid malignancies, is characterized by extremely delayed recurrence that may occur up to two decades after apparently successful initial treatment. The hormonal environment probably plays a key role in such delayed recurrence along with mediating metastasis dormancy. The estrogen receptor (ER) promotes survival and mitosis in both normal and cancerous breast epithelium, and a majority of breast cancers are ER-positive at diagnosis, although a significant minority are not (perhaps 25%). A role for hormones in breast cancer progression and recurrence is supported by the success of hormonal therapy. Efficacy of oophorectomy (removal of the ovaries) was recognized as early as 1896 [26]. A more recent study showed that treatment using the ER antagonist tamoxifen reduces annual breast cancer mortality by 31% [23], and this benefit persists for up to 15 years, with continually increasing mortality divergence between treated and untreated groups. Further support comes from the Women's Health Initiative, which in 2002 first showed that hormone replacement therapy (HRT) increases breast cancer risk. Significantly, while ER-positive tumors respond to hormonal therapy and have a better prognosis than ER-negative tumors, essentially no ER-negative tumor recurs more than five years after diagnosis, while the risk of recurrence in ER-positive tumors continues for 15 years or more [26].

Drawing on the work of Demicheli and others studying irregular breast cancer growth kinetics (see Section 7.5), Love and Niederhurber [52] endorsed a model suggesting that breast cancer is a disorder of host or environmental control, rather than simply "uncontrolled growth" that can only be stopped by annihilating tumors to the last cell. These authors also commented upon a surprising result: surgical oophorectomy in premenstrual women confers a significant survival benefit, yet this benefit appears to be restricted to women in whom surgery was performed during the luteal phase of the menstrual cycle. This observation led to the *progesterone trigger hypothesis*:

1. High systemic levels of progesterone during the luteal phase support metastasis survival and vasculature;

2. Sudden withdrawal of progesterone by luteal oophorectomy leads to micrometastasis arrest, perhaps by withdrawal of survival signals or regression of the vasculature.

Finally, we note that early pregnancy, which exposes the breast to sustained, high levels of estrogen, is strongly *protective* against breast cancer. This pattern seems puzzling if we view estrogen as a driver of cancer growth, as suggested by the adverse effects of HRT in later life. Possible explanations for the protective effect of estrogen include the following: (1) hormones of pregnancy drive a population of cells susceptible to neoplastic transformation to a mature, terminally differentiated state incompatible with malignant transformation, (2) early neoplastic cells are differentially sensitive to high levels of estrogen and are killed rather than supported, or (3) pregnancy alters sensitivity of cells to estrogen such that neoplastic cell survival cannot be supported [39].

7.7 The natural history of breast cancer and screening protocols

The impact of different screening protocols on cancer mortality has been the subject of various statistical and dynamical models. The efficacy of any cancer screening program can only be definitively established by randomized trials that track cancer *mortality*. Increases in survival time related to screening do not necessarily indicate a benefit to screening, due to *lead-time* and *length-time biases*. Lead-time is the survival time gained by detecting a disease earlier in its natural history. For example, assume a cancer is initiated at year 0 and destined to cause death at year 10. Further assume that the disease will become clinically evident at year 8, but screening detects it at year 5. In this case, screening increases survival by 3 years, even though the course of disease is unaffected by early detection. Length-time bias refers to the tendency of screening to detect diseases with longer natural histories. Figure 7.6 depicts these biases schematically. Both biases are inherent to screening and cannot be simply adjusted away. Therefore, it is difficult to know if earlier detection improves treatment efficacy without prospective trials that measure mortality.

Randomized prospective trials have, in fact, demonstrated that mammography screening reduces breast cancer mortality by perhaps 15–20% for women aged 40–74. Absolute benefit increases with patient age, and relative benefit may also reach its maximum in the 60–69 age group [61]. The appropriate use of screening remains controversial younger (aged 39–49) and older (over 74) women. For younger women, screening efficacy, while real, is very small (the number needed to screen to prevent one death is 1,904), and younger women may also be most likely to suffer false positives [61].

The two major risks associated with any screening test include false positives and overdiagnosis. Overdiagnosis, which may be considered the limiting case of length-time bias, refers to screen-detected cancers that would never have caused symptoms. Treating such a cancer can therefore only cause harm. Estimates of breast cancer overdiagnosis by mammography range from less than 1% to 30%, but most are in the 1–10% range [61]. Mammography in particular also exposes women to low-dose chest radiation.

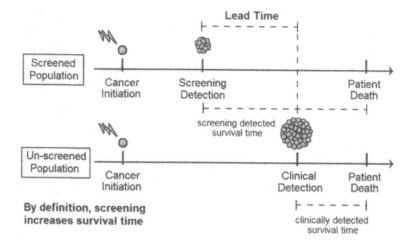

(a) Lead time bias.

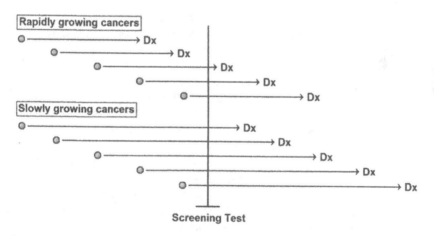

Screening detected cancers tend to be less aggressive.
Screening and clinically detected cancers are intrinsically different.

(b) Length time bias.

FIGURE 7.6: Schematic depictions of lead time and length time bias in screening.

7.7.1   Pre-clinical breast cancer and DCIS

Optimizing the value of screening is hindered by the fact that little is known about the natural history of pre-clinical breast cancer. Since the advent of mass mammography screening the incidence of ductal carcinoma in situ (DCIS) has increased dramatically. It now accounts for about 20% of new breast cancers, compared to less than 2% in the past, and this increase is largely attributable to mammography. DCIS is a neoplasm of ductal epithelial cells that has not penetrated the basement membrane to invade the surrounding stroma, although it may have spread widely within the duct lumen.

DCIS is generally believed to represent an early stage in the evolution of invasive carcinomas, and it shares many of the genetic and morphological characteristics of invasive disease. However, recent research has shown that not all cases of DCIS will progress to invasive cancer, and the proportion that do is unknown. Moreover, some more aggressive cancers may lack a recognizable in situ phase. Several studies have examined the rates of invasive carcinoma in women who received only a biopsy for DCIS and were initially misdiagnosed as having benign disease. These studies suggest that 14–53% of DCIS gives rise to invasive disease, but they had small sample sizes and considered only low-grade DCIS [25].

Welch and Black [79] reviewed seven autopsy series of women who died without evidence of breast cancer during life, and found the median prevalence of undetected DCIS to be 8.9%, although the prevalence was higher in studies that sampled more breast tissue (up to 14.7%). Furthermore, the prevalence was much higher in younger women. At the upper limit, 39% of women aged 40–49 had in situ disease, and another 12% had atypical ductal hyperplasia. This result suggests that a large pool of DCIS in younger women becomes depleted with increasing age. Depletion could occur either by spontaneous regression or progression to invasive cancer.

In accordance with Welch and Black's results, other evidence points to markedly decreasing DCIS incidence detected by screening after age 70 [24]. In addition, invasive cancers in older women tend to be lower grade with more favorable prognostic factors [24, 25]. The cause of this shift in incidence and grade remains unclear at the time of this writing.

7.7.2   CISNET program

Cancer Intervention and Surveillance Modeling Network (CISNET) is a consortium of National Cancer Institute (NCI) sponsored investigators that use statistical modeling to improve our understanding of cancer control interventions in prevention, screening, and treatment and their effects on population trends in incidence and mortality.

Given that randomized trials have established a mortality benefit to breast cancer screening, we turn our attention to optimum screening schedules. Here, modeling comes into its own because it is logistically infeasible to test all possible screening regimes using prospective trials.

Two major approaches for modeling natural history in screening trials exist: (1) stochastic state-transition models, where tumors transition through a variety of pre-clinical phases, and (2) continuous growth models, which typically assume exponential or occasionally Gompertzian growth kinetics. These two approaches may also be combined.

The CISNET Breast Cancer Program, sponsored by the National Cancer Institute, was a collaborative effort to model the effect of screening and adjuvant therapy on breast cancer mortality and incidence between 1975 and 2000 [12]. Of the seven groups, six developed models for the natural history of breast cancer, and we review them here. The other used Bayesian simulation modeling [6], and we do not discuss it. The reader may also consult [12] for a comparative review of these models.

The CISNET models employ "discrete-event microsimulation," in which individual patient life histories are simulated and the results aggregated. In general, all CISNET models have the following components:

1. A patient population component, typically divided into birth cohorts.

2. Breast cancer incidence component, where incidence rates by birth cohort are given as a model input.

3. The tumor natural history component, including clinical cancer detection.

4. The screening component, which takes into account changes over time in both mammography practice and the size threshold for tumor detection.

5. Adjuvant treatment component, where the probability of a given therapy option being used and probability of tumor cure both depend on tumor characteristics at detection. The cure probabilities typically distilled from results of a meta-analysis of adjuvant therapy efficacy by the Early Breast Cancer Trialists Collaborative Group (EBCTCG).

All CISNET models included the following fixed inputs (base case inputs) [36]:

1. Mortality in the population from causes other than breast cancer.

2. Secular (long-term) trend in breast cancer incidence.

3. Mammography screening dissemination.

4. Adjuvant therapy dissemination.

The efficacy of adjuvant therapy and the probability tumors are estrogen-receptor positive are also generally determined by empirical data. This leaves parameters governing the natural history submodel and the threshold for detection by screening as free parameters determined by calibration to data. We will return to model calibration after we address the natural history submodels in the next section.

7.7.3 Continuous growth models

Four CISNET models [36, 75, 66, 43] described tumor natural history using continuous growth formalisms. Three used exponential growth without an explicit metastasis component [75, 66], while the Wisconsin group [36] used a more sophisticated Gompertzian model that explicitly included early tumor dormancy and regression, along with a simple model for metastatic spread. The latter work is an excellent example of how a good model *confronted with data* can be used to generate predictions about unobservable early tumor history.

MISCAN-Fadia model. The MISCAN computer model [40] is a state-transition model used to study screening for a number of malignancies. For example, Boer et al. [8] used a MISCAN breast cancer model to study disease progression from disease free through DCIS and on through four invasive cancer stages (T1a, Tab, T1c, T2+). Dwell-time in each stage was exponentially distributed, and probability of detection by screening depended upon stage.

In a modification of this basic approach, Tan et al. [75] altered the natural history component of the MISCAN framework to an exponential growth model, making tumor size the primary metric (rather than stage). Furthermore, they used the notion of a "*fatal diameter*" ("Fadia"): a tumor diameter unique to each patient at which the tumor becomes fatal; this diameter is (implicitly) assumed to be the point that fatal metastasis is initiated. This model assumes an initial spherical tumor of 0.1 mm diameter. The exponential growth rate is assigned randomly according to a log-normal distribution. The fatal diameter is selected from a Weibull distribution, and the survival time after reaching the fatal diameter is log-normally distributed. The tumor may be detected by screening or clinical detection of either the primary tumor or metastases. Screening and clinical detection of the primary tumor depend on tumor diameter thresholds, while clinical detection of metastases is a function of the time since a tumor became fatal.

Metastasis is not modeled other than to assume that tumor diameter is a marker for the probability of metastatic disease, i.e., the notion of the fatal diameter. DCIS is handled by the un-modified MISCAN DCIS submodel, which tracks incidence of two DCIS subtypes: (1) tumors that ultimately regresses, and (2) those that progress to an invasive cancer. Finally, adjuvant treatment is modeled by modifying a patient's fatal diameter, implying a greater probability of cure at detection. The MISCAN-Fadia model is the simplest of the continuous growth models, and likely the least biologically reasonable.

Nevertheless, it matches SEER breast cancer incidence rates by tumor size reasonably well, with the exception of DCIS. From 1985 onward, the predicted DCIS incidence is increasingly too low. While the model mortality curve reflects the actual SEER mortality curve in shape, the model's predicted mortality is too high.

Plevritis et al. model. Plevritis et al. [66] also assumed exponential growth for their natural history model, from the onset of invasive growth at a diameter of 2 mm until clinical detection. In particular,

$$V(t) = c_0 \exp\left(\frac{t}{R}\right) \qquad (7.21)$$

where $V(t)$ is tumor volume, and the inverse growth rate R is gamma distributed. Their model for local to regional to metastatic spread is proposed in terms of hazard functions. The hazard function for the time to the onset of regional (or nodal) disease, T_N, is given by:

$$P(T_N \in [t, t + dt]|T_N \geq t) = \eta_0 dt + \eta_1 V(t)dt + o(dt). \qquad (7.22)$$

That is, there is some constant baseline hazard of regional spread, $\eta_0 dt$. The second term, $\eta_1 V(t)dt$, implies that the risk of regional spread increases in proportion to tumor volume. The $o(dt)$ term represents un-modeled higher order terms.

The hazard function for the time of metastatic disease, T_M, is very similar. The only difference is that regional disease must be established, i.e., $T_M \geq T_N$, before distant spread can occur. Therefore, this model implicitly takes the view that metastatic disease spreads from the primary tumor in an orderly manner (Halsted model). The formal hazard function is:

$$P(T_M \in [t, t + dt]|T_M \geq t, T_N = t_N) = \begin{cases} \omega_0 dt + \omega_1 V(t)dt + o(dt) & t \geq t_N, \\ 0 & t < t_N. \end{cases}$$
$$(7.23)$$

The hazard function for clinical detection within a given time interval depends only on tumor volume, and the probability of detection is linearly related to tumor volume:

$$P(T_D \in [t, t + dt]|T_D \geq t) = \gamma V(t)dt + o(dt). \qquad (7.24)$$

Finally, screening detection is modeled with a hazard function defined in terms of tumor volume intervals because screening occurs at discrete time intervals. The probability of detection by mammography is assumed to be proportional to tumor cross-sectional area, which is proportional to $V^{2/3}$. Therefore, the hazard function for the threshold volume of detection, V_{TH} is

$$P(V_{TH} \in [v, v + dv)|V_{TH} \geq v) = \lambda v^{2/3} dv + o(dv). \qquad (7.25)$$

A lower limit of detection is also imposed at a tumor diameter of 2 mm.

Adjuvant therapy efficacy is modeled by using published hazard ratio and varies by age and estrogen receptor status. The probability that a patient's tumor is ER positive is determined by SEER data.

Significantly, Plevritis et al. did not attempt to model DCIS, and did not compare their model results to stage-specific SEER incidence rates, but only to the overall mortality curve. As in the MISCAN-Fadia model, the predicted mortality curve had the same shape as the SEER curve.

The negative results of these two models suggest that that DCIS and invasive cancer behave fundamentally differently. The MISCAN-Fadia model fails to predict increasing incidence of DCIS following introduction of widespread screening. The Plevritis et al. model does not attempt to model DCIS. It matched incidence rates well but overestimated mortality, which could not be resolved even if all tumors less than 1 cm in diameter were assumed to be DCIS destined to pose no risk. However, other explanations, such as increases in treatment efficacy, may explain Plevritis et al.'s mortality overestimate, so one should not read too much into these results.

Rochester model. Briefly, the Rochester model [43] considered the probability density function of the time of tumor initiation to be determined by a two-stage stochastic model of cancer incidence. Such models are discussed in Section 7.8.2. The random variable W gives the time from tumor initiation to detection, and its distribution is determined using continuous exponential tumor growth.

Wisconsin model. The Wisconsin group [36] simulated the female population of Wisconsin between 1950 and 2000, aged 20–100 years, divided into birth cohorts. This population includes 2.95 million women. The simulation was run in 6-month cycles, with a 25 year "burn in" period between 1950 and 1975 to establish breast cancer prevalence at 1975. Results were compared to measured 1975–2000 breast cancer rates.

The Wisconsin study's natural history submodel assumes Gompertzian primary tumor growth with an initial occult mass with 2 mm diameter and an asymptotic diameter of 8 cm. The growth rate for individually simulated patients is selected from a gamma distribution, with mean and variance calibrated by data.

A nonhomogeneous Poisson process governed metastatic spread. Metastasis from primary tumor to local lymph nodes was modeled as a function of absolute tumor size and growth rate, as proposed by Shwartz [72]; that is, if $n(t)$ is the rate at which groups of lymph nodes become positive, then

$$n(t) = b_1 + b_2 V(t) + b_3(dV(t)/dt), \qquad (7.26)$$

where the b_i are all constants. In each 6-month simulation cycle, the number of new lymph node metastases is determined by random draw from a Poisson distribution with parameter $n(t)$. While the assumption that regional spread correlates with both tumor size and growth rate is intuitively reasonable, there is no mechanistic basis for this rate function. The Yorke model (see Section 7.4.1) makes metastasis a function of tumor volume and growth rate in a much more mechanistic way.

A key feature of the Wisconsin model is that it considers DCIS to be an early stage of invasive breast cancer. Based on the primary tumor size and number of positive nodes, tumors are mapped to four historical SEER stages: (1) in situ (DCIS)—tumors without positive nodes and below a critical diameter, determined to be 0.95 cm, (2) locally invasive—tumors above the threshold for in situ disease, but with negative nodes, (3) regional spread—any tumor with 1–4 positive nodes, and (4) distant spread—5 or more positive nodes. Upon reaching the distant stage, the survival time distribution is derived from SEER data.

The probability that a screening test detects a tumor is assumed to be a function of tumor diameter and varies with time to reflect changes in mammography technology. Similarly, tumors emerge clinically with some probability determined by their diameter. The probability curve for clinical surfacing was determined in model calibration.

Therapy is modeled as a cure/no-cure process. That is, either a detected tumor is immediately eliminated or there is no effect at all upon the tumor's natural course. The latter assumption is of course problematic, as non-curative therapy can still significantly extend survival time, but this is a reasonable first-approximation. Therapy is modeled as follows: baseline treatment, consisting of mastectomy with or without radiation was assumed to be applied to all patients. The probability of a cure from baseline treatment alone depends upon the stage. In addition to baseline treatment, one of five adjuvant therapies may be applied: (1) chemotherapy alone, (2) tamoxifen for 2 years, (3) tamoxifen for 5 years, (4) (cytotoxic) chemotherapy plus tamoxifen for 2 years, or (5) chemotherapy plus 5 years. The probability of each therapy is determined according to various published data and whether the tumor's ER status is known. Survival probability is modified by each treatment according to a published meta-analysis [22].

The most important result of the model was its prediction that a significant fraction of occult-onset tumors grow only to a small size and ultimately regress. This conclusion reinforces other evidence that many DCIS tumors spontaneously regress (see the introduction to this section). We outline the steps leading to this prediction [36]. The rate of occult tumor onset, i.e. the rate at which 2 mm Gompertzian tumors are initiated, is estimated from the age-period-cohort (APC) model of Holford et al. [48] (a common CISNET input). However, since clinical incidence and occult onset are not the same, as some women may die of other causes before their cancer is detected, we define the "onset proportion" as the ratio of occult onset to clinical incidence:

$$\text{onset proportion} = \frac{\text{occult onset rate}}{\text{clinical incidence rate}}. \tag{7.27}$$

We define the onset lag l as the average time between occult onset and clinical presentation. If all tumors are eventually detected, the tumor incidence rate in year Y is simply the onset rate in year $Y - l$. Taking into account the proportion of onset cancers that become incident, we have the following expression for the rate of occult onset in year Y:

$$\text{Occult onset}(Y) = \text{Incidence}(Y + l) \times \text{onset proportion.} \qquad (7.28)$$

This expression is valid only if all occult onset cancers are destined to become clinically incident. Fryback et al. [36] found that this expression does not yield the observed high rate of in situ disease following introduction of mass-screeing. It follows that not all occult onset cancers will become invasive. Therefore, we introduce a class of "limited malignancy potential" (LMP) tumors that grow to "Max LMP size," remain at this size for "LMP Dwell time," and then regress. Some fraction, the "LMP fraction," of all occult onset tumors are randomly chosen to be LMP tumors. With this new class of tumors, we derive a new expression for the rate of occult onset as a function of the predicted clinical cancer incidence. First, redefine onset proportion as the ratio of non-LMP onset to clinical incidence:

$$\text{onset proportion} = \frac{\text{non-LMP occult onset rate}}{\text{clinical incidence rate}}. \qquad (7.29)$$

Now, determine total onset rate using the following relations:

$$\text{total onset} = \text{LMP onset} + \text{non-LMP onset,} \qquad (7.30)$$

$$\text{LMP onset} = \frac{\text{LMP onset}}{1 - \text{LMP fraction}} - \text{non-LMP onset,} \qquad (7.31)$$

$$\text{non-LMP onset} = \text{Incidence}(Y + l) \times \text{onset proportion.} \qquad (7.32)$$

With some simple rearrangement we have the final expression for total occult tumor onset:

$$\text{total onset} = \text{Incidence}(Y + l) \times \text{onset proportion} \times \left(\frac{1}{1 - \text{LMP fraction}} \right). \qquad (7.33)$$

Thus, we have total onset in terms of the predicted clinical incidence rate, a model input, and two parameters with values determined by model calibration. The inclusion of a subset of small tumors destined to regress was needed to avoid depleting the pool of early tumors detected by screening in SEER data. It was also necessary to introduce two classes of hyperaggressive tumors to successfully match SEER data. In this case, some small fraction of onset tumors were assigned 4 nodes (regional stage) at 2 mm onset, while another fraction of tumors was assigned to the distant stage (5 nodes) at 2 mm onset.

Model calibration yields the following parameter values:

$$LMP\ fraction = 0.42,$$
$$Max\ LMP\ size = 1\ cm,$$
$$LMP\ Dwell\ time = 2\ years,$$
$$Percent\ 4\ nodes = 1\%,$$
$$Percent\ 5\ nodes = 2\%.$$

Thus, the model predicts that slightly less than half of all breast cancers that reach 2 mm in diameter are destined to regress by the in situ or early invasive stage of growth. This is in broad accord with studies tracking the incidence of invasive disease arising from in situ carcinomas misdiagnosed as benign [25] and autopsy studies revealing a surprisingly high prevalence of undiagnosed DCIS in younger women [79], as discussed earlier. We note that since a tumor of 2 mm diameter contains about 4.2×10^6 cells (assuming 10^6 cells per mm^3), it is unclear how modeling tumor initiation from a single cell would alter this prediction.

State-transition models. The so-called SPECTRUM model [55] and another by Lee and Zelen [51] describes tumor natural history as a series of stage transitions. Notably, the SPECTRUM model always underestimates DCIS incidence following screening, and the authors considered this indirect evidence that many DCIS never progress to invasive disease, and many screen-detected DCIS may be over-treated.

7.7.4 Conclusions and optimal screening strategies

The SPECTRUM and MISCAN-Fadia models fit clinical data well, with the exception of the increasing DCIS incidence following widespread screening. The Wisconsin model predicts that a fraction of DCIS is destined to regress. These results, along with previously discussed autopsy series [79], suggest that a significant fraction of DCIS not only never surface clinically, but fully regress.

The CISNET models were also recently applied to a more pressing clinical problem: the optimal mammography screening strategy [54]. A set of screening strategies that varied by starting and stopping age (e.g. 40–84 vs 50–74) and screening interval, either annual (A) or biennial (B), were tested using six of the CISNET models. A screening strategy is considered "inefficient" or "dominated" if it both (1) results in worse health outcomes (mortality reduction and increase in life-years), and (2) requires more resources (mammographies per 1000 women) than any other strategy considered. Across all models, eight strategies were found to be non-dominated. All but one (A40–84) used biennial screening and all but two began screening at age 50 rather than age 40.

This work further found that screening women beginning at age 40, rather than age 50, does decrease mortality, but only by about 3%. Due to the low incidence of cancer in the 40–49 age group, the absolute benefit is therefore very small, perhaps less than harm due to the screening procedure. Moreover, more resources are consumed and more false positives occur. Also, *mortality* is predicted to decrease more by screening older patients (over 74) than by screening younger women, but screening younger women should add more overall *life-years* (total years of life gained).

These results helped motivate a recent statement by the U.S. Preventive Services Task Force [77] that recommends (1) *against* routine mammography for women aged 40–49, and (2) biennial screening for women of ages 50–74. This recommendation to reduce screening has been very controversial and in some cases vehemently condemned. While certainly debatable, the recommendations are based on data, and represent a significant contribution of modeling to an important clinical problem.

7.8 Cancer progression and incidence curves

Most models we have discussed so far consider cancer growth under a constant parameter regime. Yet, cancer growth rates *evolve*. Therefore, the parameters describing tumor growth kinetics almost certainly cannot remain constant over the entire tumor history. However, simple Gompertzian models in which tumors, after initiation, progress according to constant parameter values most naturally assume cancer as an essentially "one-hit process." However, it is well known that most tumors arise from multiple "hits" to their genomes. So we have another challenge to the simple Gompertz view of tumor growth.

7.8.1 Basic multi-hit model

Cancer incidence increases dramatically with age, a fact initially regarded as inconsistent with the hypothesis that cancer arises from mutation in normal cells [62]. If clinical cancer is sparked by a single aberrant mutation, then age-specific cancer incidence rates should be constant, at least among adults if there is a significant delay between cancer initiation and clinical presentation. However, the epidemiological evidence shows clearly that age-specific incidence is not constant, an observation that motivated the first explicit multi-hit models, which we now discuss.

By the early 1950s it was well known that many carcinogenic agents are also mutagenic, and there is typically a latency period between carcinogen exposure and clinical cancer. In 1953, Nordling first proposed that malignant transformation of a single cell requires a sequence of multiple mutations [62]. He argued that if a single mutation sufficed, then cancer incidence would be the same for all ages. If two mutations were required, then the number of cells carrying a single "hit" (cancer-causing mutation) would increase in direct proportion to age, and so the incidence (requiring the second hit) would be directly proportional to age. If three hits were needed, incidence would increase as the second power of age, and so on. Nordling then examined cancer mortality data for adult men in the U.K., excluding children and the very old, and observed that mortality increased with the sixth power of age, implying that, on average, seven mutations are necessary.

Armitage and Doll [1], in 1954, reached a similar conclusion using more rigorous probabilistic reasoning. If malignant transformation requires r mutations, the ith mutation occurs with probability p_i per unit time, and if these mutations must occur in a particular order, then the probability that a cell becomes malignant (acquires the rth) in the time interval $(t, t + dt)$, is

$$\frac{p_1 p_2 ... p_r t^{r-1}}{(r-1)} dt. \tag{7.34}$$

(If the mutations can occur in any order, the following result still holds.) It follows that the cancer incidence rate, $I(t)$, is proportional to the $(r-1)$th power of time (i.e. age); i.e.,

$$I(t) \propto p_1 p_2 ... p_r t^{r-1} = k t^{r-1}, \tag{7.35}$$

where k is constant. Taking the logarithm of both sides yields

$$\log(I(t)) = \log(k) + (r-1)\log(t), \tag{7.36}$$

which implies that a log-log plot of cancer incidence versus age should be a straight line with slope $(r-1)$. Data from nine different types of cancer gave slopes between 4.97 and 6.48. The slope for colon and rectal cancer came out to be approximately five, suggesting six rate-limiting steps for this particular neoplasm.

Armitage and Doll's model has become the classic multi-hit model for cancer incidence. It assumes that intermediate mutations confer no selective advantage or functional change to the cell, and that cancer arises only after all hits have occurred. In this case, mutation order *cannot matter*, and all mutations are mathematically equivalent. However, if mutations confer differential selective advantages, then mutation order does matter because selection dynamically alters the pool of cells which may be mutated. Therefore, there is a slight conflict between Armitage and Doll's original analysis, which assumed a particular order is necessary, and the actual model which implies that order does not matter.

We can also recast the basic multi-hit model as a simple system of differential equations, as done by Frank [31]. Letting $x_i(t)$ represent the number of cells with i mutations at time t, Frank proposed the following simple model:

$$\frac{dx_0}{dt} = -u_0 x_0, \tag{7.37}$$

$$\frac{dx_i}{dt} = u_{i-1} x_{i-1} - u_i x_i, \quad i = 2, ..., r-1, \tag{7.38}$$

$$\frac{dx_r}{dt} = u_{r-1} x_r, \tag{7.39}$$

where u_i is the rate at which cells with i mutations mutate. This formulation helps point out some issues with rate parameters. It is natural to assume that transition rates should be the same, at least in first approximation, since one may expect genes to mutate at roughly the same rate. However, the basic multi-hit model requires that mutations are unordered (see discussion above). Since the number of genes available for mutation decreases with each hit, transition rates must vary among stages.

If the order of mutations does matter, then transition rates will be constant (at least under the assumption that mutations never occur out of sequence). This may be the better assumption, at least in some cases. For example, colon carcinoma seems to exhibit a preferred progression of genetic changes during transition from dysplasia to clinical cancer. Interestingly, the slope of the log-log plot is constant whether or not the u_i vary, implying that incidence curves give no information on whether mutations are ordered or not.

This discussion highlights the complexities introduced when mutations confer selective advantage. In such cases, one must be very careful to define the mutation mechanism and transition rates among mutant states correctly. Moreover, if mutation order is indeed unimportant, the rate at which precancerous lesions evolve toward malignancy may reach its maximum at an intermediate number of mutations, but slow as more hits are acquired.

7.8.2   Two-hit models

We now turn to another model by Armitage and Doll, this one from 1957 [2]. Here the authors do consider the role of selection in a pre-cancerous tumors. They showed that such selection can yield cancer incidence proportional to the sixth power of age with only two genetic hits required for cancer. The model considers the influence of two carcinogenic agents. The first acts as an initiator, and cells acquire the initial mutation at a rate proportional to dose. These precancerous cells undergo exponential clonal expansion and are susceptible to malignant transformation by the second agent. Let d_1 and d_2 be the doses of initiating and transforming agents, respectively, and n_t represent the number of precancerous cells at time t after exposure to the first dose. Then

$$n_t \propto d_1 e^{kt}, \tag{7.40}$$

$$I(t) \propto d_2 n_t = d_1 d_2 e^{kt}. \tag{7.41}$$

Relation (7.40) expresses the exponential expansion of premalignant cells following exposure to the initiating agent. Using probabilistic arguments, Armitage and Doll arrived at the following equation for cancer incidence:

$$I(t) = Np_1 \left(1 - \exp\left[\frac{-p_2}{k} \left(e^{kt} - 1 \right) \right] \right), \tag{7.42}$$

where N is the number of persons at risk for the first mutation and, as before, $I(t)$ is cancer incidence at time t. Here, p_1 is the probability per unit time that a normal cell undergoes the first mutation and is proportional to d_1; p_2 is the per unit time probability of the second mutation and is likewise proportional to d_2.

We arrive at expression (7.42) by the following arguments. Suppose a clone has undergone the initiation step at time $t - T$, but has not transformed by time t. Then the clone transforms to malignancy with probability $p_2 \exp(kT)$ per unit time. The probability that the second change occurs in the small time interval $(t, t + dt)$, but not before, is therefore approximately

$$= p_2 e^{kT} dt \times P(\text{no change occurred in } (t - T, t)),$$
$$= p_2 e^{kT} dt \times P(\text{no change occurred in } (0, T)). \tag{7.43}$$

Transitions from the intermediate class to malignancy is governed by a Poisson process with a non-constant rate proportional to the number of intermediate cells. Therefore, the transition rate parameter on the interval $(0, T)$ is $\lambda = p_2 \exp(kT)$. It follows from probability theory that:

$$P(\text{no change occurred in } (0, T)) = \exp\left(-\int_0^T \lambda \, dt \right) = \exp\left[\frac{p_2}{k} \left(1 - e^{kT} \right) \right]. \tag{7.44}$$

The probability of the second change in $(t, t + dt)$ is

$$p_2 e^{kT} dt \exp\left[\frac{p_2}{k} \left(1 - e^{kT} \right) \right]. \tag{7.45}$$

Now, we need the probability that the first change occurs in the interval $(t - T, t - T + dT)$. The time to the first mutation is exponentially distributed with parameter p_1, so

$$P(\text{1st change in } (t - T, t - T + dT)) = 1 - \exp(-p_1 dT). \tag{7.46}$$

It follows from the Taylor series of the exponential that, if $p_1 dT$ is small, we can approximate $1 - \exp(-p_1 dT)$ as

$$p_1 dT. \tag{7.47}$$

The probability we seek is therefore the product of equations (7.45) and (7.47) (given with some rearrangement):

$$p_1 p_2 \exp\left(\frac{p_2}{k} \right) \exp\left(kT - \frac{p_2}{k} e^{kT} \right) dt \, dT. \tag{7.48}$$

All that remains is to integrate over time between the first and second mutations, T, to get the final probability of a second mutation in the interval $(t, t + dt)$:

$$p_1 p_2 e^{p_2/k} dt \int_0^t \exp\left(kT - \frac{p_2}{k} e^{kT}\right) dT = p_1 \left(1 - \exp\left[\frac{-p_2}{k}\left(e^{kt} - 1\right)\right]\right) dt.$$
$$(7.49)$$

We recover relation (7.42) by multiplying the right-hand side of equation (7.49) by N, the number of persons at risk.

This model successfully matched mortality data for stomach, intestine, colorectal, and pancreatic cancer for persons aged 25–74 years. In many hormone-dependent cancers, such as breast and ovarian cancer, the slope of the incidence curve decreases in later life. Armitage and Doll suggest that this model could be made to match these cancers as well if hormones were promote clonal expansion of singly mutated cells.

From equation (7.42), we can make several predictions concerning cancer incidence. First, it should increase linearly with exposure to the initiating carcinogen and increase exponentially with exposure to the transforming carcinogen. Furthermore, as $t \to \infty$, we have $I \to Np_1$. Thus, the model predicts that all age-incidence curves should eventually become constant, which accords well with the observation that cancer incidence curves tend to flatten with age.

Finally, we note that as with the basic multi-hit model, we may also reformulate this model in terms of differential equations:

$$\frac{dx_0}{dt} = -p_1 x_0, \tag{7.50}$$

$$\frac{dx_1}{dt} = p_1 x_0 - p_2 x_1 + k x_1, \tag{7.51}$$

$$\frac{dx_2}{dt} = p_2 x_1. \tag{7.52}$$

In this case, $I(t) \propto dx_2/dt$.

Retinoblastoma, a rare childhood tumor of the eye, has been examined in the context of two-hit models by Kundson and colleagues, beginning with a seminal paper in 1971 [50]. This tumor often shows a hereditary pattern, but can also arise spontaneously. In either case, both copies of a key tumor-suppressor gene, called Rb, must be inactivated for cancer to occur. Incidence data and probabilistic arguments led Knudson et al. to hypothesize that hereditary retinoblastoma arises when one of the Rb genes is already mutated in the germ line (lineage of cells that produces gametes). Therefore, they hypothesize that only one additional hit is necessary to spark the cancer. In contrast, spontaneous cases would require both alleles to be inactivated, implying a two-hit incidence pattern. Beyond being among the first cancers to which the two-hit etiology has been applied, retinoblastoma also provided the

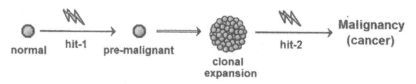

FIGURE 7.7: Simple schematic for the two-hit model, or two-stage clonal expansion model.

first evidence for tumor suppressor genes. Frank also recently revisited this hypothesis using a simple mathematical model [32].

Following Knudson et al.'s lead, Moolgavkar, Knudson, and others have developed many two-hit stochastic models for tumor growth [59, 58]. In these models, healthy stem cells transition via mutation to intermediate cells with a slight proliferative advantage. This expanding clone may then undergo a second mutation to become malignant. Like the model of Armitage and Doll, tumor initiators cause the first transition, while tumor promoters increase the proliferation rate of intermediate cells without inducing mutation [58]. This two-hit framework, sometimes referred to as the two-stage clonal expansion model (TSCE), has been widely used to model carcinogenesis. The TSCE is essentially equivalent to a common conceptual framework called the initiation-promotion-progression "model."

7.8.3 The case of colorectal cancer

While two-hit or initiation-promotion-progression models are popular, they are almost certainly too simple. Most adult carcinomas require multiple genetic modifications. Colorectal cancer, for which the key molecular events driving progression are reasonably well defined, provides an example.

The morphologic progression from early adenoma to invasive cancer is driven by genetic and genomic instability that arises early in this tumor's evolution; it characterizes nearly all colorectal carcinomas and adenomas. Two major forms of instability are observed: (1) *chromosomal instability* (CIN) which leads to aneuploidy, present in about 85% of cases, and (2) *microsatellite instability* (MSI), associated with defective DNA mismatch repair (MMR) and present in 15% of cases. These two phenomena tend to be mutually exclusive [81], suggesting that genetic stability plays an important functional role in carcinogenesis irrespective of its molecular basis.

A set of key genetic changes that occur in cancers arising by CIN, termed the "traditional pathway," were characterized in 1988 by Vogelstein, Fearon, and colleagues [78, 27] based on molecular analysis of 172 colorectal cancers. Morphologically, these lesions typically progress through the following stages: (1) normal epithelium, (2) aberrant crypt focus (ACF), (3) early adenoma, (4) late adenoma, (5) invasive cancer, and finally (6) metastatic disease.

Genetically, the earliest change is either loss of the *APC* gene on chromosome 5q or a mutation of the K-*ras* gene, which gives rise to a dysplastic ACF. While the K-*ras* mutation may actually occur first [81], sources traditionally list APC as the initiating mutation [27]. Inactivation of APC leads to loss of proliferation suppression by the Wnt signaling pathway and may contribute to the development of CIN. Further progression to invasive cancer is associated with mutations in the K-*ras* oncogene, loss of 18q, causing inactivation of *DCC*, *SMAD2*, and *SMAD4* tumor suppressor genes, and loss of 17q, which causes *p53* tumor suppressor inactivation (Figure 7.8). This set of genomic alterations is unlikely to be sufficient to cause cancer. A number of other genetic and epigenetic alterations are also associated with cancer progression, and non-invasive adenomas with all four changes exist [27]. While there is a preferred order for these changes, their net accumulation is likely more important than the order [27]. The apparent ordering may be a function of the selective advantage or genomic instability conferred by individual mutations.

Most tumors do not actually harbor all canonical changes of the traditional pathway [81], suggesting that they arise through functional changes in phenotype. The traditional pathway may (partially) represent only one of many ways to generate the same phenotype. This view accords well with the hallmarks of cancer, which include a limited number of rate-limiting steps generating malignant phenotypes. But a given phenotype may be conferred by a variety of genotypes.

Thus, from the biology we can see that a two-hit model clearly cannot describe colon cancer, and some form of a multi-hit model is required. Recall from above that Armitage and Doll's 1954 work suggested that six mutations are necessary for colon cancer. However, the fact that adenomas arise at all and accumulate genetic changes as they grow argues for a selective advantage conferred to pre-cancerous clones. It is generally accepted that adenomas advance by successive rounds of clonal expansion, where more fit clones outcompete the existing adenoma population [27]. Therefore, the basic multi-hit model without clonal expansion cannot be applied either. A clonal expansion model similar to that proposed by Frank [31] may be more appropriate. By modifying mutation rates it could also be made to handle genetic instability.

Several recent mathematical models have been used to study colon cancer [53, 65, 57]. Nowak et al. [65] studied a simple mutation network to determine the role of CIN in *APC* inactivation in early colon cancer. Later, Michor et al. [57] used a somewhat similar model to study CIN's role in tumor suppressor inactivation in a more abstract setting. While Nowak et al. used stochastic methods to study their model which we do not cover here, we briefly outline the studied mutation network, parametrization, and model conclusions.

It is frequently the case that one copy of *APC* is inactivated by a point mutation in the gene, while the second is lost via a gross chromosomal deletion, a so-called **loss-of-heterozygosity** (LOH) event. Chromosomal instability increases the rate of LOH by orders of magnitude; therefore, induction of CIN may be the rate-limiting step in tumor evolution, rather than the LOH event itself. Nowak et al.'s model considered a population of colon crypt cells in two basic classes: with and without CIN. Let x_i, $i \in \{0, 1, 2\}$ be the mass of cells in the first class (without CIN) with i inactivated copies of *APC*. Similarly, y_0, y_1, and y_2 be CIN cells with 0, 1, and 2 inactivated copies of *APC*. They assume that active *APC* genes in all cells become inactivated by point mutations at rate $u \approx 10^{-7}$ per cell division. Transition from $i = 0$ to $i = 1$ therefore occurs at rate $2u$ (to account for the two ways in which this transition can occur). The second APC gene becomes inactivated either by a second point mutation at rate u, or by LOH, assumed to occur in non-CIN cells at rate p_0. LOH in CIN cells arises at a much higher rate, $p > p_0$, estimated to be about 0.01 per cell division. Finally, they allow caretaker genes (the products of which maintain genomic stability) to mutate, resulting in a transition from non-CIN to CIN. Assuming n_c caretaker genes exist, then the transition $x_i \rightarrow y_i$ occurs at rate $2n_c u$.

Based on their simulations, Nowak et al. [65] concluded tha CIN typically arises early in tumorigenesis. However, since CIN alone decreases cell fitness, then it probably follows inactivation of the first *APC* gene, while the second most often is deleted via LOH. However, if CIN is neutral or advantageous, then CIN may be the initiating event in tumorigenesis. In either case, the model suggests that CIN is an early and rate-limiting event in tumorigenesis. Michor et al. [57] reach essentially the same conclusion.

Luebeck and Moolgavkar [53] extended 2-hit colon cancer incidence models to include 3, 4 or 5 premalignant steps. The four-stage model best matched SEER data on colon cancer incidence. Moreover, their best-fit parameter estimates suggested that colon cancers are initiated by two rare genetic events (e.g. mutations) followed by a single high-frequency event, giving rise to clonal expansion of the premalignant population from which the final event sparks malignancy.

This prediction coincides with Nowak et al.'s [65] suggestion that a point mutation in *APC* is followed by induction of CIN. The two slow steps predicted by Luebeck and Moolgavkar may be mutation of one APC and onset of CIN. Following CIN, the tumor likely rapidly inactivates the second APC and begins clonal expansion. A final *p53* mutation, which is likely to occur in short order following CIN, may be the Luebeck and Moolgavkar's fourth rate-limiting step. We now have a plausible reconciliation between the biological necessity that more (and probably far more) than four mutations are necessary to yield invasive cancer and the epidemiological agreement between a four-stage model and colon cancer incidence. This view also suggests that early tumor formation may be dominated by rare, stochastic genetic events; however, following CIN and the onset of rapid genome derangement, growth may be modeled reasonably well by deterministic kinetics.

FIGURE 7.8:   Traditional pathway of colorectal tumorigenesis. Depicted are the canonical genetic hits associated with morphological progression.

7.8.4   Multiple clonal expansions

One important aspect of tumorigenesis that we have not yet encountered in tumor models is multiple rounds of clonal expansion. Such a behavior was considered by Frank [31]. He modified the continuous version of the basic multi-hit model (see above) to consider logistic growth of each clone individually. Clones grew independently with a limiting size that depended only on the clone's size, not the tumor's, an assumption that can be challenged on biological grounds.

Nevertheless, this model yields theoretical cancer incidence curves that match several observed phenomenon. In particular, it predicts the flattening log-log incidence curve that occurs in advanced age. Frank termed this phenomenon a decrease in the "age-specific acceleration" of cancer.

7.8.5   Smoking and lung cancer incidence

We close by mentioning the case of lung cancer and smoking. Cigarette smoke is generally believed to act as both tumor initiator and promoter. That is, it directly induces genetic mutations and promotes a host environment permissive of neoplasia. Smoking's enhancement of cancer risk is well documented, and its effect on risk for ex-smokers has been tracked in large data-sets that provide excellent opportunities to test the predictive ability of multi-stage models.

After cessation of smoking, the excess risk for lung cancer declines fairly rapidly, and continues to decline for about two decades. However, the absolute cancer risk declines slightly and remains fairly static, or "frozen," over time.

The basic multi-hit model was first applied to this problem by Doll in 1978 [21], who concluded that smoking affects the first stage and a late, perhaps the last, stage. (Later work generally assumes it affects the penultimate stage.) These conclusions are supported by data. That smoking affects an early step in tumor progression follows from the fact that, when cancer incidence is plotted against time of exposure, both smokers and non-smokers show the same power relationship. That smoking accelerates a later step is implied by the observed rapid reduction in cancer risk after smoking cessation. Finally, Doll observed a quadratic relationship between the dose rate (i.e., cigarettes smoked per day) and incidence, again suggesting that two stages are affected.

In related work, Brown and Chu [9] used a multi-hit model to estimate the rates of the two events affected by smoking. Their results predicted that cigarette smoking increases the rate of the late event by twice as much as it does the early event. They concluded that the greater part of total cancer risk is due to smoking's effect on the penultimate event.

Importantly, the multi-hit models of smoking's effect on cancer risk tend to predict, contra the data, that when a person stops smoking, the excess lung cancer risk continues to rise. This is discussed extensively by Freedman and Navidi [35]. They predicted the existence of a mechanism to repair smoking-induced lesions as a possible explanation for the reduction in excess risk and the failure of multi-hit models [35].

More recently, several two-stage clonal expansion models [45, 44, 19] have been applied to the problems of lung cancer incidence and smoking. All have concluded that smoking acts primarily as a tumor promoter. It may also be an initiator, and initiation and promotion may interact synergistically to increase risk [44].

7.8.6 Summary

While we have largely focused on kinetic models in this chapter, we have introduced multi-hit epidemiological models because of their historical importance on understanding the etiology of cancer, and because they address a key facet of the natural history of cancer. We are also motivated to examine such models because the genetic changes that occur in cancer etiology must necessarily affect tumor growth dynamics.

Very different genetic hit models can give similar cancer incidence curves, implying that while incidence curves have been instrumental to understanding the genetic basis for cancer, they cannot tell us the whole story. It remains an open question how mutations give rise to gross tumor growth kinetics. It is generally accepted that repeated rounds of clonal expansion drive both genetic and morphologic evolution, at least in the case of colon cancer. But aside from an intermediate period of exponential growth for two-stage models, the genetic hit models do not address growth kinetics. The models of Nowak et al. [65] and Luebeck and Moolgavkar [53] suggest that kinetic models may describe tumor growth reasonably well following CIN, after which rare genetic events may no longer be rate-limiting (although further genetic changes would continue to be necessary). Nevertheless, kinetic models of gross tumor growth have yet to be systematically unified with models of molecular etiology of cancer.

7.9　Exercises

Exercise 7.1:　Show that if

$$N(t) = N_0 \exp\left(k\left(1 - e^{-\alpha t}\right)\right), \tag{7.53}$$

then

$$k = \ln\left(\frac{N_\infty}{N(0)}\right),$$

where the asymptotic tumor size is N_∞. Assume the tumor is detected at size N_D at time T_D. Show that

$$T_D = -\frac{1}{\alpha} \ln\left(\frac{-\ln(N_D/N_\infty)}{k}\right). \tag{7.54}$$

Exercise 7.2:　Let p be a constant. Show that the solution of

$$\frac{d\mu}{dN} = \frac{\mu}{N} + p\left(1 - \frac{\mu}{N}\right). \tag{7.55}$$

is

$$\mu(N) = N\left(1 - N(0)^p N^{-p}\left[1 - \frac{\mu(0)}{N(0)}\right]\right).$$

Exercise 7.3:　The Yorke model is stochastic and too complex to be easily amenable to analysis. Therefore, to gain insight implement and study the basic model computationally:

1. Implement the basic model for primary tumor growth and metastasis seeding. Generate the distant metastasis free survival curve for a set of simulated patients, assuming no local recurrence occurs. For a more challenging extension, add local recurrence.

   (a) We assume that the metastases have a different growth rate than the primary tumor, randomly drawn from the same distribution. Also, the model construction necessitates that all metastases have the same growth rate. Are these realistic assumptions?

   (b) From Figure 7.3(a), we see that the DMFS curve approaches an asymptote. This is obviously not realistic. What model assumption could be responsible for this behavior, and how might it affect predictions?

Exercise 7.4:　In the clinical medicine literature, patient survival data is widely reported and displayed using *Kaplan-Meier survival curves*. Such a curve gives the cumulative probability of survival at any time, and can be approximated (but not replicated) by the DFMS curves of the basic model. We note that Kaplan-Meier plots are valuable because they can incorporate *censoring*, i.e. patient loss. To generate such curves using a model it is not enough to know the probability of death at a given time: the individual times of death must be known exactly. Implement a simulated version of the Yorke model and use it to generate a Kaplan-Meier curve. A very brief pseudocode follows:

1. Pre-define the simulation time as (0, T_max).
2. For all N simulated patients:
   1. Randomly determine primary tumor growth parameter.
   2. Calculate time to detection, T_D.
   3. Divide the pre-clinical history, t = (0, T_D) into Q discrete steps of length T_s.
   4. At every time-step, calculate the number of metastatically-capable cells, m(t), as a function of the number of primary tumor cells:

      m(t) = (1 - n(t)^(-alpha)) * n(t). Also:

      A. The number of mets successfully seed during each time-step is Poisson-distributed with parameter: (eta) * m(t) * T_s.
      B. Randomly seed metastases with unique, random growth rates. Calculate the time to detection and time to death for each met. If this is the minimum, store this value.
      C. Store the time of met seeding and growth rate for all mets.
   5. Use the data from (C) to determine the total metastatic burden at all times. Use this to determine if death occurs at an earlier time than previously determined.
   6. Have local recurrence randomly occur with some probability.
   7. If recurrence happens, follow a similar algorithm to that above.
3. Discretize the time domain (0, T_max) and, from the individual patient data, store the total number of patients still alive and recurrence-free at each time-point.
4. Plot the fraction of patients still alive at each time-point as a stair-plot, giving a Kaplan-Meier plot.

   1. Use the simulated model to determine how sensitive survival (in terms of both the shape of the cumulative survival and long-term survival) is to the following:

      (a) Tumor size at diagnosis.

      (b) Intrinsic metastatic potential, as measured by α and η.

      (c) The probability of local recurrence.

      (d) The residual tumor size following local treatment.

      (e) Tumor growth rate (parameter b).

   2. Following 1(e), all else being equal, do faster-growing tumors have a better or worse prognosis than slow-growing tumors? Note that, clinically, faster tumor growth is clearly correlated with poorer survival. What does this suggest about the validity of the model?

   3. The degree to which disease free survival (DFS) (i.e. the minimum of DMFS or time to local recurrence) correlates with survival after recurrence (SAR) is controversial. How well do these metrics correlate under different probabilities of local recurrence?

Exercise 7.5: The current model assumes that it takes only one genetic hit for cells to become metastatically capable. Biologically, what capabilities must a cell acquire to enter the bloodstream and survive to found a distant colony? Are these likely to be conferred by a single hit? Develop an alternative model in which two or more hits are required for metastatic capability.

Exercise 7.6: Implement a cancer-screening program under the Yorke model. That is, for every patient, during the pre-clinical period, $t < T_D$, at regular intervals (say, every year) check to see if the tumor has passed some more smaller threshold size, $\hat{T}_D$, and if it has, initiate treatment. Characterize the effect of screening on both short-term and long-term survival under this model.

References

[1] Armitage P, Doll R: The age distribution of cancer and a multi-stage theory of carcinogenesis. *Br J Cancer* 1954, 8:1–12.

[2] Armitage P, Doll R: A two-stage theory of carcinogenesis in relation to the age distribution of human cancer. *Br J Cancer* 1957, 11:161–169.

[3] Barbolosi D, Benabdallah A, Hubert F, Verga F: Mathematical and numerical analysis for a model of growing metastatic tumors. *Math Biosci* 2009, 218:1–14.

[4] Bartoszyński R, Edler L, Hanin L, Kopp-Schneider A, Pavlova L, Tsodikov A, Zorin A, Yakovlev AY: Modeling cancer detection: Tumor size as a source of information on unobservable stages of carcinogenesis. *Math Biosci* 2001, 171:113–142.

[5] Baum M, Demicheli R, Hrushesky W, Retsky M: Does surgery unfavourably perturb the "natural history" of early breast cancer by accelerating the appearance of distant metastases? *Eur J Cancer* 2005, 41:508–515.

[6] Berry DA, Inoue L, Shen Y, Venier J, Cohen D, Bondy M, Theriault R, Munsell MF: Modeling the impact of treatment and screening on U.S. breast cancer mortality: A Bayesian approach. *J Natl Cancer Inst Monogr* 2006, 36:30–36.

[7] Bloom HJ, Richardson WW, Harries EJ: Natural history of untreated breast cancer (1805-1933). Comparison of untreated and treated cases according to histological grade of malignancy. *Br Med J* 1962, 2:213–221.

[8] Boer R, de Koning HJ, van der Maas PJ: A longer breast carcinoma screening interval for women age older than 65 years? *Cancer* 1999, 86:1506–1510.

[9] Brown CC, Chu KC: Use of multistage models to infer stage affected by carcinogenic exposure: Example of lung cancer and cigarette smoking. *J Chronic Dis* 1987, 40:171S-179S.

[10] Chen LL, Beck C: A superstatistical model of metastasis and cancer survival. *Physica A: Stat Mech Appl* 2008, 387:3162-3172.

[11] Clarke M, Collins R, Darby S, Davies C, Elphinstone P, Evans E, Godwin J, Gray R, Hicks C, James S, MacKinnon E, McGale P, McHugh T, Peto R, Taylor C, Wang Y; Early Breast Cancer Trialists' Collaborative Group (EBCTCG): Effects of radiotherapy and of differences in the extent of surgery for early breast cancer on local recurrence and 15-year survival: An overview of the randomised trials. *Lancet* 2005, 366:2087-2106.

[12] Clarke LD, Plevritis SK, Boer R, Cronin KA, Feuer EJ: A comparative review of CISNET breast models used to analyze U.S. breast cancer incidence and mortality trends. *J Natl Cancer Inst Monogr* 2006, 36:96-105.

[13] Demicheli R: Growth of testicular neoplasm lung metastases: Tumor-specific relation between two Gompertzian parameters. *Eur J Cancer* 1980, 16:1603-1608.

[14] Demicheli R, Abbattista A, Miceli R, Valagussa P, Bonadonna G: Time distribution of the recurrence risk for breast cancer patients undergoing mastectomy: Further support about the concept of tumor dormancy. *Breast Cancer Res Treat* 1996, 41:177-185.

[15] Demicheli R, Foroni R, Ingrosso A, Pratesi G, Soranzo C, Tortoreto M: An exponential-Gompertzian description of LoVo cell tumor growth from in vivo and in vitro data. *Cancer Res* 1989, 49:6543-6546.

[16] Demicheli R, Retsky MW, Hrushesky WJ, Baum M: Tumor dormancy and surgery-driven interruption of dormancy in breast cancer: Learning from failures. *Nat Clin Pract Oncol* 2007, 4:699-710.

[17] Demicheli R, Retsky MW, Swartzendruber DE, Bonadonna G: Proposal for a new model of breast cancer metastatic development. *Ann Oncol* 1997, 8:1075-1080.

[18] Demicheli R, Valagussa P, Bonadonna G: Does surgery modify growth kinetics of breast cancer micrometastases? *Br J Cancer* 2001, 85:490-492.

[19] Deng L, Kimmel M, Foy M, Spitz M, Wei Q, Gorlova O: Estimation of the effects of smoking and DNA repair capacity on coefficients of a carcinogenesis model for lung cancer. *Int J Cancer* 2009, 124:2152-2158.

[20] Devys A, Goudon T, Lafitte P: A model describing the growth and the size distribution of multiple metastatic tumors. *Disc Cont Dyn Sys B* 2009, 12:731–767.

[21] Doll R: An epidemiological perspective of the biology of cancer. *Cancer Res* 1978, 38:3573–3583.

[22] Early Breast Cancer Trialists' Collaborative Group (EBCTCG): Systemic treatment of early breast cancer by hormonal, cytotoxic, or immune therapy. 133 randomised trials involving 31,000 recurrences and 24,000 deaths among 75,000 women. *Lancet* 1992, 339:1–15.

[23] Early Breast Cancer Trialists' Collaborative Group (EBCTCG): Effects of chemotherapy and hormonal therapy for early breast cancer on recurrence and 15-year survival: An overview of the randomised trials. *Lancet* 2005, 65:1687–1717.

[24] Erbas B, Amos A, Fletcher A, Kavanagh AM, Gertig DM: Incidence of invasive breast cancer and ductal carcinoma in situ in a screening program by age: Should older women continue screening? *Cancer Epidemiol Biomarkers Prev* 2004, 13:1569–1573.

[25] Erbas B, Provenzano E, Armes J, Gertig D: The natural history of ductal carcinoma in situ of the breast: A review. *Breast Cancer Res Treat* 2006, 97:135–144.

[26] Eroles P, Bosch A, Bermejo B, Lluch A: Mechanisms of resistance to hormonal treatment in breast cancer. *Clin Transl Oncol* 2010, 12:246–252.

[27] Fearon ER, Vogelstein B: A genetic model for colorectal tumorigenesis. *Cell* 1990, 61:759–767.

[28] Fisher B: The evolution of paradigms for the management of breast cancer: A personal perspective. *Cancer Res* 1992, 52:2371–2383.

[29] Fisher B, Slack N, Katrych D, Wolmark N: Ten year follow-up results of patients with carcinoma of the breast in a co-operative clinical trial evaluating surgical adjuvant chemotherapy. *Surg Gynecol Obstet* 1975, 140:528–534.

[30] Folkman J: Tumor angiogenesis: Therapeutic implications. *N Engl J Med* 1971, 285:1182–1186.

[31] Frank SA: Age-specific acceleration of cancer. *Curr Biol* 2004, 14:242–246.

[32] Frank SA: Age-specific incidence of inherited versus sporadic cancers: A test of the multistage theory of carcinogenesis. *Proc Natl Acad Sci USA* 2005, 102:1071–1075.

[33] Franks SJ, Byrne HM, Mudhar HS, Underwood JC, Lewis CE: Mathematical modelling of comedo ductal carcinoma in situ of the breast. *Math Med Biol* 2003, 20:277–308.

[34] Franks SJ, Byrne HM, Underwood JC, Lewis CE: Biological inferences from a mathematical model of comedo ductal carcinoma in situ of the breast. *J Theor Biol* 2005, 232:523–543.

[35] Freedman DA, Navidi WC: Ex-smokers and the multistage model for lung cancer. *Epidemiology* 1990, 1:21–29.

[36] Fryback DG, Stout NK, Rosenberg MA, Trentham-Dietz A, Kuruchittham V, Remington PL: The Wisconsin breast cancer epidemiology simulation model. *J Natl Cancer Inst Monogr* 2006, 36:37–47.

[37] Goldie JH, Coldman AJ: A mathematic model for relating the drug sensitivity of tumors to their spontaneous mutation rate. *Cancer Treat Rep* 1979, 63:1727–1733.

[38] Goldie JH, Coldman AJ: Quantitative model for multiple levels of drug resistance in clinical tumors. *Cancer Treat Rep* 1983, 67:923–931.

[39] Guzman RC, Yang J, Rajkumar L, Thordarson G, Chen X, Nandi S: Hormonal prevention of breast cancer: Mimicking the protective effect of pregnancy. *Proc Natl Acad Sci USA* 1999, 96:2520–2525.

[40] Habbema JD, van Oortmarssen GJ, Lubbe JT, van der Maas PJ: The MISCAN simulation program for the evaluation of screening for disease. *Comput Methods Programs Biomed* 1985, 20:79–93.

[41] Halsted WS: I. The results of operations for the cure of cancer of the breast performed at the Johns Hopkins Hospital from June, 1889, to January, 1894. *Ann Surg* 1894, 20:497–555.

[42] Hanin L, Rose J, Zaider M: A stochastic model for the sizes of detectable metastases. *J Theor Biol* 2006, 243:407–417.

[43] Hanin LG, Miller A, Zorin AV, Yakovlev AY: The University of Rochester model of breast cancer detection and survival. *J Natl Cancer Inst Monogr* 2006, 36:66–78.

[44] Hazelton WD, Clements MS, Moolgavkar SH: Multistage carcinogenesis and lung cancer mortality in three cohorts. *Cancer Epidemiol Biomarkers Prev* 2005, 14:1171–1181.

[45] Heidenreich WF, Wellmann J, Jacob P, Wichmann HE: Mechanistic modelling in large case-control studies of lung cancer risk from smoking. *Stat Med* 2002, 21:3055–3070.

[46] Hellman S, Harris JR: The appropriate breast cancer paradigm. *Cancer Res* 1987, 47:339–342.

[47] Heuser L, Spratt JS, Polk HC Jr: Growth rates of primary breast cancers. *Cancer* 1979, 43:1888–1894.

[48] Holford TR, Cronin KA, Mariotto AB, Feuer EJ: Changing patterns in breast cancer incidence trends. *J Natl Cancer Inst Monogr* 2006, 36:19–25.

[49] Iwata K, Kawasaki K, Shigesada N: A dynamical model for the growth and size distribution of multiple metastatic tumors. *J Theor Biol* 2000, 203:177–186.

[50] Knudson AG Jr: Mutation and cancer: Statistical study of retinoblastoma. *Proc Natl Acad Sci USA* 1971, 68:820–823.

[51] Lee S, Zelen M: A stochastic model for predicting the mortality of breast cancer. *J Natl Cancer Inst Monogr* 2006, 36:79–86.

[52] Love RR, Niederhuber JE: Models of breast cancer growth and investigations of adjuvant surgical oophorectomy. *Ann Surg Oncol* 2004, 11:818–828.

[53] Luebeck EG, Moolgavkar SH: Multistage carcinogenesis and the incidence of colorectal cancer. *Proc Natl Acad Sci USA* 2002, 99:15095–15100.

[54] Mandelblatt JS, Cronin KA, Bailey S, Berry DA, de Koning HJ, Draisma G, Huang H, Lee SJ, Munsell M, Plevritis SK, Ravdin P, Schechter CB, Sigal B, Stoto MA, Stout NK, van Ravesteyn NT, Venier J, Zelen M, Feuer EJ; Breast Cancer Working Group of the Cancer Intervention and Surveillance Modeling Network. Effects of mammography screening under different screening schedules: Model estimates of potential benefits and harms. *Ann Intern Med* 2009, 151:738–747.

[55] Mandelblatt JS, Schechter CB, Lawrence W, Yi B, Cullen J: The SPECTRUM population model of the impact of screening and treatment on U.S. breast cancer trends from 1975 to 2000: Principles and practice of the model methods. *J Natl Cancer Inst Monogr* 2006, 36:47–55.

[56] Marks LB, Prosnitz LR: Postoperative radiotherapy for lung cancer: The breast cancer story all over again? *Int J Radiat Oncol Biol Phys* 2000, 48:625–627.

[57] Michor F, Iwasa Y, Vogelstein B, Lengauer C, Nowak MA: Can chromosomal instability initiate tumorigenesis? *Semin Cancer Biol* 2005, 15:43–49.

[58] Moolgavkar SH: Biologically motivated two-stage model for cancer risk assessment. *Toxicol Lett* 1988, 43:139–150.

[59] Moolgavkar SH, Knudson AG Jr: Mutation and cancer: A model for human carcinogenesis. *J Natl Cancer Inst* 1981, 66:1037–1052.

[60] Nagy JD: Competition and natural selection in a mathematical model of cancer. *Bull Math Biol* 2004, 66:663–687.

[61] Nelson HD, Tyne K, Naik A, Bougatsos C, Chan BK, Humphrey L; U.S. Preventive Services Task Force: Screening for breast cancer: An update for the U.S. Preventive Services Task Force. *Ann Intern Med* 2009, 151:727-737, W237–242.

[62] Nordling CO: A new theory on cancer-inducing mechanism. *Br J Cancer* 1953, 7:68–72.

[63] Norton L: A Gompertzian model of human breast cancer growth. *Cancer Res* 1988, 48:7067–7071.

[64] Norton L, Simon R: The Norton-Simon hypothesis revisited. *Cancer Treat Rep* 1986, 70:163–169.

[65] Nowak MA, Komarova NL, Sengupta A, Jallepalli PV, Shih IeM, Vogelstein B, Lengauer C: The role of chromosomal instability in tumor initiation. *Proc Natl Acad Sci USA* 2002, 99:16226–16231.

[66] Plevritis SK, Sigal BM, Salzman P, Rosenberg J, Glynn P: A stochastic simulation model of U.S. breast cancer mortality trends from 1975 to 2000. *J Natl Cancer Inst Monogr* 2006, 36:86–95.

[67] Punglia RS, Morrow M, Winer EP, Harris JR: Local therapy and survival in breast cancer. *N Engl J Med* 2007, 356:2399–2405.

[68] Retsky M: New concepts in breast cancer emerge from analyzing clinical data using numerical algorithms. *Int J Environ Res Public Health* 2009, 6:329–348.

[69] Retsky MW, Demicheli R, Swartzendruber DE, Bame PD, Wardwell RH, Bonadonna G, Speer JF, Valagussa P: Computer simulation of a breast cancer metastasis model. *Breast Cancer Res Treat* 1997, 45:193–202.

[70] Retsky M, Swartzendruber D, Wardwell R, Bame P, Petrosky V: Re: Larry Norton, a Gompertzian model of human breast cancer growth. *Cancer Res* 1989, 49:6443–6444.

[71] Retsky MW, Wardwell RH, Swartzendruber DE, Headley DL: Prospective computerized simulation of breast cancer: Comparison of computer predictions with nine sets of biological and clinical data. *Cancer Res* 1987, 47:4982–4987.

[72] Shwartz M: A mathematical model used to analyze breast cancer screening strategies. *Operations Research* 1978, 26:937–955.

[73] Skipper HE: Kinetics of mammary tumor cell growth and implications for therapy. *Cancer* 1971, 28:1479–1499.

[74] Speer JF, Petrosky VE, Retsky MW, Wardwell RH: A stochastic numerical model of breast cancer growth that simulates clinical data. *Cancer Res* 1984, 44:4124–4130.

[75] Tan SY, van Oortmarssen GJ, de Koning HJ, Boer R, Habbema JD: The MISCAN-Fadia continuous tumor growth model for breast cancer. *J Natl Cancer Inst Monogr* 2006, 36:56–65.

[76] Thames HD, Buchholz TA, Smith CD: Frequency of first metastatic events in breast cancer: Implications for sequencing of systemic and local-regional treatment. *J Clin Oncol* 1999, 17:2649–2658.

[77] US Preventive Services Task Force: Screening for breast cancer: U.S. Preventive Services Task Force recommendation statement. *Ann Intern Med* 2009, 151:716–726, W-236.

[78] Vogelstein B, Fearon ER, Hamilton SR, Kern SE, Preisinger AC, Leppert M, Nakamura Y, White R, Smits AM, Bos JL: Genetic alterations during colorectal-tumor development. *N Engl J Med* 1988, 319:525–532.

[79] Welch HG, Black WC: Using autopsy series to estimate the disease "reservoir" for ductal carcinoma in situ of the breast: How much more breast cancer can we find? *Ann Intern Med* 1997, 127:1023–1028.

[80] Withers HR, Lee SP: Modeling growth kinetics and statistical distribution of oligometastases. *Semin Radiat Oncol* 2006, 16:111–119.

[81] Worthley DL, Whitehall VL, Spring KJ, Leggett BA: Colorectal carcinogenesis: Road maps to cancer. *World J Gastroenterol* 2007, 13:3784–3791.

[82] Yorke ED, Fuks Z, Norton L, Whitmore W, Ling CC: Modeling the development of metastases from primary and locally recurrent tumors: Comparison with a clinical data base for prostatic cancer. *Cancer Res* 1993, 53:2987–2993.

[83] Zhdanov VP: Stochastic model of the formation of cancer metastases via cancer stem cells. *Eur Biophys J* 2008, 37:1329–1334.

Chapter 8

Evolutionary Ecology of Cancer

8.1 Introduction

In Chapters 2 and 3 we explored the foundations of two major themes in mathematical oncology—one stemming from the pioneering work of Laird and her application of the Gompertz formalism to tumor growth, and the other originating with Greenspan's extension of the PDE approach taken simultaneously by Burton, Thomlinson, and Gray. In this chapter we continue our exploration of these themes in a more modern setting. However, since both themes unleashed floods of research with rivulets in many different directions, we limit ourselves to theoretical attacks on how two important characteristics of cancer arise—necrosis and tumor cell diversity.

Modern cancer research has tended to focus on genomics. However, an adequate theory of cancer must recognize that tumor behavior is at least as much an ecological and evolutionary problem as a molecular one. Natural selection has long been recognized as the ultimate driver of cancer progression and pathogenesis (see [52] for a recent review; see also [133]). In early stages of tumor progression, heterogeneous populations of malignant and healthy cells compete for available resources. Tumor cell clones that have acquired, via mutation and epigenetic effects, malignant "hallmark" phenotypes [56, 57] gain proliferative and (or) survival advantages relative to other lineages in their tumor microenvironment. Eventually the hallmark-carrying mutant clones come to dominate the tumor and destroy tissue homeostasis. If this interpretation is correct, then the mechanism causing malignancy—heritable variation conferring advantages to particular clonal lineages—is precisely evolution by natural selection.

Host physiology and the tumor microenvironment are critical because they largely generate the selection pressures acting within the tumor at any given time. Therefore, any accurate theory must also include interactions among a variety of genetically distinct tumor cell types, perhaps genetically altered stromal cells, and unmutated healthy cells, both peritumoral and distant. Although these interactions certainly are influenced by genomes, all genomes, both cancerous and healthy, interacting within the tumor's "ecosystem" are involved.

Although cancer theory derived from molecular and cellular biology largely ignores evolutionary and ecological relationships, the same cannot be said for theory developed by mathematical and computational biologists, who for the last thirty years have produced an enormous variety of mathematical models of malignant neoplasia. Given the complexities of the problem, most of this work has concentrated on simulations and computational treatments and therefore lies somewhat outside the scope of this text. Therefore, in this chapter we focus on seminal dynamical models that can be used as a springboard into the rapidly expanding study of evolutionary oncology, or the evolutionary ecology of tumors. Also, although evolution by natural selection is well-recognized as an important process during tumor treatment, we explore those relationships in the treatment chapters.

8.2   Necrosis: What causes the tumor ecosystem to collapse?

Malignant tumors in vivo and, as we saw in Chapter 3, multicell spheroids in vitro often develop regions of necrosis in which large portions of the tissue dies (Fig. 8.1). Necrosis in multicell spheroids (MCSs) is typically ascribed to "lack of nutrient," as Burton and others originally hypothesized (see Chapter 3). Nutrient depletion occurs in the spheroid core as cells continuously consume whatever the nutrient is, but its only source is the spheroid's surface. As we mentioned in Chapter 2, tumors in vivo avoid this problem by enticing new blood vessels to grow within them through a process called angiogenesis. Nevertheless, tumors in vivo still can develop necrosis (Fig. 8.1). Various hypotheses have been presented in the mathematical and theoretical literature to explain in vivo necrosis. Possible immediate causes include deficiencies in oxygen, glucose, or perhaps other nutrients, such as phosphorus [38, 69] or iron [34, 59, 72, 107]. Some researchers suggest that inhibitory chemicals, which could be metabolic waste or other compounds, might play a role. Still others implicate mechanical destruction of cells. At one level removed from the immediate cause, if nutrient deficiency, toxin production, or both are to blame, local ischemia is almost certainly involved. But then what causes the ischemia—inefficient neoangiogenesis (the tumor "outgrowing" its blood supply), blood vessel collapse, variation in hematocrit distribution in a microvascular net, or some combination of the three? All of these hypotheses have been the subjects of mathematical investigations.

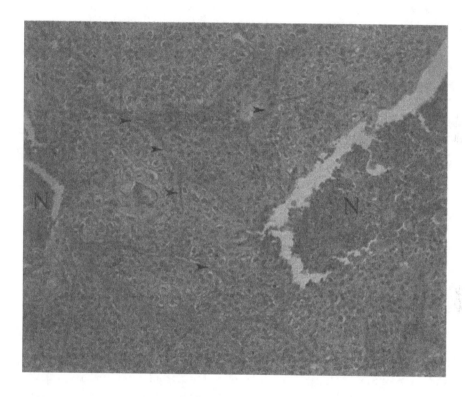

FIGURE 8.1: Squamous cell lung cancer, H. & E. stain, 100× original magnification. Regions labeled "N" are necrotic, and arrows point to examples of extracellular matrix (ECM). Dark dots in the necrotic regions are mostly immune cell nuclei, primarily of neutrophils. Note how cancer cells form deranged sheets reminiscent of epithelium but with a highly disturbed architecture.

8.2.1 Necrosis in multicell spheroids

The obvious hypothesis explaining the characteristic histological pattern in MCSs—that cells in the interior die or stop reproducing because they suffer a profound lack of nutrient caused by diffusion limitation and competition (Fig. 3.1)—is compelling but probably too simple. The Greenspan model introduced in Chapter 3, for example, introduced the notion that necrosis could be caused by accumulation of some toxin within the spheroid. This model has been instrumental in refining the diffusion-limitation hypothesis. In parallel with these studies, another formalism developed from the Thomlinson and Gray work that also has been used to probe diffusion limitation. This set of models focuses on tumor cords, essentially spheroids turned inside-out. Below we outline some of the main results from both types of models.

Among the more influential research threads using multicell spheroids to study necrosis is a series of papers by Helen Byrne and Mark Chaplain [26, 27]. These models represent a small spheroid, which Byrne and Chaplain interpret as a tiny in vivo tumor. The first of these models [26] assumes that no necrotic core exists and takes the following form under the quasi-steady-state approximation of diffusive equilibrium for both the nutrient and inhibitor:

$$
\begin{cases}
R^2 \dfrac{dR}{dt} = \displaystyle\int_0^{R(t)} S(\sigma, \beta) r^2 \, dr, \\[2ex]
0 = \dfrac{D_1}{r^2} \dfrac{\partial}{\partial r}\left(r^2 \dfrac{\partial \sigma}{\partial r} \right) + \Gamma(\sigma_B - \sigma) - \lambda \sigma - g_1(\sigma, \beta), \\[2ex]
0 = \dfrac{D_2}{r^2} \dfrac{\partial}{\partial r}\left(r^2 \dfrac{\partial \beta}{\partial r} \right) - g_2(\sigma, \beta),
\end{cases}
\tag{8.1}
$$

where R is the radius of the spheroid; $S(\sigma, \beta)$ is cell proliferation rate; D_1 and D_2 are diffusivities of nutrient and inhibitor, respectively; Γ measures vascular delivery of nutrient; σ_B is concentration of nutrient in the blood plasma; g_1 and g_2 represent sources and sinks of nutrient and inhibitor, respectively, within the spheroid; and all other notation is consistent with that in Chapter 3. Byrne and Chaplain subsequently apply Buron's original hypothesis that necrosis occurs whenever the nutrient concentration falls below some critical value to obtain this model [27] :

$$
\begin{cases}
R^2 \dfrac{dR}{dt} = \displaystyle\int_0^{R(t)} [S(\sigma, \beta) H(r - r_i) - N(\sigma, \beta) H(r_i - r)] r^2 \, dr, \\[2ex]
0 = \dfrac{D_1}{r^2} \dfrac{\partial}{\partial r}\left(r^2 \dfrac{\partial \sigma}{\partial r} \right) - [\Gamma \sigma + \gamma_1 \beta] H(r - r_i), \\[2ex]
0 = \dfrac{D_2}{r^2} \dfrac{\partial}{\partial r}\left(r^2 \dfrac{\partial \beta}{\partial r} \right) - \gamma_2 \beta H(r - r_i),
\end{cases}
\tag{8.2}
$$

with N the rate at which necrotic cells disintegrate and γ_1 and γ_2 the rates at which the inhibitor decays within the proliferative and necrotic regions, respectively. All other notation is as before, with one exception—in model (8.2), Γ represents the consumption rate of nutrient by living cells. Furthermore, nutrient is delivered to the interior by diffusion from the media or interstitium, not through an interior vascular network as in (8.1). There is no quiescent layer in either model.

Among the advances introduced by models (8.1) and (8.2) is a more realistic action for the inhibitor than that proposed by Greenspan. Instead of causing quiescence, the inhibitor is hypothesized to increase cell mortality within nonnecrotic regions of the spheroid, which Byrne and Chaplain equate to apoptosis. Therefore, $S(\sigma, \beta)$ is interpreted as pointwise differences between births and deaths and generally increases with σ and decreases with β. This hypothesis allows richer dynamics than Greenspan's model. Of particular importance, model (8.1) shows that spheroids can reach a steady state without necrosis. (See [32, 42] for more details.) In fact, a sufficiently large apoptosis rate can cause complete spheroid regression without development of a necrotic core in both models. Model (8.2) also predicts that spheroids with a necrotic core arise only if loss to apoptosis is less significant than loss to necrosis and if the external oxygen concentration is not too large. More precisely, they show that for a particular realization of g_1 and g_2, the width of the proliferating rim is proportional to $\sqrt{\sigma_\infty - \sigma_l}$, where σ_∞ is the nutrient concentration in the media and σ_l is defined above.

More recently, Davide Ambrosi and Francesco Mollica [6, 7] modeled nutrient deficiency in multicell spheroids cultured either free in suspension or embedded in agarose. These models introduce mechanical stress generated within the tumor and externally through the agarose gel under the assumption of an elastic spheroid. As in the previous models, nutrient is assumed to diffuse into the spheroid from the media very rapidly relative to cell proliferation rates. Since Ambrosi and Mollica imagine the nutrient to be a storable form of energy to power cell proliferation, one can interpret it as glucose. They assume that glucose determines reproductive potential of cells within the spheroid. In particular, if we let n be the nutrient concentration and g be the growth potential (birth rate minus death rate) at a certain point within the spheroid, then g is an increasing linear function of glucose concentration such that $g(n, \cdot) < 0$ for $0 \le n < n_0 < \infty$, reflecting the dominance of deaths over births in low-glucose environments. In addition, g decreases with a measure of stress on the cells.

The complexity of Ambrosi and Mollica's formalism takes a complete description of their model outside the scope of this chapter. However, their results are of interest. Numerical investigations show that the spheroid naturally develops an outer proliferative rim with a core dominated by lack of nutrient. Predictions about necrosis per se cannot be made from their analysis, because they chose $n_0 = 0$. However, their formalism hints at the possibility of combining the nutrient-deficiency and mechanical-deformation hypotheses (see Section 8.2.4) into a single model, which promises an incisive instrument to tease these two hypotheses apart.

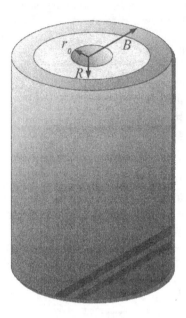

FIGURE 8.2:   Idealized section of a tumor cord. The middle tube represents a microvessel with radius r_0. Surrounding the vessel is a sleeve of living tumor tissue with outer radius R that is further surrounded by a necrotic rind of radius B. (Based on Fig. 1 of [16, p. 163]).

8.2.2   Necrosis in tumor cords

The tumor cord is a concept introduced around the same time that theorists started modeling multicell spheroids [61, 129]. In essence, the tumor cord turns the spheroid inside-out, placing the source of nutrient in the center, and transforms it from a sphere to a cylinder. The cord itself is a sleeve of tumor tissue surrounding a microvessel, which supplies nutrients and waste removal services. The outer portion of the cord is often necrotic (Fig. 8.2). Tumor cords can be observed in certain regions of certain tumors (see [64] for example) but not all (see Fig. 8.1 for example).

Recently, Alessandro Bertuzzi, Alberto Gandolfi, and their colleagues [15, 17, 18] (reviewed in [16]; see also [37]) have studied tumor cords theoretically. In their models, we imagine a rigid-walled capillary of radius r_0 surrounded by a cylinder of tissue. The maximum (fixed) width of the cord is R. In later work, they also introduce a necrotic rind of outer radius B. As in the spheroid models of Chapter 3, surface tension and incompressibility of cells and interstitium maintain the cord's shape, and the tissue can exist as a mosaic of both proliferative and quiescent cells. The rate at which cells become quiescent decreases with local nutrient concentration. However, a distinct annulus of quiescent cells can arise when nutrient concentration falls below a prescribed threshold. In some of these models, the active cell population is structured by age. Within the tissues, both living and necrotic, the volume is entirely exhausted by three components: living cells (ν_p), necrotic cells (ν_n), and extracellular space (ν_e). Cell packing is assumed to be homogeneous (ν_e and $\nu_p + \nu_n + \nu_e$ are constants).

These assumptions lead Bertuzzi et al. [15] to a model that really consists of two submodels for the cell dynamics. The first, expressed generally as

$$\begin{cases} \dfrac{\partial \nu_p}{\partial t} + \nabla \cdot (\mathbf{u}\nu_p) = \chi(\sigma)\nu_p - [\mu(\sigma) + \mu_c(c,\sigma) + \mu_r(\sigma,t)]\nu_p, \\[2mm] \dfrac{\partial \nu_n}{\partial t} + \nabla \cdot (\mathbf{u}\nu_n) = [\mu(\sigma) + \mu_c(c,\sigma) + \mu_r(\sigma,t)]\nu_p - \mu_n\nu_n, \end{cases} \tag{8.3}$$

describes dynamics in the region of living tissue ($r_0 < r < R$). The second,

$$\frac{\partial \nu_n}{\partial t} + \nabla \cdot (\mathbf{u}\nu_n) = -\tilde{\mu}_n\nu_n, \tag{8.4}$$

valid for $R < r < B$, models the necrotic region, because no living cells are present there ($\nu_p \equiv 0$). The vector $\mathbf{u}$ represents a cell velocity field that arises as cells push on one another as they reproduce or are squeezed together as some die. If one assumes that this motion is confined to the radial plane in a perfect cylinder, then $\mathbf{u} = u(r)$, where r is the radial position. In addition, $\chi(\sigma)$ is the per-capita proliferation rate, which depends on nutrient concentration σ; $\mu(\sigma)$ is the "natural" mortality rate; and μ_n and $\tilde{\mu}_n$ represent the disintegration rate of necrotic cells in living tissue and the necrotic rind, respectively. The additional death terms $\mu_c(c,t)$ and $\mu_r(\sigma,t)$ denote death rates from chemotherapeutics and radiation treatment, respectively, where c is the drug concentration. The dependence of μ_r on σ arises because the authors assume the nutrient is oxygen, and radiation-induced mortality is well known to depend on local O_2 concentration [61].

Nutrient is assumed to move entirely by diffusion on a much faster time scale than cell velocity. Therefore, they assume that

$$\Delta\sigma = f(\sigma)\nu, \tag{8.5}$$

where f depends on the diffusivity of O_2 and the rate at which the tissue consumes it. Nutrient enters the cord only across the microvessel wall at a constant rate, reflecting blood O_2 homeostasis. They also impose a no-flux condition for O_2 at the outer cord boundary. As with the spheroid models, these authors assume that all cells become necrotic whenever O_2 concentration falls below some threshold σ_n. Cells enter a reversible quiescent state, modeled in $\chi(\sigma)$, if O_2 drops below another threshold, $\sigma_q > \sigma_n$.

Interestingly, this model predicts that the boundary between necrotic and living tissue cannot always be identified as the point at which σ drops below

σ_n. For example, suppose the cord sits at its steady-state radius with the demarcation of its necrotic region r_n defined by $\sigma(r_n) = \sigma_n$. Suppose further that a chemotherapeutic attack kills a large number of tumor cells. Afterward, competition for oxygen among survivors transiently slackens, and the cord begins to grow, pushing the necrotic region outward. Then for a short period the boundary of the necrotic region is determined by history and not by nutrient availability. One can therefore in principle use this model to predict the transient dynamics of a tumor cord following cytotoxic treatment as a way to test the hypothesis that necrosis is caused by a lack of O_2 or, with proper modification, some other nutrient. More mundane phenomena, in particular the size of the viable and necrotic sleeves as a function of O_2 delivery, can be used for a similar purpose, at least in principle.

8.2.3   Diffusion limitation in ductal carcinoma in situ

Whole autochthonous tumors are often much more difficult to model than laboratory systems, such as multicell spheroids, or special in vivo systems, such as tumor cords or explants, because their geometries are usually much more irregular. However, breast ductal carcinoma in situ (DCIS) is something of an exception and so has attracted the attention of mathematical oncologists [41, 137]. By definition, this lesion is confined to the lumenal side of the duct's basement membrane, which means it is usually forced to grow in a cylindrical shape around 700 μm in diameter on average [41]. Unlike tumor cords, however, nutrients are delivered to the cancer cells via diffusion from the external surface of the cylinder rather than a central blood vessel. (See also Chapter 7 for more on DCIS.)

A model of early DCIS by Susan Franks et al. [41], although including no explicit mechanism of necrosis, helps explain why some such lesions have a necrotic interior that others lack. In their investigation, Franks et al. imagine a tumor growing within a rigid-walled cylinder representing a milk duct. Tumor cell proliferation therefore generates pressures that force cells to move with velocity $\mathbf{v}(\mathbf{x})$ at point $\mathbf{x}$. The portion of the model describing tumor volume and nutrient concentration takes the following form:

$$
\begin{cases}
\dfrac{\partial n}{\partial t} + \nabla \cdot (n\mathbf{v}) = D_n \nabla^2 n + (k_m(c) - k_d(c))n, \\[2mm]
\dfrac{\partial m}{\partial t} + \nabla \cdot (m\mathbf{v}) = D_m \nabla^2 m + k_d(c)n, \\[2mm]
\dfrac{\partial \rho}{\partial t} + \nabla \cdot (\rho\mathbf{v}) = D_\rho \nabla^2 \rho, \\[2mm]
\dfrac{\partial c}{\partial t} + \nabla \cdot (c\mathbf{v}) = D_c \nabla^2 c + -\beta k_m(c)n,
\end{cases}
\tag{8.6}
$$

where n and m are densities of living and dead cells, respectively; ρ is the density of interstitial fluid; and c is nutrient concentration. Functions k_m

and k_d represent per-capita births and deaths, respectively. Generally, k_m increases with c, and k_d decreases with c. Diffusion coefficients are represented as D_i, $i \in \{n, m, \rho, c\}$, and β is the amount of nutrient required to produce a new cell. Dead cells never disintegrate, and living, nondividing cells do not consume nutrient. By assuming that the tumor mass is exhausted by living cells, dead cells, and interstitial material, they show that

$$\nabla \cdot \mathbf{v} = k_m(c)n. \tag{8.7}$$

They then complete the model by using Stokes's law to derive expressions for intratumoral pressure.

This model produced no necrosis, because the duct diameter was so small that nutrient concentration favored proliferation over mortality everywhere in the tumor interior. However, the authors point out that allowing the duct wall to distend will decrease nutrient concentration in the tumor core, perhaps to the point where necrosis develops, as in comedocarcinoma [73], for example. The model could be extended to allow one to probe this scenario empirically, either in animal models or human histopathology samples, as a test of the nutrient-deficiency hypothesis.

8.2.4 Necrosis caused by mechanical disruption of cells

In distinct contrast to the nutrient-limitation hypothesis, Colin Please et al. [109, 110] hypothesize that necrotic regions form because cells are torn from their anchors to the extracellular matrix (ECM) and each other by pressures within the tumor. This mortality could arise either from literal destruction of the plasma membrane or apoptosis caused by loss of cell-ECM or cell-cell contact. The basic models [109] assume that the tumor's interior consists of two "phases"—cells and interstitial fluid. (ECM is not explicitly modeled in these early explorations.) Please et al. assume that cell interiors are composed of the same material as the interstitial fluid; therefore, fluid moves between phases, entering cells through the process of proliferation and reentering the interstitium as dead cells disintegrate. The models are thereby controlled primarily by conservation of this fluid. The requirement for fluid conservation produces pressures within the tumor that cause both cell and fluid movement. For example, as cells proliferate, fluid entering the cell phase produces an "outward" pressure, pushing the cell phase outward. Cells move freely in response to this force, because unlike the previous models, cells in this system do not adhere to each other. On the other hand, pressures in the fluid phase force interstitial fluid to move among the cells as if the tumor were a porous medium. Therefore, two pressures must enter the model: (1) the pressure on the interstitial fluid (P_e) and (2) the pressure exerted cell-to-cell via the ECM scaffold (P_c). The intracellular pressure is not modeled.

If in any region within the tumor $P_c > P_e$, then the cells in that region feel a compressive force through the surrounding ECM. If the inequality is reversed, then cells are assumed to be torn apart as tension forces rip them from the ECM. Although Please et al. hypothesize that this form of cell destruction occurs only to physiologically stressed cells, as in hypoxia, in the model any cell in such an environment is destroyed. This, then, is the mechanism of necrosis under investigation.

Although nutrient deficiency does not cause necrosis here, nutrients still play a role. As before, the nutrient moves primarily by diffusion on a much faster time scale than cell kinetics, so the nutrient concentration, denoted $C(\mathbf{x})$, is assumed to be in a quasi-steady state. Please et al. assume that cells proliferate at a rate proportional to the nutrient concentration and die at a constant per-capita rate; that is, at $\mathbf{x}$ the per-capita growth rate density is $dC(\mathbf{x}) - e$. If we let k be the constant permeability of fluid through the interstitium and ϕ be the proportion of the tumor volume taken up by interstitial space, also assumed to be constant throughout the tumor, then the above assumptions can be modeled as follows [110]:

$$k(1 - \phi)^2 \nabla^2 P_c = e - dC. \tag{8.8}$$

Consider the application of this model to a multicell spheroid. Again we assume perfect spherical symmetry and a nutrient diffusing into the spheroid from the external media (see [110] for the detailed boundary conditions). Then once again dynamics vary only over the radial position, so we can replace $\mathbf{x}$ with r defined in Section 8.2.1. In this case, the O_2 concentration obeys the following relation:

$$C(r, t) = \frac{R_0[\sinh(r - R_i) + R_i \cosh(r - R_i)]}{r[\sinh(R_0 - R_i) + R_i \cosh(R_0 - R_i)]}, \tag{8.9}$$

with R_0 and R_i defined as in Section 8.2.1 except that the necrosis condition is now $P_c < P_e$. In addition, the tumor and necrotic radii must satisfy the following conditions:

$$R_0^2 \frac{dR_0}{dt} = \int_{R_i}^{R_0} r^2(C(r, t) - \alpha) \, dr, \tag{8.10}$$

$$0 = \int_{R_i}^{R_0} \left(r - \frac{r^2}{R_0}\right)(C(r, t) - \alpha) \, dr, \tag{8.11}$$

where $\alpha = e/dC(R_0)$.

Superficially this model predicts observed spheroid behavior. Starting with a small spheroid, the radius grows exponentially until the moment a necrotic core begins to develop in the center. At that time, it enters a "linear" growth phase with a necrotic core eventually growing at the same rate, producing the proper histology. The depth of the proliferative layer will in general differ from that predicted by nutrient-deficiency models and can therefore be used to contrast the two hypotheses empirically.

Unfortunately, this model makes a disturbing prediction—in the absence of surface tension, the spheroid grows without bound. In fact, this result is general across choices of cell-growth models as long as cell proliferation is nondecreasing with oxygen concentration. However, Please et al. show in [110] that one can relax the assumption of inviscid cells and allow a (small) surface tension that can halt runaway growth. Doing so requires only the addition of the term $2\Gamma/R_0$ to equation (8.11), where Γ is a measure of surface tension.

Working from an extension [71] of the previous model, C. Y. Chen et al. [30] investigate the mechanical disruption hypothesis in spheroids growing in agarose gels. The gel is assumed to be elastic and therefore exerts pressure on the spheroid as it grows. Once again the tumor consists of cellular and extracellular fluid phases permeated by a nutrient, all of which obey the following relations:

$$
\begin{cases}
\dfrac{\partial \psi}{\partial t} + \nabla \cdot (\psi \mathbf{U}_c) = \psi S(C), \\[2mm]
\dfrac{\partial (1 - \psi)}{\partial t} + \nabla \cdot [(1 - \psi)\mathbf{U}_e] = -\psi S(C), \\[2mm]
D\nabla^2 C = \psi \Sigma(C),
\end{cases}
\tag{8.12}
$$

where ψ is the cellular volume fraction within the tumor ($\equiv 1 - \phi$ for ϕ defined above); $\mathbf{U}_c$ and $\mathbf{U}_e$ represent velocity fields for cell and extracellular fluid, respectively; S denotes cell-growth rate; D is oxygen diffusivity through the tumor; Σ represents rate of nutrient consumption; and all other notation is defined earlier in this section. Following [71], Chen et al. include in the force balance equations hydrodynamic drag, hydrostatic forces in the interstitial fluid, and forces among cells transmitted by an ECM scaffold. In the nonnecrotic region, by definition $P_c > P_e$ and cells are maximally packed such that $\psi = \psi_0$, ψ_0 a constant; however, in the necrotic core, $P_e = P_c$ but $\psi \leq \psi_0$.

The model becomes quite tractable if one limits the investigation to a perfect sphere, defines $\Sigma(C) = 1$, and sets

$$
S(C) = \begin{cases} 1 & \text{if } C > \alpha, \\ -\rho & \text{if } C \leq \alpha, \end{cases}
\tag{8.13}
$$

where α represents a lower O_2 threshold below which cell death predominates and ρ is a positive constant representing sensitivity of cells to nutrient deficiency. With this definition of the growth function, spheroids obeying model (8.12) can develop three histologically distinct regions: a necrotic core, a middle annulus characterized by cell mortality dominating proliferation, and an outer annulus of proliferative tissue. As in [110], this model predicts that spheroids suspended in liquid media (zero gel stiffness) obtain the traditional histology, including a necrotic core, but contrary to observation tend to grow without bound. In a gel, however, spheroids always asymptotically approach a limited size, with or without a necrotic core.

For our purposes the most important prediction made by this model involves the relationship of necrosis to gel stiffness. In a very rough sense, spheroids in stiffer gels tend to be smaller at their steady-state size, with a lower likelihood of becoming necrotic than spheroids in more elastic gels. Even if necrosis does develop, it tends to arise later in stiffer gels. Apparently, stiff gels squeeze fluid out of the spheroid while favoring cell compression, so the necrosis conditions are less likely to be met. Therefore, this model encourages one to test the mechanical disruption hypothesis against nutrient deficiency by varying gel stiffness and nutrient availability using a fully crossed, factorial experimental design. (See [62] for example.)

8.2.5 Necrosis from local acidosis

A series of investigations by Robert Gatenby and his colleagues focusing on how malignant neoplasms invade surrounding tissue has also produced an explanation of necrosis that harkens back to the inhibitors hypothesized by Greenspan, Byrne, and Chaplain. In this case, Gatenby and his colleagues identify the inhibitor as acid. Most malignant tumors acidify their local environments because parenchyma cells metabolize glucose via glycolysis and fermentation, which produces lactate that cells then secrete [47, 49, 114] (see Section 8.3.2.2 for an elaboration of this idea). Gatenby et al. [45, 46, 106] suggest that this acidification selects for tumor cells able to withstand acidosis, allowing them to outcompete and therefore invade adjacent healthy tissue. In one model of this hypothesis [45, 46], one represents the densities of cancer and healthy cells with $N_1(\mathbf{x}, t)$ and $N_2(\mathbf{x}, t)$, respectively, at point $\mathbf{x}$ in the tumoral or peritumoral environment at time t. The excess hydrogen ion or lactate concentration is denoted $L(\mathbf{x}, t)$. With this notation, Gatenby et al.'s model becomes

$$\begin{cases} \dfrac{\partial N_1}{\partial t} = r_1 N_1 \left(1 - \dfrac{N_1}{K_1} - \alpha_{12} \dfrac{N_2}{K_2} \right) - d_1 L N_1, \\[2mm] \dfrac{\partial N_2}{\partial t} = r_2 N_2 \left(1 - \dfrac{N_2}{K_2} - \alpha_{21} \dfrac{N_1}{K_1} \right) - d_2 L N_2 + \nabla \cdot [D_{N_2} \nabla N_2], \quad (8.14) \\[2mm] \dfrac{\partial L}{\partial t} = r_3 N_2 - d_3 L + D_L \nabla^2 L, \end{cases}$$

where d_1 and d_2 are excess death rates of the two cell types due to local acidosis; $D_{N_2}(N_1, N_2)$ represents cancer cell motility, modeled in particular as either $D_2(1 - N_1/K_1)$ or $D_2(1 - N_1/K_1 - N_2/K_2)$; r_3 is the per-cancer-cell H^+ secretion rate; d_3 is the rate at which hydrogen ions wash out in blood or are absorbed by physiological buffers; and D_L is the acid diffusivity through tumor tissue. A cellular automaton (CA) analog of this system, with an addition of glucose delivery through a vascular net, has also been studied [46, 106].

Both model (8.14) and its CA analog support the notion that acid secretion facilitates invasion, even in small tumors. These models also suggest a relationship between the morphology along the tumor edge and invasiveness; namely, a gap between tumor and healthy tissue tends to form in more aggressively invasive cancers. More important for our current purposes is the observation of necrosis in the CA model. Under certain circumstances, Aalpen Patel et al. [106] show that areas within the tumor can become so acidic that all cells are destroyed, yielding a region of necrosis.

Although this mechanism is distinctly different from all others presented above, it may be hard to tease apart from the nutrient limitation hypothesis for the following reason. Cancer cells might evolve to rely on glycolysis instead of the tricarboxylic acid cycle precisely because of nutrient limitation in nascent tumors [47, 49]. So areas where nutrients are limited are precisely those areas where selection favors cells that acidify the environment, resulting in a spatial correlation between nutrient deficiency and acidosis. However, the acidification hypothesis predicts that in older tumors at least regions with a higher density of parenchyma cells and therefore regions of high acid secretion, should be more prone to necrosis than regions with a more mixed histology, even if nutrient delivery does not vary between the areas. Models such as (8.14) and those presented in Sections 8.2.1, 8.2.2 and 8.2.6 may be employed to refine this prediction into something empirically testable.

8.2.6 Necrosis due to local ischemia

The irregular pattern of necrosis often observed in real tumors begs for more complex hypotheses than those presented above. In particular, how necrosis-inducing conditions arise in larger, irregularly shaped tumors in the face of vascularization needs explanation. If necrosis is caused by nutrient limitation, for example, then what determines where it occurs within a vascularized tumor? If, on the other hand, mechanical disruption of cells causes necrosis, then can one predict where such necrotic regions will crop up within a growing tumor given its location within the body?

Little theoretical work has been done on the mechanical-disruption hypothesis in an in vivo setting, probably for two reasons. The first is the obvious complexity involved; the region of the body itself would have to be modeled, including organ shape and tissue compositions. The importance of these factors is highlighted by the observation that multicell spheroid behavior depends in part on the stiffness of the media (see Section 8.2.4). Second, the hypothesis itself is relatively young and so has not been fully analyzed beyond the simplified geometry of a spheroid.

On the other hand, three distinct variations of the nutrient-limitation hypothesis have been proposed to explain necrosis in vascularized tumors. All three point to local ischemia (lack of blood delivery at the tissue level) as the culprit, but they disagree on what causes the ischemia. One, commonly cited in oncology texts, suggests that tumors "outgrow their blood supply"; that is, parenchyma growth exceeds vascular growth in some region within the tumor, resulting in a local ischemic necrosis. The viability of this hypothesis is questionable, because the net proliferation rate appears to depend on local perfusion, so it is unclear how the parenchyma could overshoot its local "carrying capacity" so wildly. Nevertheless, such a mechanism could account for some subtle oscillations in growth rate [95] (see Section 8.4 below). In the second variation, local ischemia is caused by compressive pressure within the tumor, which collapses tumoral blood vessels. This compression is thought to arise in part through high fluid pressure in the interstitium [89], although bulk pressure from cells probably plays a dominant role [66]. Finally, the third variation places the blame on irregular distributions of blood flow and hematocrit that can arise even within a highly organized microvascular net. Here, we focus only on models of these last two variations.

One version of the collapsed blood vessel hypothesis has been modeled by Mollica et al. [89] at the level of a single microvessel. In this model we imagine a capillary of length L situated within a tumor. The interstitial fluid pressure, π_i, is assumed to be constant, so the pressure acting on the capillary at location x along its length at time t, denoted $p(x,t)$, is $\pi(x,t) - \pi_i$, where π represents the vessel pressure. If $p < 0$, then the capillary feels a compressive force and will begin to collapse. Sufficient compression causes the capillary to buckle, resulting in almost complete cessation of blood flow. On the other hand, the capillary is assumed to have some elasticity, so it may dilate in response to the distending force the capillary feels when $p > 0$.

If we let $u(x,t)$ be the displacement of the capillary wall from its average width h_0, then the main model equations take the following form:

$$\begin{cases} -\dfrac{T}{w}\dfrac{\partial^2 u}{\partial x^2} + \Phi(u) - p + c\dfrac{\partial u}{\partial t} + \rho H\dfrac{\partial^2 u}{\partial t} = 0, \\[2mm] -\dfrac{\partial}{\partial x}\left((h_0 + u)^3 \dfrac{\partial p}{\partial x}\right) + \dfrac{3kp}{w\delta} + \dfrac{3\mu}{w}\dfrac{\partial u}{\partial t} = 0, \end{cases} \tag{8.15}$$

where T represents capillary wall tension, w normalizes the vessel cross section so changes in its area can be equated to changes in the height of a rectangle of equal area, $\Phi(u)$ represents the capillary wall stiffness, c is a drag coefficient, ρ represents interstitial fluid density, H is the virtual mass coefficient, δ is the capillary wall's thickness, k denotes the permeability of the capillary wall to serum, and μ is blood viscosity. At the venous and arterial ends, the capillary is held at fixed width h_0, and the entrance (arterial, π_a) and exit (venous, π_v) pressures are also fixed, with $\pi_a > \pi_v$.

Numerical investigation of this model uncovered a regime in which blood flow through the vessel cycles on a 100-millisecond time scale. The cycles appear to be chaotic and persistent, indicating a continuous "pulsing" of blood through the capillary independent of the cardiac cycle, which was not modeled. The original purpose of the model was to investigate observed variation in blood flow rate and direction in actual tumors. Mollica et al. note that their results, while intriguing, fail to explain observed behavior because the oscillations occur much too rapidly. However, they suggest that a network of such capillaries and relaxation of certain simplifying assumptions may yield more realistic behavior. A similar cautionary remark applies to the use of this model to study necrosis. The basic premise is promising, but a more coarse-grained scale is probably required. Nevertheless, this model is an important mechanistic attack on the problem.

Although it is well known that tumors frequently suffer disrupted circulation [66, 114], vessel collapse from compressive pressure is not the only possible mechanism. Tomas Alarcón et al. [3, 4] investigated an alternative in which ischemia arises as a result of capillary accommodation responses and heterogeneous distribution of hematocrit throughout a tumor. Using a "hybrid" cellular automaton model, so called because it includes a traditional diffusion formalism along with the CA mechanism, they studied the effects of heterogeneous blood distribution on competition between tumor cells and healthy cells. The setting is a prescribed, two-dimensional vascular net overlaying a 60-by-60 pixel CA grid. Each pixel in the grid represents one (biological) cell, so the domain is about 1200 μm^2, assuming an average cell diameter of 20 μm. The vascular bed is a regular "hexagonal" net with anastamoses every 80 to 90 μm or so. This arrangement results in a maximum avascular interval of 160 μm along the vertical axis and 240 μm along the horizontal.

The vessels themselves are not inert tubes. Rather, they change diameter in response to changes in transmural pressure (pressure across the vessel wall), sheer stress, effective oxygen delivery and intrinsic mechanisms. In short, if we let $R_{i,t}$ be the diameter of the ith capillary section at time step t, then accommodation dynamics of the capillary are described by the following equation, which because of the CA formalism is expressed in discrete time steps of length Δt:

$$\frac{R_{t+\Delta t} - R_t}{\Delta t} = R \left[\ln \left(\frac{\tau_w}{\tau(P)} \right) + k_m \ln \left(\frac{\dot{Q}_r}{H\dot{Q}} + 1 \right) - k_s \right], \qquad (8.16)$$

where the i subscripts have been dropped, τ_w is the sheer stress along the capillary wall, $\dot{Q}$ is whole blood flux, $\dot{Q}_r$ measures constant oxygen demand of cells serviced by the capillary, k_m measures how sensitive the capillary response is to discrepancies between O_2 demand and O_2 delivery, and k_s represents an innate tendency of the capillary to shrink in the absence of other modifiers. In addition, the authors assume that capillaries homeostatically regulate sheer stress around a set point that can vary with transmural pressure, P; that set point is represented by $\tau(P)$ and was determined empirically. The variable H is a measure of red blood cell count or, alternatively, moles of O_2 and can be thought of as proportional to the hematocrit, the red cell volume fraction of whole blood.

The most important aspect of this paper is the recognition that hematocrit tends not to remain homogeneous in a microvascular net; therefore, if hematocrit in one region of a tumor became very low, nutrient delivery would be impaired and necrosis might result. The mechanism causing hematocrit inhomogeneity is the tendency of erythrocytes to disproportionately enter branches with larger flow rates per unit area in the branch's cross section. To model this phenomenon, Alarcón et al. assume that at a vessel bifurcation, erythrocytes prefer to enter branches with a larger flow rate. In fact, if the difference in flow rates is large enough, all erythrocytes enter the faster branch.

Results of this model showed that both blood volume and hematocrit can vary wildly throughout a microvascular net. Unfortunately, Alarcón et al. did not use this model to investigate it as a mechanism of necrosis, but certainly this intriguing hypothesis is worth following up.

8.3　What causes cell diversity within malignant neoplasia?

Along with necrosis, cell diversity is another biologically and clinically significant feature of malignant neoplasia, the explanation of which benefits from an ecological perspective. This cellular diversity exists on two levels. The first, which for convenience we will call "Type I" diversity, includes all types of cells within malignant tumors, including parenchyma cells, "healthy" cells of the reactive stroma—primarily fibroblasts and myofibroblasts—and cells of the circulatory infrastructure—blood and lymph endothelial cells, pericytes, smooth muscle cells, and a few others. In addition, immune reactive cells, including lymphocytes, macrophages, and neutrophils, are also present (Fig. 8.1).

The second type of diversity, "Type II," is variation in parenchyma cell anatomy and physiology. Anatomical variation alone is typically referred to as cellular pleomorphism, or more specifically nuclear pleomorphism if one focuses on that organelle. But cancer cells also vary in genomic architecture and general physiology. Any variation primarily generated by genetic differences tends to be called clonal variation. Although mathematical attacks on the causes of necrosis have a deeper history than studies of tumor cell diversity, theories for both types have appeared, as reviewed below.

8.3.1　Causes of Type I diversity

Cancer can be understood as a result of natural selection favoring certain cell lineages that one can describe as "selfish 'cheats' that exhibit antisocial characteristics" [100, p. 493]. In the short term, selection favors aggressive mutant cells over "healthy," cooperating cells at the expense of integrated tissue architecture. Tissue integration breaks down because mutant cells enter a competition for resources that otherwise would not exist among cooperating, genetically similar clones. Since this destruction of tissue architecture, which defines malignancy, arises through disrupted relationships among all cell types within the lesion, Type I diversity has become a major focus of theoretical oncology. However, despite efforts to model angiogenesis (see [9, 28, 29, 74, 75, 82, 108, 119] for reviews) and interactions between parenchyma and ECM [29], the most modern empirical research of phenomena at this level of diversity (which can be thoroughly explored in [19, 33, 65, 76, 80, 105, 113, 115, 117, 134]) has attracted surprisingly little attention from the theoretical oncology community.

Despite the relative paucity of effort directed at Type I diversity, at least three hypotheses explaining how it arises can be derived from existing research. The first suggests that invading parenchyma cells cannot entirely outcompete the original healthy population, leaving remnants of the healthy population in pockets or spread evenly throughout the tumor. The second supposes that tumor tissue invades surrounding healthy tissue with fingerlike projections, like the fungiform invasion described in pathology texts, caused by known reaction-diffusion mechanisms. Finally, the third hypothesis is an extension of the first. Complex interactions among parenchyma, healthy, and immune cells within the lesion cause Turing-like patterns to arise in which densities of the various cell types vary throughout the tumor, with some areas inhabited primarily by parenchyma, others by normal cells. Each of these ideas is explored more fully below.

8.3.1.1 Incomplete competitive exclusion

Perhaps the most straightforward way to represent competition among cell types characteristic of malignancy is a direct application of the Lotka-Volterra model, as was made by Gatenby [44]:

$$
\begin{cases}
\dfrac{dN}{dt} = r_N N \left(1 - \dfrac{N + \alpha T}{K_N}\right), \\[3mm]
\dfrac{dT}{dt} = r_T T \left(1 - \dfrac{T + \beta T}{K_T}\right),
\end{cases}
\tag{8.17}
$$

where $N(t)$ and $T(t)$ represent the number of healthy cells and tumor cells, respectively; K_i is the "carrying capacity" for a body made exclusively of cell type i; α measures the competitive impact of tumor cells on healthy cells; β is the competitive impact of healthy cells on tumor cells; and r_i is the intrinsic rate of increase for cell type i. The dynamics of this model are well understood [63, 93]; the novelty is Gatenby's interpretation of the behaviors. He views parameter regions that allow an attracting interior (nonboundary) fixed point as benign neoplasia. Malignancy is recognized as an attracting fixed point on the boundary for which $N = 0$ and $T = K_T$. Later, Gatenby et al. [48] used this system in a general reaction-diffusion model,

$$
\frac{\partial \mathbf{n}}{\partial t} = \mathbf{f}(\mathbf{n}) + D\nabla^2\mathbf{n},
\tag{8.18}
$$

where $n(x, t)$ is a vector of cell population densities for all cell types at point x and time t, D is a matrix of cell motility coefficients, and $f(n)$ is a generalization of model (8.17) that includes an arbitrary number of cell types competing at a point in space. In their application, Gatenby et al. limit the model to one space dimension and two competing species, again cancer versus healthy cells, so $f(n)$ is the right-hand side of model (8.17).

Again, model (8.18) is well studied [94] and known to admit travelling-wave solutions with well-characterized velocities, interpreted by Gatenby et al. as tumor invasion of surrounding tissue. In particular, if

$$\alpha K_T > K_N \tag{8.19}$$

and

$$\beta K_N < K_T, \tag{8.20}$$

then the tumor will invade surrounding healthy tissue, completely replacing it, at a speed no less than

$$2\sqrt{r_T D_T \left(1 - \frac{\beta K_N}{K_T}\right)}. \tag{8.21}$$

However, if the inequality in equation (8.20) is reversed, then the tumor still invades but does not entirely eliminate surrounding healthy tissue. This result is interpreted as desmoplasia, a mixture of cancerous and noncancerous cells within a tumor. Therefore, this model explains tumor cell diversity as incomplete competitive exclusion.

Model (8.18) makes some interesting practical predictions about how tumors will respond to treatment. Most basically, any successful treatment must reverse both inequalities (8.19) and (8.20). If the treatment is successful and scar-forming tissue has essentially the same properties as the original healthy tissue destroyed by the tumor, then the lesion will scar over at a minimum speed given by an expression formally equivalent to (8.21), with subscripts switched and α replacing β. Also, as Gatenby et al. point out, cytotoxic therapy can kill tumor cells directly and may blunt the tumor population's intrinsic rate of increase r_T. In neither case will it have an effect on the asymptotic behavior of the tumor—r_T does not determine the stability properties of the steady states—unless the tumor is entirely eradicated. This observation may help explain why cytotoxic therapy often fails.

Gatenby et al. use this insight to identify other parameters that might make more promising targets for therapy. In fact, this model supports attacks on a potential target already identified—tumor vasculature [40]. In this context, an attack on tumor angiogenesis at the least reduces K_T, which will tend to reverse both inequalities (8.19) and (8.20). If all else remains equal, angiogenesis inhibition could therefore cause stability of the boundary equilibria to switch, in which case the tumor would regress *without further cytotoxic treatment*. In essence, the body itself would destroy the tumor by outcompeting it. However, whatever effect such a treatment has on K_T, it must not equally degrade K_N, as is clear from relations (8.19) and (8.20).

These inequalities also suggest that one might profitably attack the tumor by altering the competitive relationship between cancerous and healthy cells since decreasing α and increasing β will also tend to favor healthy cells. Gatenby et al. suggest that techniques to decrease tumor cell nutrient uptake and perhaps increase healthy cell uptake might work. This idea is in line with results obtained by [69]. Gatenby et al. also recommend looking for ways to decrease protease expression and acid secretion as ways to decrease α. They also suggest one might consider trying to increase K_N by somehow attenuating contact inhibition among normal cells.

8.3.1.2 Fungiform invasion

As an alternative to the incomplete-competitive-exclusion hypothesis, a model by Shusaku Tohya et al. [126] suggests that nutrient dynamics within the tumor drives Type I diversity. This model was originally designed to explore the irregular penetration of dermis by nodular lesions of basal cell carcinoma (BCC), a largely curable form of skin cancer. In this model, we look at a cross section through a BCC lesion perpendicular to the skin. All dynamics occur on the plane of the cut, so it is convenient to let x and y be the dimensions parallel and perpendicular to the skin's surface, respectively. Also, let $0 \leq x \leq X$ and $0 \leq y \leq Y$. Nutrient is delivered to tumor cells by a capillary that lies along the basal edge of the tumor, so the nutrient along the line $y = Y$ for all allowable x is fixed at n_0. Cells take up this nutrient, metabolize it, and use it for both movement and growth. If we let $n(x, y, t)$ and $c(x, y, t)$ be the nutrient concentration and cancer cell density, respectively, at point (x, y) and time t, then Tohya et al.'s model becomes

$$\begin{cases} \dfrac{\partial n}{\partial t} = D_n \nabla^2 n - knc, \\[2mm] \dfrac{\partial c}{\partial t} = \nabla(D_c(n, c)\nabla c) + \theta f(n, c), \end{cases} \tag{8.22}$$

where D_n is the diffusivity of nutrient; k is the base rate at which cells uptake and metabolize nutrient; $D_c(n, c) = \sigma nc$, σ constant, expresses cancer cell motility; and θ measures how efficiently cells convert nutrient into new growth. Initially the lesion starts as a flat layer of cells of fixed thickness y_0, with the remaining space $y_0 < y < Y$ for all allowable x considered to be normal dermal tissue.

Despite this model's formal simplicity, it produces an intriguing hypothesis. Under certain conditions, in particular when $n_0\sqrt{\theta/D_n}$ is sufficiently small and with the proper initial conditions, the lesion extends tumorous "fingers" into the dermis. If sectioned in any way other than exactly perpendicular to the skin surface, a microscopic examination of such tissue would look like islands of normal tissue within a sea of cancer tissue, or vice versa. Alternatively, it is not hard to imagine a more realistic extension of model (8.22) in which these fingers grow together, engulfing islands of healthy tissue and yielding a realistic histology. As it stands, the simulations make certain predictions about the width of the tumorous fingers and the rate at which they grow as functions of parameters that may help guide empirical investigation.

8.3.1.3 Turing instabilities

A model by Markus Owen and Jonathan Sherratt [104] offers a distinctly different explanation of Type I diversity from either the incomplete-competitive exclusion or fungiform-invasion hypotheses. Their model is a spatially explicit description of macrophage-tumor interactions that includes dynamics of a chemical regulator secreted by cancer cells that both attracts and activates macrophages. Macrophages are seen as able to bind to parenchyma cells to form a parenchyma-macrophage complex. Such complexes then fall apart, yielding an intact macrophage and unmodeled debris. If we let

$$l(\mathbf{x}, t) = \text{macrophage density at spatial point } \mathbf{x} \text{ at time } t,$$
$$m(\mathbf{x}, t) = \text{cancer (parenchyma) cell density,}$$
$$n(\mathbf{x}, t) = \text{healthy cell density,}$$
$$f(\mathbf{x}, t) = \text{concentration of the chemical regulator, and}$$
$$c(\mathbf{x}, t) = \text{parenchyma cell-macrophage complex density,}$$

then the following is a nondimensional version of Owen and Sherratt's model:

$$
\begin{cases}
\dfrac{\partial l}{\partial t} = D_l \nabla^2 l - \chi_l \nabla(l \nabla f) + \dfrac{\alpha f l (N+1)}{N+l+m+n} + I(1 + \sigma f) \\
\qquad - k_1 f l m + k_2 c - \delta_l l, \\[2mm]
\dfrac{\partial m}{\partial t} = D_m \nabla^2 m + \dfrac{\xi m (N+1)}{N+l+m+n} - m - k_1 f l m, \\[2mm]
\dfrac{\partial n}{\partial t} = D_n \nabla^2 m + \dfrac{n(N+1)}{N+l+m+n} - n, \\[2mm]
\dfrac{\partial f}{\partial t} = D_f \nabla^2 f + \beta m - \delta_f f, \\[2mm]
\dfrac{\partial c}{\partial t} = D_c \nabla^2 c + k_1 f l m - k_2 c - \delta_c c.
\end{cases}
\tag{8.23}
$$

Everything in this model moves about by simple diffusion with diffusion constants D_i, $i \in \{l, m, n, f, c\}$, and macrophages tend to migrate up the chemical regulator gradient with basic motility χ_l. All cell proliferation terms have the form

$$
\frac{\psi(N+1)}{N+l+m+n},
\tag{8.24}
$$

with ψ some simple function of assorted dependent variables and N a parameter that describes sensitivity of cells to crowding. The healthy cell population reaches equilibrium whenever $l+m+n = 1$ in the absence of diffusion. Therefore, the variables are scaled such that when total cell density is unity, a sort of "carrying capacity" for healthy cells is reached. Note that in this model crowding inhibits only proliferation, not mortality. The remaining parameters include the rates at which macrophages proliferate in response to the chemical regulator (α), macrophages leave blood vessels to enter the tumor interstitium (I), blood-borne macrophages enter the tumor interstitium in response to the chemical regulator (σ), the macrophage-tumor cell complex forms (k_1) and dissociates (k_2), free macrophages disappear (δ_l), the chemical regulator is secreted by cancer cells (β), the chemical regulator decays (δ_c), and macrophage-cancer cell complexes dissociate. Finally, $\xi > 1$ measures the proliferative advantage cancer cells enjoy over healthy cells.

If one simplifies this model by turning chemotaxis off ($\chi_l = 0$), then numerical investigation reveals two interesting regimes. The first represents a smooth wave-front of cancer tissue infiltrating and completely eliminating surrounding healthy cells. In this regime, the wave speed is approximately $2[D_m(\xi - 1)]^{\frac{1}{2}}$ per time, to first order. Their parameter estimates applied to this formula indicate that a 1 mm diameter tumor would take on the order of 100 days to grow.

The second regime is dynamically more surprising and shows the potential importance of immune attack on Type I diversity within a tumor. If the chemical regulator diffuses sufficiently well (D_f is large enough), then behind the invasion front a Turing pattern of alternating regions of high and low cancer cell density develops. In one dimension, the healthy cell density becomes very ragged as it decays outward in a pattern reminiscent of actual cancerous lesions. These patterns form because local areas in which cancer cell density, and therefore chemical regulator production, is high cause a sharp chemical gradient to form. Since both the gradient and diffusion constant are large, most of the chemical regulator moves out of areas of high cancer cell density. As a result, macrophages, following the chemical regulator, tend to cluster in areas of relatively low cancer cell density. Since these areas contain very few cancer cells, and therefore produce very little chemical regulator of their own, the gradient is maintained as long as the chemical regulator decay rate is sufficiently high. These patterns were observed in both one- and two-dimensional solutions.

Allowing macrophages to migrate up the chemical regulator gradient (allowing $\chi_l > 0$) stabilizes these Turing patterns in the following sense: chemotaxis tends to increase the critical value of D_f above which these patterns form. However, chemotaxis also appears to favor even wilder behavior once D_f gets high enough. The pattern following the invasion wavefront, which before was more or less regularly repeating regions of high and low cancer cell density, can become highly irregular, exhibiting the mixed histology often characteristic of malignant neoplasia.

8.3.2 Causes of Type II diversity

As already mentioned, diversity within malignant neoplasms is not limited to differences between parenchyma and a few "healthy" cell types. Even among parenchyma, cellular pleomorphism and clonal variation are common features [83] and may predict prognosis in some cases [81], although the amount of variation itself varies among tumors and even within the same tumor over time, typically declining as the tumor ages [77]. The question we now address is, how does anatomical and physiological pleomorphism arise? At least three hypotheses exist [83]. First, as already discussed, a number of different aspects of the tumor microenvironment—nutrient concentration, hydrostatic and mechanical pressure, among other things—vary both temporally and spatially. Since cells are physiologically plastic, they can change their behavior and even form to accommodate the demands of their local environment. Therefore, pleomorphism may represent nothing more than accommodation of cells to different environmental conditions. We have already seen an example of this idea in the quiescent layer of some multicell spheroids. Also, it is well established that cancer cells change phenotype between differentiated epithelia-like cells and less differentiated mesenchyal cells [111]. This epithelial-mesenchymal transition (EMT) appears to be controlled in part by epigenetic mechanisms [123]. Cancer cells may also be phenotypically plastic with respect to their metabolic response to variation in oxygen tension [67].

A second hypothesis focuses on what evolutionary biologists would call mutation pressure. Parenchyma cells have long been known to exhibit striking genetic variation caused in part by dysfunction of their DNA-maintenance machinery [78, 91]. From this observation an explanation of cancer cell variation almost immediately follows—pleomorphism is in fact clonal variation driven by genetic polymorphism caused by the rapid accumulation of mutations among cancer cells. This hypothesis has been applied directly to explain variations in proliferation rate, invasion and metastasis potential, anaplasia (lack of differentiation), and senescence, in addition to cellular and nuclear anatomy (reviewed in [21, 22]). Although elegant, this hypothesis rarely has been used in the mathematical oncology literature (see [120] for an exception).

The third hypothesis is really an extension of this genetic polymorphism idea but adds natural selection. The history of mutations among cancer cells, while important, is still insufficient to explain the pattern of pleomorphism within any given tumor. One must also know how natural selection then sifted through the mutations to understand fully the diversity and frequency of parenchyma cell phenotypes. The role that natural selection plays depends critically on the functional nature of the pleomorphism. That is, are tumors integrated tissues, with a variety of cell types working together for their mutual benefit? Or are tumors a collection of uncooperative cell types competing for scarce resources? If the latter, then pleomorphism is a manifestation of niche segregation, and damaging or removing one cell type should have little effect on overall tumor growth. If the former, then pleomorphism is an adaptation of the tumor to the host; destruction of one subpopulation will cause disproportionate damage as the disruption of integrated function will ripple throughout the tumor.

Although the idea that natural selection acts within tumors is old [99, 103] and presented dogmatically in standard texts, the magnitude of selection's impact still demands evaluation. For example, if mutation rates are very high and environmental conditions extremely spatially and temporally variable, no consistent selection pressures will exist, thereby minimizing natural selection's role. Therefore, one should maintain natural selection and genetic polymorphism as distinct hypotheses.

For the remainder of this section we focus on the natural-selection hypothesis and ask, what traits does selection favor in the competition among parenchyma cell types? Certainly growth rate is an obvious candidate, but others have also been proposed, including efficient nutrient use and decreased dependence on oxygen. Below we review models of each of these suggestions.

8.3.2.1 Natural selection favoring proliferation rate and efficient nutrient use

As with Type 1 diversity, the earliest models of pleomorphism, by Seth Michelson et al. [86], grew from the Lotka-Volterra competition models. However, unlike Gatenby, Michelson et al. interpret the species as two different strains of cancer cells within a single tumor. In essence, the model "begins" after mutation has already created a challenger to the resident parenchyma strain. The question then is, will one strain eventually dominate or will a polymorphism result? In one variation, for example, Michelson et al. (see also Michelson and Leith [85]) allow one cell type to mutate into the other. In this model they represent the population sizes of the two cell strains as x and y and define the following model:

$$
\begin{cases}
\dfrac{dx}{dt} = r_1 x \left(1 - \dfrac{x}{K_1} - \lambda_1 y \right) - px, \\[2mm]
\dfrac{dy}{dt} = r_2 y \left(1 - \dfrac{y}{K_2} - \lambda_2 x \right) + px,
\end{cases}
\tag{8.25}
$$

with r_i and K_i the intrinsic rate of increase and carrying capacity, respectively, of cell type i; λ_i the effect of competition on strain i; and p the rate at which cell type 1 mutates into cell type 2. This model can have three fixed points: the origin, the point $(0, K_2)$, and a point in the interior representing a polymorphism. The authors use Dulac's criteria to show that no relevant limit cycles exist. The origin is never asymptotically stable if one assumes both $r_i > 0$; so, the dynamics are (almost always) well characterized. In short, if $p > r_1(1 - \lambda_1 K_2)$, then solutions always approach the boundary fixed point, representing a monomorphic y-type population. If the inequality is reversed, the population approaches a well-characterized polymorphism asymptotically. So, if the x-type population suffers either a low intrinsic reproductive rate or high mortality, then selection will favor its complete annihilation. We can also see in this model selection punishing cells that use nutrients or space inefficiently—if y-type cells are inefficient, manifested as a limited "carrying capacity" (K_2 small), then x-type cells are less likely to be completely excluded.

A more sophisticated extension to these simple competition models provided by Gatenby and Thomas Vincent [49] can be used to predict how tumor populations are likely to evolve in the face of competition for resources, in this case glucose. Consider a tumor that contains one healthy cell population and $p - 1$ subpopulations of parenchyma cell types. The number of healthy cells at time t is denoted $N_1(t)$, while $N_i(t)$, $i \in \{2, \ldots, p\}$, represents the size of the ith cancer subpopulation. Cells of all types take up and metabolize glucose, the absolute amount of which is denoted $R(t)$. Gatenby and Vincent then write the following model to represent competition for glucose within this heterogeneous tumor:

$$\begin{cases} \dfrac{\mathrm{d}N_1}{\mathrm{d}t} = \alpha_n \left(1 - \dfrac{N_1}{K_n}\right) \left(\dfrac{E_n R^2}{R_n^2 + R^2} - m_n\right) N_1, \\[2ex] \dfrac{\mathrm{d}N_i}{\mathrm{d}t} = \alpha_c \left(1 - \dfrac{S}{K_i}\right) \left(\dfrac{E_i R^2}{R_c^2 + R^2} - m_c\right) N_i, \quad i \in \{2, \ldots, p\}, \\[2ex] \dfrac{\mathrm{d}R}{\mathrm{d}t} = r - \dfrac{E_n R^2}{R_n^2 + R^2} N_1 - \sum_{i=2}^{p} \dfrac{E_i R^2}{R_c^2 + R^2} N_i, \\[2ex] r = r_e(m_n N_1 + m_c S), \\[2ex] S = \sum_{i=2}^{p} N_i, \end{cases} \qquad (8.26)$$

where α_n and α_c are intrinsic rates of increase for healthy and cancer cells (invariant across strains), respectively; K_i and E_i, $i \in \{1, \ldots, p\}$, are "carrying capacities" and maximum substrate uptake rates for all cell types, respectively; R_n and R_c measure the sensitivity of nutrient uptake to changes in nutrient concentration, and m_n and m_c represent glucose oxidized for purposes other than proliferation, which one can think of as maintenance metabolism. The function r represents glucose delivery through the blood, which increases with tumor size. Gatenby and Vincent appear to assume that microvessel density varies in proportion to glucose demand for maintenance metabolism, so they modify the basic glucose delivery rate, r_e, by the weighted average of basic glucose demand.

In this model the parameters assumed to be under selection are K and E and are considered to be random variables. They assume that normal cells' carrying capacities and basic nutrient uptake rates distribute normally around means μ_K and μ_E, with variance σ_n^2 for both distributions. Similarly, parameter values for cancer cells are normally distributed with means ν_K

and ν_E and variance σ_c for both. To determine how the population evolves, Gatenby and Vincent exploit a method involving fitness-generating functions that essentially allows them to write an expression for the fitness of all possible cell types for any given population composition. With this adaptive landscape, they can then write a differential equation for the change in population size for any strategy in any population. With such an equation one can find evolutionary equilibria, equivalent to evolutionary stable strategies [50, 84], by finding the population configuration at which the fitnesses of all cell types are zero.

The results of this model suggest that when cancer arises, glucose concentration tends to decline, because the basic metabolic rate of the tissue (tumor) increases. Because glucose becomes scarce, natural selection favors cells that can sequester and metabolize glucose efficiently (maximize both K and E). So over time the tumor is able to maintain its proliferation rate in the face of fierce competition for glucose. In addition, this model is among the first to reproduce the observed decline in tumor pleomorphism as tumors progress.

Further support for the hypothesis that selection favors efficient nutrient use comes from a model by Yang Kuang et al. [69]. In this model we imagine a tumor growing in an organ with mass $x(t)$. The tumor contains two different parenchyma cell types with masses $y_1(t)$ and $y_2(t)$. Nutrient is delivered to the cells through a dynamic vascular network with a total mass of vascular endothelial cells (VECs) of $z(t)$. Total tumor phosphorus is denoted by P and is partitioned into five different compartments: the interstitial fluid, healthy cells, cells of the first parenchyma type, cells of the second parenchyma type, and VECs. Each unit mass of healthy cells, including VECs, contains n units of phosphorus, while parenchyma cells of types 1 and 2 hold m_1 and m_2 units, respectively. Therefore, if we denote extracellular phosphorus as P_e, then $P_e = P - [n(x + z) + m_1 y_1 + m_2 y_2]$. With this notation, Kuang et al. suggest the following model (see Section 6.6.1):

$$
\begin{cases}
\dfrac{dx}{dt} = x \left(a \min\left[1, \dfrac{P_e}{fnk_h}\right] - d_x - (a - d_x)\dfrac{x + y_1 + y_2 + z}{k_h} \right), \\[2ex]
\dfrac{dy_1}{dt} = y_1 \left(b_1 \min\left(1, \dfrac{\beta_1 P_e}{fm_1 k_h}\right) \min(1, L) - d_1 - (b_1 - d_1)\dfrac{y_1 + y_2 + z}{k_t} \right), \\[2ex]
\dfrac{dy_2}{dt} = y_2 \left(b_2 \min\left(1, \dfrac{\beta_2 P_e}{fm_2 k_h}\right) \min(1, L) - d_2 - (b_2 - d_2)\dfrac{y_1 + y_2 + z}{k_t} \right), \\[2ex]
\dfrac{dz}{dt} = c \min\left(1, \dfrac{P_e}{fnk_h}\right) (y_1(t - \tau) + y_2(t - \tau)) - d_z z,
\end{cases}
$$

$$(8.27)$$

where $L = g(z - \alpha(y_1 + y_2))(y_1 + y_2)$, a, b_1, b_2, and c represent intrinsic rates of increase of all cell types; all terms d_i are basic death rates; k_h and k_t represent limiting sizes for the healthy organ and tumor, respectively; f is the intracellular fluid fraction; and L is a measure of vascular supply. In

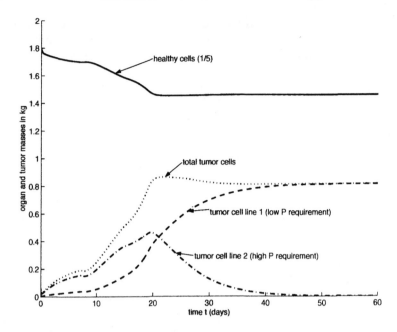

FIGURE 8.3: A numerical solution to model (8.27) with $a = 3$, $b_1 = 6$, $b_2 = 6.6$, $c = 0.05$, $d_x = 1$, $d_z = 0.2$, $d_1 = 1$, $d_2 = 1$, $f = 0.6667$, $g = 100$, $k_h = 10$, $k_t = 3$, $m_1 = 20$, $m_2 = 22$, $n = 10$, $P_e = 150$, $\alpha = 0.05$, $\beta_1 = 1$, $\beta_2 = 1$, $\tau = 7$, and $[x(0), y_1(0), y_2(0), z(0)] = [9, 0.01, 0.05, 0.001]$. This example shows that selection appears to favor neither cell type until phosphorus becomes limiting.

particular, α represents the mass of tumor cells one unit of blood vessels can just barely maintain, and g measures sensitivity of tumor tissue to lack of blood. Also, Kuang et al. assume that tumor tissue starved for blood releases an angiogenic signal. This signal is distilled by VECs from a complex mix of pro- and antiangiogenic chemical growth factors released by all cells in the tumor. Upon receipt of the signal, VECs respond by reproducing, moving toward the blood-starved region and forming new microvessels. Kuang et al. further assume that the VEC response is delayed by τ time units, representing the time needed for cells to transduce and respond to the chemical signal and complete their reproductive, motility, and differentiation programs. Finally, parameters β_i represent the effect of a drug able to modulate, generally inhibit, cell type i's ability to sequester phosphorus from the interstitium.

Model (8.27) admits two possible limiting factors: blood supply and phosphorus. However, simulations with reasonably realistic parameter values suggest that phosphorus is the key limiting factor, determining both growth rate and final tumor mass. In this case, then, what type of cell does natural selection favor—cells with a high or low phosphorus requirement (high or low

m_i)? The nutrient-use efficiency hypothesis suggests that selection will favor the most efficient type; that is, the type that minimizes m_i. However, the reality is complicated by the fact that phosphorus use relates to growth rate [38, 69, 118] in the following way. Cells require phosphorus primarily for nucleic acid, and rapidly proliferating cancer cells must synthesize large amounts of nucleic acids, primarily in the form of ribosomes [118], to build proteins needed for cell division. In fact the number of ribosomes in cancer cells appears to correlate with cancer aggressiveness (reviewed in [38]). Therefore, selection for the aggressive cell type can work directly against selection for efficient nutrient use.

Model (8.27) suggests that selection's choice between nutrient-use efficiency and aggressive proliferation depends on the state of the tumor. In particular, when tumors are small and well supplied with phosphorus and blood, the more aggressively proliferating cell type may have the advantage, growing faster than its less aggressive competitor. However, as the tumor approaches its asymptotic limit, competition for phosphorus increases as it becomes limiting. Then selection changes favor and gives the advantage to the more efficient type, which eventually drives the aggressive phenotype to extinction (Fig. 8.3). Therefore, this model predicts that as tumors age they become less aggressive and more miserly with nutrients.

8.3.2.2 Natural selection favoring insensitivity to hypoxia

As already discussed (Section 8.2.5), malignant tumors and their peritumoral environments tend to be relatively acidic, probably because tumor cells favor glycolysis and fermentation over the tricarboxilic acid cycle. Gatenby and his colleagues [47, 49, 114] suggest that natural selection provides the answer as to why. Carcinomas by definition begin within epithelial tissue. This tissue is defined by the presence of a basement membrane upon which the epithelial cells live. Usually the vasculature servicing such tissue lies on the opposite side of the basement membrane; therefore, premalignant carcinoma precursors, which by definition cannot penetrate the basement membrane, are constrained to expand away from the blood supply. (Such geometry raises doubts about the validity of tumor cell spheroids as models of nascent carcinoma.) In such a situation, cells able to produce ATP under hypoxic conditions will then be favored. Since these geometrical constraints apply to essentially all carcinomas, selection favoring glycolytic oxidation of glucose will be nearly ubiquitous [47, 49, 114].

Two recent models [43, 125] connect this hypothesis with molecular biology of cancer cells through the tumor suppressor gene *p53*, whose product, p53, the editors of *Science* declared "molecule of the year" in 1993. Among its many demonstrated functions, *p53*'s product activates the apoptosis mechanism in "stressed" cells. One form of stress to which p53 appears to respond is hypoxia. Evidence for this conclusion comes from studies of *p53*-deficient cells in culture, which commit apoptosis less frequently than intact wild-type cells in hypoxic environments [51, 116]. This observation is profoundly significant to cancer biologists, because *p53* is widely regarded as the most commonly disrupted gene among cancers as a whole, mutated in over 50% of all malignant neoplasms, and many cancerous tumors suffer regions of local hypoxia (see Section 8.2.6). If selection frequently favors cancer cells able to withstand hypoxia, as Gatenby and colleagues have suggested, perhaps *p53* disruption is a common mechanism, along with other metabolic changes, by which cells acquire the favored trait. Empirical support for this interpretation comes from observations of cell populations evolving a dysfunctional *p53* gene when exposed to hypoxic environments [92].

Selection for dysfunctional *p53* was studied quantitatively by David Gammack et al. [43]. Building on the earlier model of Kevin Thompson and Janice Royds [125], Gammack et al. model the dynamics of three quantities: the number of tumor cells with the wild-type (normal) *p53* gene ($N(t)$), the number of tumor cells with the mutated *p53* gene ($M(t)$), and molecular oxygen concentration ($C(t)$). Tumor cells of both types consume oxygen, proliferate, and die at rates dependent on O_2 concentration. Oxygen is supplied to cells in one of two ways, depending on whether the model represents cells in culture in some virtual experiment or an in vivo tumor. In the in vitro situation, O_2 is supplied exogenously in the media. We imagine the researchers of this virtual experiment varying the O_2 concentration in the following way: for a period of length δ, oxygen is maintained at physiologically normal levels (normoxia); then a period of hypoxia that lasts for τ time units follows. We imagine that the researchers control δ and τ and repeat the procedure some number of times. In the model of an in vivo tumor, cells can also be exposed to repeated rounds of normoxia and hypoxia but only if blood vessels collapse from internal pressures (see Section 8.2.6). This occurs whenever total cell numbers reach a prescribed threshold, N^*. Once $N + M = N^*$, hypoxia begins for a fixed period of time, τ, representing the time required for angiogenesis to reconstruct a sufficient vascular infrastructure.

From these assumptions, Gammack et al. build the following model:

$$
\begin{cases}
\dfrac{dN}{dt} = \dfrac{A_N C^p}{C_{N_1}^p + C^p} N - B_N \left(1 - \dfrac{\sigma_N C^q}{C_{N_2}^q + C^q}\right) N(N + M), \\[3mm]
\dfrac{dM}{dt} = \dfrac{A_M C^p}{C_{M_1}^p + C^p} M - B_M \left(1 - \dfrac{\sigma_M C^q}{C_{M_2}^q + C^q}\right) M(N + M), \\[3mm]
\dfrac{dC}{dt} = \lambda C_{ex}(t) - \Gamma_N \dfrac{A_N C^p}{C_{N_1}^p + C^p} N - \Gamma_M \dfrac{A_M C^p}{C_{M_1}^p + C^p} M - \Gamma_C C.
\end{cases}
\qquad (8.28)
$$

Parameters A_i and B_i represent maximum proliferation and mortality rates of cell type i, respectively; $C_{N_1}^p$ and $C_{M_1}^p$ measure how sensitive each cell type's reproductive response is to changes in O_2 concentration. Similarly, $C_{N_2}^p$ and $C_{M_2}^p$ measure how sensitive mortality rates are to changes in O_2 concentration. Both p and q are free parameters with no obvious physiological meaning beyond making the response to changes in oxygen tension more or less switch-like. Parameters σ_N and σ_M can be interpreted as a measure of basal mortality that occurs even in a perfect environment; that is, as the O_2 concentration gets large, mortality asymptotes at $B_i(1 - \sigma_i)$. However, actual estimates of σ from cell culture data reviewed by Gammack et al. put it very close to unity for both wild-type and *p53*-deficient cell lines, indicating that mortality in these assays was negligible when O_2 concentration was high. The function $\lambda C_{ex}(t)$ represents the rate at which O_2 is supplied to the system as described in the previous paragraph. Oxygen is consumed by cells at base rates Γ_i for cell type $i \in \{N, M\}$, and total O_2 consumption depends on the growth rate of each cell type. Finally, O_2 diffuses out of the system or is consumed by other processes at linear rate Γ_C.

Data from Thompson and Royds [125] indicate that wild-type and *p53*-deficient cell lines differ mostly in basic mortality rates and sensitivity to O_2. Roughly speaking, wild-type cells have a larger basic mortality rate ($B_N > B_M$) and suffer more from hypoxia ($C_{N_1} > C_{M_1}$ and $C_{N_2} > C_{M_2}$). Not surprisingly, *p53*-deficient cells tend to outcompete wild-type cells in the virtual experiment. In one run, for example, with an initial cell culture in which wild-type cells outnumbered *p53*-deficient mutants by orders of magnitude and periods of normoxia were only slightly longer than hypoxic periods, mutants became the dominant cell type between the fourth and fifth hypoxic episode. Of course, the length of time it takes for mutants to become dominant depends strongly on how long hypoxic and normoxic periods last. In general, Gammack et al. found that higher oxygen availability—either from longer periods of normoxia or higher O_2 concentrations during normoxic episodes— favors mutant invasion. In contrast to the nutrient-use efficiency hypothesis presented in Section 8.3.2.1, changes in the rate at which mutant cells consume oxygen, Γ_M, had very little effect; however, increasing oxygen consumption by mutant cells very slightly decreased the invasion rate, as predicted by the nutrient-use hypothesis.

Results of the in vivo case were similar. Once again, under the estimated parameters, *p53*-deficient mutants tended to invade tumors in which they were initially rare, whether solutions permitted oscillations or not. Oscillatory solutions like those in the in vitro case arose when the oxygen concentration during normoxic episodes was sufficiently high, and the rate at which mutants consumed O_2 was sufficiently low. Once again, variations in Γ_M hardly affected the mutant invasion rate, which was once again driven primarily by the duration of the hypoxic episode.

8.3.2.3 Natural selection favoring fungiform invasion

As noted in Section 8.3.1.2, malignant tumors characteristically invade surrounding tissues. To invade, carcinomas—malignant tumors originating in epithelial tissue—must degrade and rupture the epithelial basement membrane. This requires the tumor to elaborate or otherwise activate enzymes that digest connective tissue proteins in the matrix, referred to generally as matrix degrading enzymes (MDEs). There is an important side effect of matrix degradation by MDEs. As the matrix degrades, a variety of signaling molecules are solubilized that activate migration, proliferation and angiogenesis programs, among others, in cancer cells. This process often creates a kind of "positive feedback," in which an active, invasive tumor releases signals from its surrounding that make the tumor more active and invasive. Tumors differ in their tendency to invade, and some of this variation may be explained by the fact that tumor cells vary in their ability to express, either directly or indirectly, matrix degrading enzymes. The question we address here is, how does natural selection act on matrix degradation ability in carcinoma, and how does evolution of this trait affect tumor growth and morphology?

This question was addressed by Anderson and colleagues [8, 10] using a hybrid model comprising a system of coupled partial differential equations linked to a stochastic process. The PDEs describe spatial dynamics of matrix degrading enzyme concentration $(m(\mathbf{x}, t))$, density of extracellular matrix $(f(\mathbf{x}, t))$ and concentration of oxygen $(c(\mathbf{x}, t))$ at spatial point $\mathbf{x} \in \mathbb{R}^2$ and time t. The stochastic process is an individual-based model in which individual cells move, proliferate and die within the concentration fields governed by the PDEs. The PDE model takes the following form:

$$\frac{\partial m}{\partial t} = D_m \nabla^2 m + \mu N_{i.j} - \lambda m, \tag{8.29a}$$

$$\frac{\partial f}{\partial t} = -\delta m f, \tag{8.29b}$$

$$\frac{\partial c}{\partial t} = D_c \nabla^2 c - \alpha c - \gamma N_{i,j} c + \beta f. \tag{8.29c}$$

In this model, both matrix degrading enzymes and oxygen diffuse with (constant) diffusion coefficients D_m and D_c, respectively. This diffusion occurs in a continuous space that overlays a 2-D lattice of discrete positions that may contain a number of cancer cells (including 0). The number of individual cancer cells in lattice position (i, j) is $N_{i,j}$. Anderson et al. assume that matrix degrading enzymes are elaborated from individual cells at essentially the same constant rate (in this instance), namely $\mu > 0$. MDEs themselves degrade naturally with first-order kinetics at per-unit rate λ. Matrix degradation is assumed to be a mass-action process with rate δ. Oxygen is absorbed by various entities in the model domain at rates proportional to its concentration: individual cells absorb oxygen at rate γ, while other elements, which are not dynamic in this model, absorb O_2 at rate α. Finally, oxygen is released by blood vessels within the extracellular matrix. As a first approximation Anderson et al. assume that vessel density is proportional to ECM density; therefore, oxygen is released at per-unit rate β.

The rules and simulation procedures for the individual-based stochastic process are outside the scope of this book. However, three aspects of this part of the model are key. First, individual cells can move, and they move both by Fickian (random) diffusion and in a biased way up the ECM density gradient (haptotaxis). Therefore, cells tend to congregate in the most dense portions of the matrix, which of course increases the rate at which the matrix is degraded. Second, cell proliferation and death are both oxygen dependent. Finally, individual cells can vary in their motility (both random and biased), proliferation probabilities, O_2 consumption rates, and rates at which they elaborate degrading enzymes. In the model, trait variations arise via mutation, and Anderson et al. explore evolution of these traits in a variety of mutational settings. The model predicts that tumor edge morphology—the size and depth of the invasive "fingers"—depends critically on the initial matrix density distribution, but natural selection acting on the traits listed above tends to promote a more invasive morphology. Therefore, this model links natural selection to one of the key characteristics of malignancy.

8.4 Synthesis: Competition, natural selection and necrosis

Here we address the niche segregation hypothesis proposed to explain clonal diversity in tumors. A game theory model by Lars Bach et al. [13] represents an early foray in this direction. Their model, based on earlier work by Ian Tomlinson and Walter Bodmer [128, 127], assumes that two different strains of parenchyma cells exist within the tumor. One strain, denoted by A+, secretes some chemical that is beneficial to the cells in its neighborhood—say, an angiogenesis signal. The other strain, A−, does not. Otherwise, all cells are identical. In Bach et al.'s model, neighborhoods consist of three cells. It is assumed that, if only one cell in the neighborhood secretes the chemical, then the group gains no benefit because the chemical concentration remains too small to elicit the effect. However, if at least two of the three cells secrete the chemical, then all three enjoy an increased reproductive output of j units above normal. On the down side, there is a cost associated with making the chemical; those that do suffer a deduction of i to their expected reproductive output. These considerations lead to the evolutionary payoff matrix shown in Table 8.1.

A discrete-time dynamical system model of these payoffs shows that selection can allow coexistence of both strains under certain circumstances. However, niche segregation cannot explain this behavior, for the following reason. If a tumor starts with "cooperators" that secrete the angiogenesis factor and is later invaded by a mutant "defector" strain that does not, then both defector and cooperator populations can persist. However, if the roles of resident and potential invader are reversed—that is, a population of defectors is challenged by cooperators—the cooperators die out, leaving a monomorphic

TABLE 8.1: Evolutionary payoff matrix for the model by Bach et al. [13]. Strain A+ secretes an angiogenesis factor, whereas strain A− does not. Adapted from [13].

Neighborhood	A+	A−
A+, A−	$1 - i - j$	$1 + j$
A+, A−	$1 - i - j$	1
A−, A−	$1 - i$	1

defector population. One can interpret these results biologically as follows: defectors can invade a tumor full of cooperators as a sort of parasite living off of the cooperator's ability to bring in resources; however, cooperators are always hurt by the presence of defectors. Therefore, polymorphism can be explained as parasitism instead of niche segregation.

This conclusion was corroborated in a series of studies using dynamical models that are biologically analogous to the Back et al. model but formally very different [20, 95, 96, 97, 98]. The simplest of these models assumes that the tumor contains two competing clones that differ in their abilities to secrete angiogenesis signals. Unlike Bach et al.'s model, this one allows dynamic population sizes and a broader array of potential differences among cell types. Similar in construction to models by Zvia Agur, Levon Arakelyan and their colleagues [2, 11], this model tracks masses of the two clones, denoted by $x_1(t)$ and $x_2(t)$. In addition, the model also follows the total mass of immature VECs from which mature microvessels are made, $y(t)$, and the total length of microvessels within the tumor, $z(t)$. With this notation, the model becomes

$$\begin{cases} \dfrac{dx_1}{dt} = \Phi_1(v)x_1, \\[2mm] \dfrac{dx_2}{dt} = \Phi_2(v)x_2, \\[2mm] \dfrac{dy}{dt} = (\alpha H(x_1, x_2, z) - \beta)\, y, \\[2mm] \dfrac{dz}{dt} = \gamma y - \delta v z, \end{cases} \qquad (8.30)$$

where $v(t) = z/(x_1 + x_2)$, $H(x_1, x_2, z) = (x_1 h_1(v) + x_2 h_2(v))/(x_1 + x_2)$, Φ_i represents per-capita growth functions for both types of parenchyma, α is the rate at which immature VECs convert the angiogenesis signal into growth, H is the mean angiogenesis secretion rate within the tumor, h_i is the per-capita angiogenesis secretion rate for cell type i, β expresses both the VEC death and maturation rates, γ represents the rate at which maturing VECs convert themselves into new blood vessels, and δ is the base rate at which blood vessels are broken down during remodeling. In numerical analysis, the functions Φ take a form similar to that of model (8.28) without the crowding term; that is,

$$\Phi_i(v) = \frac{A_i C(v)^{p_i}}{\hat{c}_{i1}^{p_i} + C(v)^{p_i}} - B_i \left(1 - \frac{\sigma_i C(v)^{q_i}}{\hat{c}_{i2}^{q_i} + C(v)^{q_i}}\right), \qquad (8.31)$$

where $C(v)$ is the oxygen pressure and all other parameters equate to their analogues in model (8.28). The cell type-specific angiogenesis secretion rate obeys a function like the following:

$$h_i(v) = r_i C(v) e^{-\xi_i C(v)}, \qquad (8.32)$$

where r_i measures a type i cell's commitment to producing the angiogenesis signal and ξ_i expresses how sensitive this commitment is to changes in local oxygen pressure.

This model supports the hypothesis that natural selection favors aggressively proliferating cell types, at least early in tumor growth, but with a twist. If clones differ only in their basic proliferation rates such that $A_1 > A_2$, strain 1 always ends up dominating the tumor regardless of initial conditions. In fact, one can show that for a broad array of forms for the growth functions Φ, aggressively proliferating cell types are always favored. However, a tumor invaded by a more aggressive strain may end up *clinically* less aggressive. This paradox arises when the invading, aggressive cell type is sufficiently inept at producing the angiogenesis signal. Natural selection is blind to angiogenesis secretion and so always favors the aggressive invader. But because the favored clone cannot entice new blood vessels to grow very well, the tumor eventually ends up with a lower microvessel density and is therefore relatively hypoxic. In certain circumstances, this hypoxia can become so profound that the tumor regresses. One can describe such circumstances as a hypertumor—one tumor invading and destroying part of an existing tumor. And once again we see competition resulting in something akin to parasitism.

This hypertumor phenomenon hands us yet another hypothesis explaining necrosis. Superficially, one can view it as a variant of the nutrient-limitation hypothesis. However, certain predictions distinguish it from the other ideas. In particular, one will recognize a hypertumor not just as regions of nutrient deficiency but as regions of nutrient deficiency that always correlate with invading cells displaying cytological or genetic features of aggressive proliferation.

8.5   Necrosis and the evolutionary dynamics of metastatic disease

An advanced tumor is a harsh environment, as evidenced by the necrosis we have been studying in this chapter. Whatever its cause, necrosis represents ecological collapse within the tumor. In Nature, organisms living in failing ecosystems often disperse, seemingly in search of more permissive living spaces. By analogy, one might guess that metastasis is an evolutionary strategy by which cancer cells escape a failing tumor ecosystem. Support for this hypothesis comes from the observation that metastatic ability is an acquired trait—healthy epithelial cells do not normally circulate—and in general metastatic potential correlates with tumor stage and therefore likelihood of ecosystem derangement.

However, while metastasis might allow the occasional mutant cell to avoid ecological challenges in the primary tumor, this alone cannot explain why tumors tend toward greater metastatic potential with age. Selection in the primary tumor occurs exclusively within the environment of the primary tumor, unless metastatic colonies exert some unknown feedback to the tumors that spawned them. At the moment, however, it appears that the success or failure of distant metastatic seeds does not directly affect the fitness of cells confined to the primary tumor. Selection pressures within the primary lesion alone apparently favor the metastatic phenotype.

8.5.1   Pre-metastatic selection hypothesis

Thalhauser et al. [124] proposed the "pre-metastatic selection hypothesis" to address this issue. The hypothesis posits that vascular collapse within the primary tumor selects *directly* for cells with a migratory phenotype. The hypothesis was generated by a model of a collection of tumor cords (Fig. 8.2). As before (Section 8.2.2), tumor density is assumed to be both radially and axially symmetric around the central vessel, allowing the three-dimensional problem to be reduced to one dimension. The vessel wall is the innermost point of the domain at radius R_{in}, and the cord extends to radius R_{out}. Oxygen is supplied at the vessel wall and diffuses into the surrounding tissue where it is consumed by cancer cells.

The key to Thalhauser's model is the assumption that cellular migration and proliferation are separate and competing processes, following the "go-or-grow" hypothesis derived from observations of glioblastoma multiforma (GBM) tumors. A high cellular density causes proliferating cells to take on a migratory phenotype. Motile cells transition to the proliferating class when the cellular density is sufficiently low. Proliferation is governed by the local oxygen tension and cell density, and cell death occurs where oxygen tension is low. The model variables are $G(r,t)$, $M(r,t)$, and $O(r,t)$, representing the density of proliferating cells, migrating cells, and oxygen, respectively, at time

t and radial distance r. The complete model is the following:

$$\frac{\partial G}{\partial t} = A\frac{O^p}{C_1^p + O^p}\left(1 - \frac{T}{T_{max}}\right)G - B\frac{C_2^q}{C_2^q + O^q}G - \lambda_{G,M}G + \lambda_{M,G}M,$$
(8.33)

$$\frac{\partial M}{\partial t} = D_T\nabla\cdot(M\nabla M) - B\frac{C_2^q}{C_2^q + O^q}M + \lambda_{G,M}G - \lambda_{M,G}M,$$
(8.34)

$$\frac{\partial O}{\partial t} = D_{O_2}\nabla^2 O - \gamma A\frac{O^p}{C_1^p + O^p},$$
(8.35)

where

$$\lambda_{G,M} = \epsilon k\frac{T^2}{T^2 + K_M^2},$$

$$\lambda_{M,G} = k\frac{K_G^2}{T^2 + K_G^2},$$

$$T = M + G.$$

Oxygen concentration at the vessel wall is imposed by a Dirichlet boundary condition, $O(R_{in}, t) = O_0$, and all other boundary conditions are no-flux. Consistent with the go-or-grow notion, there is assumed to be a continuum of phenotypes, with "aggressive grower" at one extreme and "aggressive mover" at the other. The location of a clone on this continuum is determined by the K_G/K_M ratio and the parameter ϵ; the larger the ratio and (or) smaller the ϵ, the farther out the phenotype is in the "aggressive grower" direction of the spectrum.

In this model, aggressive movers form a tumor cord that is less dense with fewer cells at the vessel wall, while aggressive growers form dense tumor cords. A high cellular density exerts compressive force on the vessel wall, which may cause local vascular collapse, ischemia, and necrosis. Therefore, somewhat paradoxically, prevention of ischemia may select for motile phenotypes and higher metastatic potential.

Thalhauser et al. also connected this result to Nagy's [95] theoretical notion of a hypertumor (see Section 8.4). Recall that a hypertumor occurs in a vascular tumor when a clone with inferior angiogenic potential but superior growth potential invades an existing tumor. The angiogenic clones are out-competed, causing a type of "evolutionary suicide" in the non-spatial ODE version of the model. The tumor ecosystem completely collapses as the nonangiogenic clone comes to dominate the tumor. In Thalhauser's tumor cord model, aggressive growers are the counterpart to the hypertumor. It is plausible that "local hypertumors" may be responsible for the regions of necrosis frequently observed in advanced tumors.

Thalhauser et al.'s model addresses the selective forces at work in the primary tumor. However, it is also probable that metastatic cells continue to evolve once they have colonized distant sites. Here, the Halstedian curtain wall model of metastatic spread may have some utility. We can imagine that a spontaneously arising mutant in the primary tumor is capable of colonizing, say, a lymph node. Thus, the first curtain wall has been colonized. Now that a basic metastatic capability has been acquired, another mutant may arise in this colony with the ability to spread further. Thus, mutants could drive the continued distal expansion of disease, where further metastatic capability is acquired at each subsequent site. We suggest the development of a formal model for such a process as a project for the interested student.

Such evolutionary dynamics likely interact with the physical dynamics of tumor cell dissemination within the vascular and lymphatic systems to determine the overall pattern of spread in a clinical cancer.

8.5.2 Reproductive fitness and export probability

A recent series of studies by Michor, Iwasa, Dingli, Haeno, and colleagues [36, 58, 87, 88] have examined the interaction between mutation to a metastatic phenotype within the primary tumor, the mutant's reproductive fitness within the tumor, and mutant's export probability from the primary tumor. These models are generally cast as stochastic birth-death processes, and we briefly present Dingli et al.'s [36] model as an example.

Consider a quasi-steady state tumor with a total of N cells, M of which are mutants with metastatic capability and relative fitness r within the primary tumor. The model is a discrete, stochastic model. Cell divisions occur at some rate, λ, and with each replication a mutation occurs with probability μ. The probability that an existing mutant replicates depends on its relative fitness, r, and is given as:

$$p_M = \frac{rM}{rM + N - M}. \tag{8.36}$$

Note that if $r = 1$, then $p_M = M/N$. The probability that a wild-type cell replicates is $p_N = 1 - p_M$. Following replication of a mutant or a *de novo* mutation, the daughter cell metastasizes with probability q. If, after all of this, the cell population is greater than N, mortality is assumed to reduce the population back to N cells. While Dingli et al. studied this system in the context of agent-based stochastic simulations, the model can be recast as a system of ODEs (which yields similar numerical results) as follows:

$$\frac{dM}{dt} = \lambda p_M(1 - q)N + \mu\lambda p_N N - \lambda\frac{M}{N}(1 - q)p_M N - \lambda\frac{M}{N}p_N N, \tag{8.37}$$

$$\frac{dE}{dt} = \lambda Nq(p_M + \mu p_N), \tag{8.38}$$

where $E(t)$ is the total number of metastatic cells exported. If $r = 0$ (mutations are neutral) and the probability of metastasis is nonzero ($q > 0$), then this model predicts that mutants will make up only a very small fraction of the tumor and the metastatic rate will be low at equilibrium. However, when mutants have a high relative fitness, the export rate and survival advantage within the primary tumor compete. That is, a high export probability depletes the tumor of mutants and ultimately lowers metastasis rate. Somewhat surprisingly, the tumor fraction of the tumor composed of mutants at steady-state switches quite suddenly from 1 to nearly 0 as q increases.

Michor et al. [87] considered a very similar model, but without the explicit removal of mutant cells from the primary tumor by metastasis. This work concluded that the probability of establishing distant metastases is negligible for mutants with $r \leq 1$ unless the export rate, q, is very high, suggesting that most mutations that confer metastatic ability also increase the mutant's fitness within the primary tumor. Michor et al. reached essentially the same conclusion in a second paper [88], where two mutations were required to confer metastatic ability. This result relates to Thalhauser et al.'s [124] prediction that increased motility, which may predispose to to metastasis, confers a survival advantage within the primary tumor.

8.5.3 Tumor self-seeding

Finally, we briefly discuss recent experimental results on tumor re-seeding by metastatic cells. While metastasis has been traditionally viewed, in both the modeling and basic biology literature, as a unidirectional process where cells spread strictly from the primary tumor to distant sites, cells from metastatic colonies may also return to, or "self-seed," the original tumor site. Kim et al. [68] recently showed that tumor self-seeding by aggressive circulating tumor cells occurs in a variety of experimental cancer systems. Moreover, they showed that the most aggressive circulating cells preferentially reinvade the primary tumor mass. These aggressive cells were capable of increasing primary tumor growth rate by releasing cytokines that influence the tumor stroma and foster angiogenesis and invasion.

We now have a dynamical picture where the primary tumor continually sheds cells into the circulation. The primary tumor itself is fertile soil for such potentially metastatic cells, as they are already adapted to the tumor microenvironment, and the leaky tumor vasculature is a weak barrier to extravasation. The most aggressive and metastatically capable cells colonize the primary tumor site, increasing its growth rate and inducing further angiogenesis. With increasing vascularization, more cells are shed into the circulation and ever more return to the tumor. The tumor comes to be dominated by clones with increasing metastatic capability, eventually leading to distant metastasis. Distant metastatic colonies themselves may shed cells into the circulation that seed the primary tumor, further influencing its evolution.

However, the interaction between circulating tumor cells and the primary tumor is likely not so straightforward. Kim et al. [68] also speculate that a large primary tumor may filter many aggressive cells from the circulation, perhaps decreasing the overall metastatic potential of the tumor. The interaction between the vasculature, circulating cells, and primary cells could be explored in a formal setting, and we suggest it as a potential project.

8.6   Conclusion

Perhaps the most obvious conclusion one can draw from this review is that existing mathematical theory provides significant insight into the causes of necrosis, overall tumor diversity (coexistence of different developmental lineages within the tumor), clonal diversity (diversity within the malignant lineage) and metastasis. Mathematical oncology provides a much richer theory than that underlying explanations of these phenomena in standard pathology texts or even the massive compendium edited by Vincent DeVita et al. [35]. In essence, these more clinical sources either treat them as largely explained or ignore them altogether. But as insights from existing theory make clear, these phenomena are far from adequately explained.

Unfortunately most of these insights have gone largely unexploited by the empirical cancer biology community. To see this in a very profound way, compare the literature-cited sections of the basic cancer biology chapters in [35] and the historical review of mathematical oncology in [12]. The former is a compendium of modern cancer biology from a pathologist's or clinician's standpoint, and the latter is an outstanding review of the major themes in mathematical oncology. Despite their focus on the same disease, these two sources share very few citations, especially theory papers. Traditionally, each group tends to be unaware of the other's literature, although there are some notable exceptions.

Trying to establish why mathematical oncology has influenced the work of experimentalists and medical doctors so lightly is dangerous business, but ignoring it is even more dangerous. One possibility, perhaps the most obvious, may be that the two groups by and large cannot communicate because they speak different technical languages, and translators are rare. Hopefully this situation is changing as more students seek training in both advanced molecular biology and mathematics—the very students to whom this text is addressed. Nevertheless, despite its obvious appeal, this explanation cannot be the only reason for the disconnect, because although rare, very talented translators have always existed between the empirical and theoretical communities.

Another contributing factor to the lack of knowledge transfer may also be the wildly different research focuses of the two groups. As discussed in the introduction, empirical cancer biologists tend to focus on the molecular biology of cancer cells, as is obvious from a casual inspection of any empirical cancer journal from the last four decades. Mathematical oncologists, on the other hand, tend to focus on aspects of tumor ecology like those reviewed here, plus immune predation, the effects of cytotoxic chemotherapy and radiation therapy on competition among tumor clones, and other aspects of tumor ecology. So, the outlook, interests, and research tools characteristic of experimentalists and theoreticians have traditionally differed so much that there has been very little overlap in research programs. Again, a look through DeVita et al. [35] makes clear that the practice of oncology would hardly be changed if no one had ever written a mathematical model of cancer.

The question then becomes, where, if anywhere, will empirical and mathematical oncology meet? In particular, how best can mathematical oncology serve experimentalists by helping direct their work in the lab and clinic? Certainly the insights gained so far by mathematical oncology should not be abandoned, and work on specific systems, especially drug trials on cell cultures and angiogenesis inhibition, along with other recent collaborations between empiricists and theoreticians, are increasingly building momentum. However, one major question of growing importance appears to be a perfect place where the interests and tools of molecular biologists, experimental cell biologists, and mathematical biologists meet. That question is, what precisely is the role of the tumor microenvironment—including the malignant lineage(s), ECM and stromal cells, vascular, immune and peritumoral cells—in generating, promoting and maintaining malignancy? This question obviously involves molecular and cellular biology. Genes for growth factors like the various forms of VEGF, their receptors, such as flt-1, matrix metalloproteinases, and a host of other molecules, choreograph all interactions among all cell types and even nonliving elements within malignant and premalignant tumors. But these interactions are primarily ecological in nature, ultimately determining the outcome of competition, cooperation, and predation. Mathematical oncologists can attack this problem with extensions of formalisms already in place, like multicell spheroids with mixtures of cell types currently under investigation by empiricists [91]. More important, mathematical oncologists must move beyond the MCS and build formalisms representative of more realistic geometries encountered in carcinoma, especially when modeling nascent tumors. But, no matter how one attacks the problems posed by malignant neoplasia, the time has arrived to begin melding the molecular and evolutionary ecology approaches to cancer biology, and the mathematical oncology community has, above all others, the skill set to do it.

8.7 Exercises

Exercise 8.1: For the angiogenesis model (8.30), assume that $x_2(t) \equiv 0$, $x_1(0) > 0$. Further, suppose $\Phi_1(v) \in C^1$, is monotonically increasing with $\Phi_1(0) < 0$ but

$\lim_{v \to \infty} \Phi_1(v) > 0$. Also, suppose that $h_1(v) \in C^1$, is unimodal (monotonically increasing to the left and monotonically decreasing to the right of the mode), and that $h_1(0) = \lim_{v \to \infty} = 0$. Finally, let $w = y/x$ and $v = z/x$. Assume that $x(t) > 0$ for all t.

1. Interpret in words the biological meanings of the variables w and v.

2. Show that model (8.30) can be written in the form:

$$w = f_1(w, v), \qquad\qquad (8.39a)$$
$$v = f_2(w, v). \qquad\qquad (8.39b)$$

3. Show that there are two "boundary" fixed points of model (8.39): the origin (i.e., the point $(0, 0)$) and the point $(\tilde{v}, 0)$, where $\tilde{v}$ satisfies

$$\Phi_1(\tilde{v}) + \delta\tilde{v} = 0.$$

4. Show that, if there is an "interior" equilibrium (one in which $w, v > 0$), then $v = \hat{v}$, where $\hat{v}$ satisfies

$$\Phi_1(\hat{v}) = \Psi(\hat{v}), \quad \Psi(v) = \alpha h_1(v) - \beta.$$

5. Analyze the local stability of these interior fixed points, assuming any exist.

Exercise 8.2: As an advanced project, develop the model suggested in Section 8.5.1. That is, develop and analyze as far as you can a model of subsequent colonizations of successive "Halstedean curtain walls" and selective forces driving metastasis in each. What predictions does your model make about how metastatic aggressiveness will change as disease burden increases over time in a single patient?

References

[1] Adam JA: General aspects of modeling tumor growth and immune response. In: *A Survey of Models for Tumor-Immune System Dynamics.* Adam JA, Bellomo N, eds. Berlin: Birkhäuser, 1997, 15–87.

[2] Agur Z, Arakelyan L, Daugulis P, Ginosar Y: Hopf point analysis for angiogenesis models. *Disc Cont Dyn Sys B* 2004, 4:29–38.

[3] Alarcón T, Byrne HM, Maini PK: A cellular automaton model for tumor growth in inhomogeneous environment. *J Theor Biol* 2003, 225:257–274.

[4] Alarcón T, Byrne HM, Maini PK: Towards whole-organ modelling of tumour growth. *Prog Biophys Mol Biol* 2004, 85:451–472.

[5] Alberts B, Bray D, Lewis J, Raff M, Roberts K, Watson J: *The Molecular Biology of the Cell,* 3rd ed. New York: Garland, 1994.

[6] Ambrosi D, Mollica, F: On the mechanics of a growing tumor. *Int J Eng Sci* 2002, 40:1297–1316.

[7] Ambrosi D, Mollica F: The role of stress in the growth of a multicell spheroid. *J Math Biol* 2004, 48:477–499.

[8] Anderson ARA: A hybrid mathematical model of solid tumour invasion: The importance of cell adhesion. *Math Med Biol* 2005, 22:163–186.

[9] Anderson ARA, Chaplain MAJ: Continuous and discrete mathematical models of tumor-induced angiogenesis. *Bull Math Biol* 1998, 60:857–900.

[10] Anderson ARA, Weaver AM, Cummings PT, Quaranta V: Tumor morphology and phenotypic evolution driven by selective pressure from the microenvironment. *Cell* 2006, 127:905–915.

[11] Arakelyan L, Merbl Y, Daugulis P, Ginosar Y, Vainstein V, Selitser V, Kogan Y, Harpak H, Agur Z: Multi-scale analysis of angiogenic dynamics and therapy. In: *Cancer Modeling and Simulation*, Preziosi, ed. Boca Raton, Fl: Chapman and Hall/CRC, 2003, pp.185–219.

[12] Araujo RP, McElwain DL: A history of the study of solid tumor growth: The contribution of mathematical modelling. *Bull Math Biol* 2004, 66:1039–1091.

[13] Bach LA, Bentzen SM, Alsner J, Christiansen FB: An evolutionary-game model of tumor-cell interactions: Possible relevance to gene therapy. *Eur J Cancer* 2001, 37:2116–2120.

[14] Barcellos-Hoff MH, Ravani SA: Irradiated mammary gland stroma promotes the expression of tumorigenic potential by unirradiated epithelial cells. *Cancer Res* 2000, 60:1254–1260.

[15] Bertuzzi A, D'Onofrio A, Fasano A, Gandolfi A: Regression and regrowth of tumour cords following single-dose anticancer treatment. *Bull Math Biol* 2003, 65:903–931.

[16] Bertuzzi A, D'Onofrio A, Fasano A, Gandolfi A: Modelling cell populations with spatial structure: Steady state and treatment-induced evolution of tumor cords. *Disc Cont Dyn Sys B* 2004, 4:161–186.

[17] Bertuzzi A, Fasano A, Gandolfi A, Marangi D: Cell kinetics in tumour cords studied by a model with variable cell cycle length. *Math Biosci* 2002, 177 & 178:103–125.

[18] Bertuzzi A, Gandolfi A: Cell kinetics in a tumor cord. *J Theor Biol* 2000, 204:587–599.

[19] Bhowmick NA, Neilson EG, Moses HL: Stromal fibroblasts in cancer initiation and progression. *Nature* 2004, 432:332–337.

[20] Bickel ST, Juliano JD, Nagy JD: Evolution of proliferation and the angiogenic switch in tumors with high clonal diversity. *PLoS ONE* 2014, 9:e91992.

[21] Bignold LP: The mutator phenotype theory can explain the complex morphology and behaviour of cancers. *Cell Mol Life Sci* 2002, 59:950–958.

[22] Bignold LP: The mutator phenotype theory of carcinogenesis and the complex histopathology of tumours: Support for the theory from the independent occurrence of nuclear abnormality, loss of specialization and invasiveness among occasional neoplastic lesions. *Cell Mol Life Sci* 2003 60:883–891.

[23] Britto Garcia S, Novelli M, Wright NA: The clonal origin and clonal evolution of epithelial tumours. *Int J Exp Pathol* 2000, 81:89–116.

[24] Burton AC: Rate of growth of solid tumours as a problem of diffusion. *Growth* 1996, 30:157–176.

[25] Byrne HM: Modelling avascular tumor growth. In: *Cancer Modelling and Simulation*. Preziosi L, ed., Boca Raton, FL: Chapman and Hall/CRC, 2003.

[26] Byrne HM, Chaplain MAJ: Growth of nonnecrotic tumors in the presence and absence of inhibitors. *Math Biosci* 1995, 130:151–181.

[27] Byrne HM, Chaplain MAJ: Growth of necrotic tumors in the presence and absence of inhibitors. *Math Biosci* 1996, 135:187–216.

[28] Chaplain MAJ: Mathematical modelling of angiogenesis. *J Neurooncol* 2000, 50:37–51.

[29] Chaplain MAJ, Anderson ARA: Mathematical modelling of tumor invasion. *In:* Cancer Modeling and Simulation. Preziosi, ed., Boca Raton, FL: Chapman and Hall/CRC, 2003. pp.269–297.

[30] Chen CY, Byrne HM, King RJ: The influence of growth-induced stress from the surrounding medium on the development of multicell spheroids. *J Math Biol* 2001, 43:191–220.

[31] Cotran RS, Kumar V, Collins T: *Robbins' Pathologic Basis of Disease*, 6th ed. Philadelphia: Saunders, 1999

[32] Cristini V, Lowengrub J, Nie Q: Nonlinear simulation of tumor growth. *J Math Biol* 2003, 46:191–224.

[33] De Wever O, Mareel M: Role of tissue stroma in cancer cell invasion. *J Pathol* 2003, 200:429–447.

[34] Deugnier Y: Iron and liver cancer. *Alcohol* 2003, 30:145–150.

[35] DeVita VT, Hellman S, Rosenberg SA (eds.): *Cancer: Principles and Practice of Oncology*, 5th ed. Philadelphia: Lippencott Raven, 1997.

[36] Dingli D, Michor F, Antal T, Pacheco JM: The emergence of tumor metastases. *Cancer Biol Ther* 2007, 6:383–90.

[37] Dyson J, Villella-Bressan R, Webb G: The steady state of a maturity structured tumor cord cell population. *Disc Cont Dyn Sys B* 2004, 4:115–134.

[38] Elser JJ, Nagy JD, Kuang Y: Biological stoichiometry: An ecological perspective on tumor dynamics. *Biosci* 2003, 53:1112–1120.

[39] Evan GI, Vousden KH: Proliferation, cell cycle and apoptosis in cancer. *Nature* 2001 411:342–348.

[40] Folkman J, Hahnfeldt P, Hlatky L: Cancer: Looking outside the genome. *Nat Rev Mol Cell Biol* 2000, 1:76–79.

[41] Franks SJ, Byrne HM, King JR, Underwood JCE, Lewis CE: Modelling the early growth of ductal carcinoma in situ of the breast. *J Math Biol* 2003, 47:424–452.

[42] Friedman A, Reitich F: Analysis of a mathematical model for the growth of tumors. *J Math Biol* 1999, 38:262–284.

[43] Gammack D, Byrne HM, Lewis CE: Estimating the selective advantage of mutant p53 tumour cells to repeated rounds of hypoxia. *Bull Math Biol* 2001, 63:135–166.

[44] Gatenby RA: Models of tumor-host interaction as competing populations: Implications for tumor biology and treatment. *J Theor Biol* 1995, 176:447–455.

[45] Gatenby RA, Gawlinski ET: A reaction-diffusion model of cancer invasion. *Cancer Res* 1996, 56:5745–5753.

[46] Gatenby RA, Gawlinski ET: The glycolytic phenotype in carcinogenesis and tumor invasion: Insights through mathematical models. *Cancer Res* 2003, 63:3847–3854.

[47] Gatenby RA, Gillies RJ: Why do cancers have high aerobic glycolysis? *Nat Rev Cancer* 2004, 4:891–899.

[48] Gatenby RA, Maini PK, Gawlinski ET: Analysis of tumor as an inverse problem provides a novel theoretical framework for understanding tumor biology and therapy. *Appl Math Lett* 2002, 15:339–345.

[49] Gatenby RA, Vincent TL: An evolutionary model of carcinogenesis. *Cancer Res* 2003, 63:6212–6220.

[50] Geritz SAH, Kisdi E, Meszéna G, Metz JAJ: Evolutionarily singular strategies and the adaptive growth and branching of the evolutionary tree. *Evol Ecol* 1998, 12:35–57.

[51] Graeber TL, Osmanian C, Jacks T, Housman DE, Koch CJ, Lowe SW, Giaccia AJ: Hypoxia-mediated selection of cells with diminished apoptotic potential in solid tumours. *Nature* 1996, 379:88–91.

[52] Greaves M, Maley CC: Clonal evolution in cancer. *Nature* 2012, 481:306–313.

[53] Greenspan HP: Models for the growth of a solid tumor by diffusion. *Stud Appl Math* 1972, 52:317–340.

[54] Greenspan HP: On the growth and stability of cell cultures and solid tumors. *J Theor Biol* 1976, 56:229–242.

[55] Hahn WC, Weinberg RA: Modelling the molecular circuitry of cancer. *Nature Rev Cancer* 2002, 2:331–341.

[56] Hanahan D, Weinberg RA: The hallmarks of cancer. *Cell* 2000, 100: 57–70.

[57] Hanahan D, Weinberg RA: The hallmarks of cancer: The next generation. *Cell* 2011, 144:646–674.

[58] Haeno H, Michor F: The evolution of tumor metastases during clonal expansion. *J Theor Biol* 2009, doi:10.1016/j.jtbi.2009.11.005

[59] Hann HW, Stahlhut MW, Hann CL: Effect of iron and desferoxamine on cell growth and in vitro ferritin synthesis in human hepatoma cell lines. *Hepatology* 1990, 11:566–569.

[60] Hayward SW, Wang Y, Cao M, Hom YK, Zhang B, Grossfeld GD, Sudilovski D,Cunha GR: Malignant transformation in a nontumorigenic human prostatic epithelial cell line. *Cancer Res* 2001, 61:8135–8142.

[61] Hellman S: Principles of cancer managemant: Radiation therapy. In: *Cancer: Principles and Practice of Oncology.* DeVita VT, Hellman S, Rosenberg SA, eds., Philadelphia: Lippencott-Raven, 1997. ch. 16, 307–332.

[62] Helmlinger G, Netti PA, Lichtenbeld HC, Melder RJ, Jain RK: Solid stress inhibits the growth of multicellular tumor spheroids. *Nature Biotech* 1997, 5:778–783.

[63] Hirsch MW, Smale S, Devaney RL: *Differential Equations, Dynamical Systems, and an Introduction to Chaos*, 2nd ed. Amsterdam: Elsevier, 2004.

[64] Holash J, Maisonpierre PC, Compton D, Boland P, Alexander CR, Zagzag D, Yancopolous GD, Weigand SJ: Vessel cooperation, regression and growth in tumors mediated by angiopoietins and VEGF. *Science* 1998, 221:1994–1998.

[65] Ingber DE: Cancer as a disease of epithelial-mesenchymal interactions and extracellular matrix regulation. *Differentiation* 2002, 70:547–560.

[66] Jain RK: Normalization of tumor vasculature: An emerging concept in antiangiogenic therapy. *Science* 2005, 307:58–62.

[67] Kianercy A, Veltri R, Pienta KJ: Critical transitions in a game theoretic model of tumour metabolism. *Interface Focus* 2014, 4:20140014.

[68] Kim MY, Oskarsson T, Acharyya S, Nguyen DX, Zhang XH, Norton L, Massagué J: Tumor self-seeding by circulating cancer cells. *Cell* 2009, 139:1315–26.

[69] Kuang Y, Nagy JD, Elser JJ: Biological stoichiometry of tumor dynamics: Mathematical models and analysis. *Disc Cont Dyn Sys B* 2004, 4:221–240.

[70] Kunz-Schughart LA: Multicell tumor spheroids: Intermediates between monolayer culture and in vivo tumor. *Cell Biol Int* 1999, 23:157–161.

[71] Landman K, Please CP: Tumor dynamics and necrosis: Surface tension and stability. *IMA J Math Appl Med Biol* 2001, 18:131–158.

[72] Le NT, Richardson DR: The role of iron in cell cycle progression and the proliferation of neoplastic cells. *Biochim Biophys Acta* 2002, 1603:31–46.

[73] Lester SC, Cotran RS: The breast. In: *Robbins' Pathologic Basis of Disease*, 6th ed. Cotran RS, Kumar V, Collins T, eds., . Philadelphia: W.B. Saunders, 1999, ch. 25, 1093–1119.

[74] Levine HA, Pamuk S, Sleeman BD, Nilsen-Hamilton N: Mathematical modeling of capillary formation and development in tumor angiogenesis: Penetration into the stroma. *Bull Math Biol* 2001, 63:801–863.

[75] Levine HA, Sleeman BD: Modelling tumor-induced angiogenesis. In: *Cancer Modelling and Simulation*. Preziosi L, ed., Boca Raton, FL: Chapman & Hall/CRC, 2003.

[76] Liotta LA, Kohn EC: The microenvironment of the tumour-host interface. *Nature* 2001, 411375–379.

[77] Loeb LA: A mutator phenotype in cancer. *Cancer Res* 2001, 61:3230–3239.

[78] Loeb LA, Loeb KR, Anderson JP: Multiple mutations and cancer. *Proc Natl Acad Sci USA* 2003, 100:776–781.

[79] Lundberg AS, Weinberg RA: Control of the cell cycle and apoptosis. *Eur J Cancer* 1999, 35:1886–1894.

[80] Lynch CC, Matrisian LM: Matrix metalloproteinases in tumor-host cell communication. *Differentiation* 2002, 70:561–573.

[81] Maley CC, Galipeau PC, Finley JC, Wongsurawat VJ, Li X, Sanchez CA, Paulson TG, Blunt PL, Risques R-A, Rabinovitch PS, Reid BJ: Genetic clonal diversity predicts progression to esophageal adenocarcinoma. *Nat Gen* 2006, 38:468–473.

[82] Mantzaris NV, Webb S, Othmer HG: Mathematical modeling of tumor-induced angiogenesis. *J Math Biol* 2004, 49:111–187.

[83] Marusyk A, Polyak K: Tumor heterogeneity: Causes and consequences. *Biochim Biophys Acta* 2010, 1805:105–117.

[84] Maynard Smith J, Price GR: Logic of animal conflict. *Nature* 1873, 246:15–18.

[85] Michelson S, Leith JT: Positive feedback and angiogenesis in tumor growth control. *Bull Math Biol* 1997, 59:233–254.

[86] Michelson S, Miller BE, Glicksman AS, Leith JT: Tumor micro-ecology and competitive interactions. *J Theor Biol* 1987, 128:233–246.

[87] Michor F, Iwasa Y: Dynamics of metastasis suppressor gene inactivation. *J Theor Biol* 2006, 241:676–89.

[88] Michor F, Nowak MA, Iwasa Y: Stochastic dynamics of metastasis formation. *J Theor Biol* 2006, 240:521–30.

[89] Mollica F, Jain RK, Netti PA: A model for temporal heterogeneities of tumor blood flow. *Microvasc Res* 2003, 65:56–60.

[90] Mueller-Klieser W: Three-dimensional cell cultures: From molecular mechanisms to clinical applications. *Am J Physiol* 1997, 273:C1109–C1123.

[91] Mueller-Klieser W: Tumor biology and experimental therapeutics. *Crit Rev Oncol Hematol* 2000, 36:123–139.

[92] Murphy BJ: Regulation of malignant progression by the hypoxia-sensitive transcription factors HIF-1α and MTF-1. *Comp Biochem Physiol B* 2004, 139:495–507.

[93] Murray JD: *Mathematical Biology I: An Introduction*, 2nd ed. Berlin: Springer, 2002.

[94] Murray JD: *Mathematical Biology II: Spatial Models and Biomedical Applications*, 2nd ed. Berlin: Springer, 2002.

[95] Nagy JD: Competition and natural selection in a mathematical model of cancer. *Bull Math Biol* 2004, 66:663–687.

[96] Nagy JD: The ecology and evolutionary biology of cancer: A review of mathematical models of necrosis and tumor cell diversity. *Math Biosci Eng* 2005, 2:381–418.

[97] Nagy JD, Victor EM, Cropper JH: Why don't all whales have cancer: A novel hypothesis resolving Peto's Paradox. *Int Comp Biol* 2007, 47:317–328.

[98] Nagy JD, Armbruster D: Evolution of uncontrolled proliferation and the angiogenic switch in cancer. *Math Biosci Eng* 2012, 9:843–876.

[99] Nowell PC: The clonal evolution of tumor cell populations. *Science* 1976, 194:23–28.

[100] Nunney L: Lineage selection and the evolution of multistage carcinogenesis. *Proc R Soc Lond B* 1999, 266:493–498.

[101] Ohuchida K, Mizumoto K, Murakami M, QianL-Q, Sato N, Nagai E, Matsumoto K, Nakamura T, Tanaka M: Radiation to stromal fibroblasts increases invasiveness of pancreatic cancer cells through tumour-stromal interactions. *Cancer Res* 2004, 64:3215–3222.

[102] Olumi AF, Grossfeld GD, Hayward SW, Carroll PR, Tlsty TD, Cunha GR: Carcinoma-associated fibroblasts direct tumour progression of initiated human prostatic epithelium. *Cancer Res* 1999, 59:5002–5011.

[103] Ono S: Genetic implication of karyological instability of malignant somatic cells. *Physiol Rev* 1971, 51:496–526.

[104] Owen MR, Sherratt JA: Mathematical modelling of macrophage dynamics in tumours. *Math Mod Methods Appl Sci* 1999, 9:513–539.

[105] Park CC, Bissel MJ, Barcellos-Hoff H: The influence of the microenvironment on the malignant phenotype. *Mol Med Today* 2000, 6:324–329.

[106] Patel AA, Gawlinski ET, Lemieux SK, Gatenby RA: A cellular automaton model of early tumor growth and invasion: The effects of native tissue vascularity and increased anaerobic tumor metabolism. *J Theor Biol* 2001, 213:315–331.

[107] Pescramona GP, Scalerano M, Delsanto PP, Condat CA: Non-linear model of cancer growth and metastasis: A limiting nutrient as a major determinant of tumor shape and diffusion. *Med Hyp* 1999, 53:497–503.

[108] Plank MJ, Sleeman BD: Lattice and non-latice models of tumour angiogenesis. *J Math Biol* 2004, 66:1785–1819.

[109] Please CP, Pettet G, McElwain DLS: A new approach to modelling the formation of necrotic regions in tumours. *Appl Math Lett* 1998, 11:89–94.

[110] Please CP, Pettet G, McElwain DLS: Avascular tumour dynamics and necrosis. *Math Mod Meth Appl Sci* 1999, 9:569–579.

[111] Polyak K, Weinberg WA: Transitions between epithelial and mesenchymal states: Acquisition of malignant and stem cell traits. *Nat Rev Cancer* 2009, 9:265–273.

[112] Ponder BAJ: Cancer genetics. *Nature* 2001, 411:336–341.

[113] Quaranta V: Motility cues in the tumor microenvironment. *Differentiation* 2002, 70:590–598.

[114] Raghunand N, Gatenby RA, Gillies RJ: Microenvironmental and cellular consequences of altered blood flow in tumours. *Br J Radiol* 2003, 76:11–22.

[115] Roskelley CD, Bissel MJ: The dominance of the microenvironment in breast and ovarian cancer. *Cancer Biol* 2002, 12:97–104.

[116] Royds JA, Downer SK, Qwarnstrom EE, Lewis CE: Responses of tumour cells to hypoxia: Role of p53 and NFκB. *Mol Pathol* 1998, 51:55–61.

[117] Shiomi T, Okada Y: MT1-MMP and MMP-7 in invasion and metastasis of human cancers. *Cancer Metastasis Rev* 2003, 22:145–152.

[118] Sterner RW, Elser JJ: *Ecological Stoichiometry: The Biology of Elements from Molecules to the Biosphere*. Princeton, NJ: Princeton University Press, 2002.

[119] Stoll BR, Migliorini C, Kadambi A, Munn LL, Jain RK: A mathematical model of the contribution of endothelial progenitor cells to angiogenesis in tumors: Implications for antiangiogenic therapy. *Blood* 2003, 102:2555–2561.

[120] Subramanian B, Axelrod DE: Progression of heterogeneous breast tumors. *J Theor Biol* 2001, 210:107–119.

[121] Sutherland RM: Importance of critical metabolites and cellular interactions in the biology of microregions of tumors. *Cancer* 1986, 58:668–1680.

[122] Sutherland RM, McCredie JA, Inch WR: Growth of multicell spheroids in tissue culture as a model of nodular carcinomas. *J Natl Cancer Inst* 1971, 46:113–120.

[123] Tam WL, Weinberg RA: The epigenetics of epithelial-mesenchymal plasticity in cancer. *Nat Med* 2013, 19:1438–1449.

[124] Thalhauser CJ, Sankar T, Preul MC, Kuang Y: Explicit separation of growth and motility in a new tumor cord model. *Bull Math Biol* 2009, 71:585–601.

[125] Thompson KE, Royds JA: Hypoxia and reoxygenation: A pressure for mutant p53 cell selection and tumor progression. *Bull Math Biol* 1999, 61:759–778.

[126] Tohya S, Mochizuki A, Imayama S, Iwasa Y: On rugged shape of skin tumor (basal cell carcinoma). *J Theor Biol* 1998, 194:65–78.

[127] Tomlinson IP, Bodmer WF: Modelling the consequences of interactions between tumour cells. *Br J Cancer* 1997, 75:157–160.

[128] Tomlinson, IPM: Game-theory models in interactions between tumor cells. *Eur J Cancer* 1997, 33:1495–1500.

[129] Tomlinson RH, Gray LH: The histological structure of some human lung cancers and possible implications for radiotherapy. *Br J Cancer* 1955, 9:539–549.

[130] Valk-Lingbeek ME, Bruggeman SWM, van Lohuizen M: Stem cells and cancer: The polycomb connection. *Cell* 2004, 118:409–418.

[131] Vogelstein B, Fearon ER, Hamilton SR, Preisinger AC, Willard HF, Michelson AM, Orkin SH: Clonal analysis using recombinant DNA probes from the X-chromosome. *Cancer Res* 1987, 47:4806–4813.

[132] Weinberg RA: *One Renegade Cell.* New York: Basic Books, 1998.

[133] Weinberg, RA: *The Biology of Cancer.* New York: Garland, 2007.

[134] Wernert N: The multiple roles of tumour stroma. *Virchows Arch* 1997, 430:433–443.

[135] Wernert N, Löcherbach C, Wellman A, Behrens P, Hügel A: Presence of genetic alterations in microdissected stroma of human colon and breast cancers. *Anticancer Res* 2001, 21:2259–2264.

[136] Wogan GN, Hecht SS, Felton JS, Conney AH, Loeb LA: Environmental and chemical carcinogenesis. *Sem Cancer Biol* 2004, 14:473–486.

[137] Xu Y: A free boundary problem model of ductal carcinoma in situ. *Disc Cont Dyn Sys B* 2004, 4:337–348.

Chapter 9

Models of Chemotherapy

Modern chemotherapy and antibiotics arose in the same era, between the two world wars through World War II. The sulfonamides were introduced in 1935, and penicillin followed in 1940. The era immediately before and after World War II was revolutionary in the treatment of infectious disease.[1] Rapidly fading was the therapeutic nihilism of the previous era, and many once deadly diseases were now easily curable. As Papac points out in her review of the beginnings of cancer chemotherapy [57], there was a (sadly doomed) hope that the "magic bullet" concept enjoying so much success in antimicrobial therapy would be applicable to cancers. Despite (and partly because of) their success, from the moment of their introduction, both antibiotic and anticancer agents have been plagued by the evolution of resistance. Understanding the evolution of resistance in cancer is a central goal of this chapter.

Cancers are resistant to systemic chemotherapy for at least three major reasons. First, as tumors advance, their growth fraction decreases. Therefore, since chemotherapy preferentially targets rapidly dividing cells, kinetic resistance to treatment tends to develop. Second, drug penetrance into the tumor microenvironment often declines due to high interstitial fluid pressure, drug wash-out, impaired blood flow and large inter-capillary distances within advanced tumors. Finally, and most importantly, heritable resistance to essentially any drug can be conferred by random genetic mutations followed by natural selection, which readily results in tumors (or microbial populations) refractory to treatment.

The "Norton-Simon" and "Goldie-Coldman" hypotheses have been influential in understanding kinetic and genetic resistance, respectively, and optimizing treatment regimens to minimize resistance. The Norton-Simon hypothesis, which describes tumor growth using the Gompertz model, has led to the notion of dose-density in chemotherapy scheduling. The Goldie-Coldman model considers the acquisition of drug resistance via random mutations in individual cells. We devote a significant portion of this chapter to these historically important models and several extensions and modifications of them. In addition, we look at antimicrobial chemotherapy, including the mutant se-

---

[1]It is worth reminding the reader that, despite the dramatic success of antibiotics, the greater part of the decrease in mortality from infectious disease during the 20th century was due to changes in public health unrelated to therapeutic intervention; see Chapter 1.

lection window hypothesis and the role of synergy in selecting for resistance in multi-drug regimens.

In this chapter we also introduce pharmacokinetics and pharmacodynamics models, which describe distribution of drugs in the body and their physiologic effects. These highly mathematical fields have long been informed by dynamical models of the physiologic systems studied. Pharmacokinetic models, some more complex than others, are used to study the distribution into the body of essentially all drugs. Dose-response curves, which describe the relationship between drug dose, exposure time, and cytotoxicity to cancer cells, can be understood using simple dynamical models. More complex dynamical models have also been proposed to describe the uptake and intracellular action of antineoplastic agents. Several mathematical models specifying a given spatial geometry have been used to study the delivery of chemotherapy to tumors, the tumor spheroid and tumor cord being the two most common geometries used.

Our goals for this chapter are to (1) develop a basic mathematical basis for commonly observed dose-response curves for anticancer cancer agents, and to further study representative models of *in vitro* uptake and the intracellular dynamics, (2) develop the basic principles of pharmacokinetics and drug transport within the body and the spatial environment of the tumor, (3) optimize chemotherapy regimes to overcome kinetic resistance, (4) model the evolution of drug resistance in neoplastic cells.

9.1 Dose-response curves in chemotherapy

In Chapter 11 we largely focus on cellular response to radiation and how repair processes give rise to the observed dose-response curves. Here, in the context of chemotherapy, the situation is complicated by the dynamics of agent delivery and its intracellular processing, which we briefly discuss.

9.1.1 Simple models

The log-kill model for cancer cells was first established by Skipper and colleagues [62, 63] in the 1960s and early 1970s, who studied experimental tumors and found that the cell-kill of exponentially expanding tumor cell populations increased logarithmically with dose, giving the log-kill or log-linear model. That is, the surviving cell fraction, S, is given by

$$S = \exp(-kD), \qquad (9.1)$$

where D is the total dose and k is a scaling parameter. An extension of this includes dependence upon drug concentration, C, and the time of drug exposure, T:

$$S = \exp(-kC \times T). \tag{9.2}$$

The quantity $C \times T$ is the area under the curve (AUC) for the drug concentration vs. time curve. (Note that this notation is used even for non-constant drug concentrations.) For specific agents, survival relationships are generally determined by in vitro assays. There are a number of such assays, but the standard is the clonogenic assay. That is, an assay is performed where two cell populations of equal initial size are grown under conditions for exponential expansion. One population, N_T, is exposed to the agent being tested, and the other, N_C, serves as the untreated control. The surviving fraction S after time T is given as

$$S = \frac{N_T(T)}{N_C(T)}. \tag{9.3}$$

This is taken as a surrogate for the expected cell kill, $K = 1 - S$. However, even for cell-cycle nonspecific drugs, cell kill is usually greater for actively proliferating than quiescent cells. The latter can make up a substantial portion of an in vivo tumor, and in vitro cytotoxicity does not directly predict in vivo toxicity, but the two are generally assumed to be correlated.

The log-linear model for the AUC can successfully predict in vitro cell survival and clinical response for a number of agents, and is generally considered to apply well to most cell cycle phase-non-specific (PNS) drugs. Several representative dose-response curves from Drewinko et al. [18] that can be described to some degree using the log-kill model are shown in Figure 9.1.

For many cell cycle phase-specific (PS) drugs, cytotoxicity is not a simple function of AUC. Time of exposure may be more important than drug concentration, and plateaus in cytotoxicity are observed with increasing drug concentrations. One modification of the AUC model that accounts for differential importance of concentration and time is the pharmacodynamic model applied to chemotherapy by Adams [1]:

$$C^n \times T = k, \tag{9.4}$$

where n and k are empirically determined parameters. When $n = 1$, drug concentration and time are equally important and the model is equivalent to the AUC model. If $n < 1$, cell kill increases less with increasing dose than predicted by the AUC, and concentration is, in a sense, less important than exposure time. If $n > 1$, cell kill is greater than predicted by the AUC for a given concentration, and drug concentration or dose is more important than exposure time. Broadly speaking, this model can predict whether it is better to maximize drug concentration or exposure time for a given antineoplastic agent. Several other groups have also used a Hill function to model cell survival, e.g. [22, 23, 46].

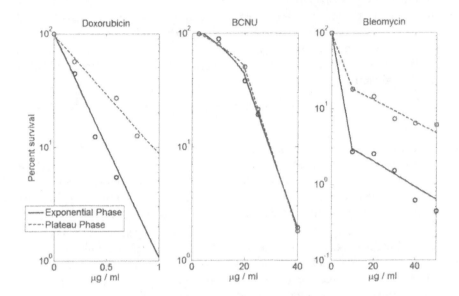

FIGURE 9.1: Dose-response curves for three anticancer agents incubated in agent-containing medium for 1 hour. The data is from Drewinko et al. [18], and response curves for both exponential growth phase (proliferating) and plateau phase (non-proliferating) cells are shown. Note that proliferating cells are usually, but not always, more sensitive. The log-kill model applies directly for the first agent, doxorubicin. The curve for BCNU displays a "shoulder" region, where the surviving fraction is linear in AUC. This is generally interpreted as signifying recovery by some cells from sublethal damage, and the shown curve-fit is a superposition of a linear and exponential model. The curve for bleomycin is biphasic and is described by a double-exponential model.

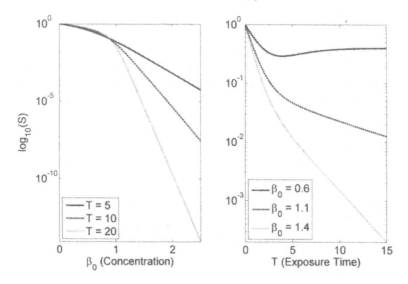

FIGURE 9.2: Dose-response curves when the growth dynamics of the clonogenic assay are logistic (see Exercise 9.2). Constant parameter values are $K = 10$, $\alpha = 1$, and $N_0 = 1$. They are ad-hoc and for demonstrative purposes only.

9.1.2 Concentration, time, and cyotoxicity plateaus

While the simple log-linear model describes some data sets well, many dose-response curves have a two-phase appearance in which cytotoxicity plateaus with either dose or exposure time. Such curves are sometimes described as concave upward. Some authors have interpreted a concave upward curve as indicating the existence of a resistant sub-population of cells. There are several mechanisms that can explain a two-phase response (cytotoxicity plateau) for large drug doses for both cell cycle phase-specific (PS) and non-specific (PNS) drugs. In summary, saturation effects in either drug transport or saturation of the drug target can cause a plateau for either drug class. The instability of some PNS drugs in medium explains plateaus for these agents. Toxicity in PS drugs can be explained as a direct consequence of cell cycle specificity. We defer further discussion of cell-cycle specific chemotherapy to Section 9.6.

9.1.3 Shoulder region

The shoulder region seen in chemotherapy dose-response curves is similar to that seen in radiotherapy dose-response curves. Such a curve is predicted by models of DNA double-strand break interaction and repair in radiotherapy, and it has been generally assumed that a similar mechanism is at work in chemotherapy curves.

9.1.4 Pharmacodynamics for antimicrobials

We briefly mention dose-response pharmacodynamic models, which have been mostly applied to antimicrobial agents; the reader may consult Mueller et al. [52] for a more substantive treatment. A common measure of drug efficacy remains the *minimum inhibitory concentration* (MIC). The MIC is usually defined as the lowest concentration that completely inhibits visible bacterial growth in vitro after a 24-hour incubation period [52].

The efficacy of an in vivo course of therapy is often estimated using the serum concentration of the drug and the MIC as determined in vitro. The most commonly used metrics are the *time above the MIC* (T > MIC), the ratio of *peak serum concentration to MIC* (C_{peak} / MIC), and the *serum AUC over the MIC* (AUC / MIC) [52]. Such approaches are problematic for many reasons; e.g., serum concentration does not necessarily reflect drug concentration at the infection site, and using the MIC alone falsely assumes an effect/no-effect phenomenon for microbes exposed to a drug.

The cytotoxic response saturates with drug concentration for many antimicrobials, leading to the fairly commonly used E_{max} model for drug effect. For exponentially growing bacteria, $N(t)$, drug-induced death is modeled by a Hill function and we have the simple equation:

$$\frac{dN}{dt} = N\left(k_0 - k_{max}\frac{C_t}{EC_{50} + C_t}\right), \tag{9.5}$$

where C_t is the drug concentration at a given time, k_0 is the unperturbed growth rate, k_{max} is the maximum killing rate, and EC_{50} is the drug concentration of half-maximal activity. Example E_{max} models may be found in [41, 71]. These models are also interesting because they combine non-trivial pharmacokinetics for the experimentally determined drug distribution within the body with a pharmacodynamic E_{max} model.

9.2 Models for in vitro drug uptake and cytotoxicity

Drug transport across the cell membrane generally occurs by three different mechanisms: (1) passive diffusion, (2) facilitated diffusion, (3) active carrier-mediated transport. At least some passive diffusion occurs for most small drugs, and it is the only means of transport for many anticancer agents. For passive diffusion, the flux across the membrane is proportional to the concentration difference across the membrane and the membrane surface area:

$$J = PA(S_E - S_I), \tag{9.6}$$

where A is the surface area, S_E is the extracellular concentration, S_I is the intracellular concentration, and P is a permeability constant. Passive diffusion is also generally taken to imply that uptake and efflux are linearly related to extracellular and intracellular drug concentration, respectively. The ratio of drug concentration in the intracellular water to extracellular drug concentration never exceeds unity for drugs that are transported only by diffusion. Note that this does not imply that the overall intracellular:extracellular drug ratio never exceeds unity, as intracellular drug can bind extensively to cellular components.

Many drugs are incidentally similar to natural substances that are transported by membrane-bound carriers, and active transport is the dominant mechanism for trans-membrane transport for a number of anticancer agents.

Once inside the cell, the drug must exert its effect by interacting with intracellular elements, such as DNA. It also may be metabolized or actively excreted by the cell, and damage may be repaired by cellular mechanisms.

9.2.1 Models for cistplatin uptake and intracellular pharmacokinetics

We briefly mention cisplatin as a model drug to study the mechanistic basis for dose-response curves. Cisplatin forms mono- and bifunctional DNA adducts that cross-link the cellular DNA, and the kinetics of such DNA interaction have been quantified. Binding to intracellular thiols detoxifies cisplatin, and DNA repair mechanisms can repair cisplatin:DNA lesions.

Sadowitz et al. [60] developed a simple model to describe cisplatin uptake and intracellular detoxification, and it considers the following behaviors: (1) uptake from the extracellular medium to the cytoplasm, (2) transfer of cisplatin from cytoplasm to the nucleus, (3) reaction of cistplatin with cytoplasmic thiols, (4) binding of cisplatin to the DNA to form DNA adducts, and (5) repair of platinum:DNA adducts.

The variables considered, all in units of concentration (μM), are cytoplasmic cisplatin, $C(t)$, nuclear cisplatin, $N(t)$, cytoplasmic thiol (sulfhydryl), $S(t)$, free nuclear DNA, $D(t)$, and cisplatin:DNA adducts, $A(t)$.

Transport across the cell membrane is assumed to occur by simple diffusion, and transfer between the cytoplasmic and nuclear compartments is similarly assumed to be a diffusion process. The DNA and nuclear cisplatin interact according to second-order kinetics, as do cytoplasmic cisplatin and thiols. Thiols also enter the cell by simple diffusion, and DNA is repaired according to first-order kinetics. These considerations yield the full model:

$$\frac{dC}{dt} = k_a(C_e - C) - k_b CS - k_c(C - N), \tag{9.7}$$

$$\frac{dN}{dt} = k_c(C - N) - k_d ND, \tag{9.8}$$

$$\frac{dS}{dt} = -k_b CS + k_f(S_e - S), \tag{9.9}$$

$$\frac{dD}{dt} = -k_d ND + k_e A, \tag{9.10}$$

$$\frac{dA}{dt} = k_d ND - k_e A, \tag{9.11}$$

where C_e and S_e are the constant extracellular cisplatin and thiol concentrations, respectively, and all other parameter meanings are straightforward to interpret. Assuming that the free DNA concentration is nearly constant, the term $k_d ND$ for cisplatin-DNA interaction can be replaced by $k_d D_0$. We present this model as an example of applying basic chemical kinetic concepts to the topic and do not comment on it further. El-Kareh and Secomb [22] also proposed a simple model for cisplatin uptake and DNA binding by first-order kinetics.

9.2.2   Paclitaxel uptake and intracellular pharmacokinetics

We study a model of paclitaxel uptake and intracellular pharmacokinetics developed by Kuh and colleagues [44] that successfully described paclitaxel in vitro dynamics. Paclitaxel is a mitotic inhibitor that binds to tubulin and induces polymerization of microtubules. Paclitaxel inhibits microtubule dynamic remodeling, and at high doses it increases cellular microtubule mass. It is highly protein bound in the extracellular space, and within the cell it is almost entirely bound to cellular components. Overall, the model accounts for the following behaviors.

1. Transport across the cell membrane of free drug by simple diffusion.

2. Saturable binding of paclitaxel to extracellular proteins within the medium.

3. Saturable and non-saturable binding of paclitaxel to cellular components.

4. Time and concentration dependent changes in microtubule mass. Such microtubules serve as cellular paclitaxel binding sites, and this increase in microtubule mass increases cellular saturable binding.

5. Time and drug concentration dependent changes in cell number.

The model derivation is somewhat non-standard and clever. The experiment considered 10^6 cells incubated in 1 ml of medium. Six drug populations are explicitly considered in the model derivation: the total amount of intracellular drug ($A_{T,c}$), the total amount of drug in the medium ($A_{T,m}$), the total concentration of intracellular drug ($C_{T,c}$), the total concentration of drug in medium ($C_{T,m}$), and the concentrations of free (versus bound) drug within the cell ($C_{F,c}$) and medium ($C_{F,m}$).

The total amounts of cellular and extracellular drug are simply

$$A_{T,c} = V_c \times C_{T,c}, \tag{9.12}$$

$$A_{T,m} = V_m \times C_{T,m}, \tag{9.13}$$

where V_c is the total cellular volume, and V_m is the volume of the medium. Differentiating and applying the product rule implies that

$$\frac{dA_{T,c}}{dt} = V_c \frac{dC_{T,c}}{dt} + C_{T,c} \frac{dV_c}{dt}, \tag{9.14}$$

$$\frac{dA_{T,m}}{dt} = V_m \frac{dC_{T,m}}{dt} + C_{T,m} \frac{dV_m}{dt}. \tag{9.15}$$

It is assumed that the volume of the medium stays constant, so the second term for extracellular drug can be neglected. The change in both drug amounts can also be expressed as a function of the concentration differences of free drug across the membrane. Therefore, altogether we have that changes in the total drug amounts can be expressed as follows:

$$\frac{dA_{T,c}}{dt} = V_c \frac{dC_{T,c}}{dt} + C_{T,c} \frac{dV_c}{dt} = CL_f N (C_{T,m} - C_{T,c}), \tag{9.16}$$

$$\frac{dA_{T,c}}{dt} = V_m \frac{dC_{T,c}}{dt} = CL_f N (C_{T,c} - C_{T,m}), \tag{9.17}$$

where CL_f is the clearance (i.e., diffusion) rate of free drug across the cell membrane per cell, and N is the total number of cells.

The total number of cells, and hence the total cellular volume, V_c, is not constant, as the cell population in this experimental setup either expands or contracts exponentially. At low paclitaxel concentrations there is a small growth rate overall, while at higher concentrations there is net death. Instead of making the change in N a function of cellular drug concentration, Khu et al. determined clonogenic expansion as an empirical function of drug concentration in the medium. That is,

$$N(t) = N(0)e^{k_N t}, \tag{9.18}$$

where the parameter k_N is determined empirically as a function of the initial drug concentration in the medium; i.e., $k_N = k_N(C_{total,m}(0))$. Also, total cellular volume can be expressed as

$$V_c = V_0 N = V_0 N(0)e^{k_N t}, \tag{9.19}$$

where V_0 is the volume of a single cell.

Next we consider drug binding to cellular components and proteins. Instead of explicitly modeling the time-dependent binding of intracellular drug to cellular components and extracellular drug to proteins, the authors use the steady-state expressions for total drug concentrations as a function of free drug. For saturable binding of a ligand to a protein, the concentration of bound protein:ligand can be expressed as a function of free ligand in general terms. Specifically,

$$B\frac{C}{K+C},\tag{9.20}$$

where B is the number of binding sites, C is the concentration of free ligand, and K is a constant. This follows from standard chemical kinetics analysis. For non-saturable binding, Khu et al. assumed that the bound drug is a linear function of free drug. It follows that total drug concentrations can be expressed as

$$C_{T,c} = C_{F,c} + B_c\frac{C_{F,c}}{K_{d,c}+C_{F,c}} + NSB \times C_{F,c},\tag{9.21}$$

$$C_{T,m} = C_{F,m} + B_m\frac{C_{F,m}}{K_{d,m}+C_{F,m}},\tag{9.22}$$

where $B_c(t)$ and B_m are the total number of intracellular and extracellular saturable binding sites, respectively. Note that $B_c(t)$ is a function of time, and NSB is the proportionality constant for non-saturable binding sites within the cell; there is no non-saturable binding in medium.

These equations are quadratic in $C_{F,c}$ and $C_{F,m}$. Some algebra yields the (positive) solutions for $C_{F,c}$ and $C_{F,m}$ in terms of $C_{T,c}$, $C_{T,m}$, and several parameters:

$$C_{F,c} = \frac{-B + (B^2 + 4(1+NSB)K_{d,c}C_{T,c})^{\frac{1}{2}}}{2(1+NSB)},\tag{9.23}$$

$$C_{F,m} = \frac{-A + (A^2 + 4K_{d,m}C_{T,m})^{\frac{1}{2}}}{2},\tag{9.24}$$

$$A = K_{d,m} + B_m - C_{T,m},\tag{9.25}$$

$$B = (1+NSB)K_{d,c} + B_c(t) - C_{T,c}.\tag{9.26}$$

We now discuss the time dependency of $B_c(t)$. Paclitaxel binds saturably to tubulin and microtubules. However, such binding also increases the stability of microtubules and, at high drug concentrations, increases the mass of microtubules. Therefore, the number of intracellular binding sites available to paclitaxel increases with intracellular paclitaxel binding. This feedback loop was not directly modeled by Khu et al., but the time-dependent increase in microtubule mass was instead determined as an empirical function of the initial extracellular drug concentration:

$$B_c(t) = B_{c0}(1 + k_{B,c}t), \tag{9.27}$$

$$k_{B,c} = k_{B,c}(C_{T,m}(0)). \tag{9.28}$$

The number of binding sites increases linearly in time with slope $k_{B,c}$, which is determined empirically as a function of the initial drug concentration in the medium.

Finally, using equations (9.14) and (9.15), solving in terms of $dC_{T,c}/dt$ and $dC_{T,m}/dt$ and substituting the algebraically determined solutions for $C_{F,c}$ and $C_{F,m}$ yields a final system of 2 differential equations:

$$\frac{dC_{T,c}}{dt} = (C_{F,m} - C_{F,c})\frac{CL_f}{V_0} - k_N C_{T,c}, \tag{9.29}$$

$$\frac{dC_{T,m}}{dt} = (C_{F,c} - C_{F,m})\frac{CL_f N_0 e^{k_N t}}{V_m}. \tag{9.30}$$

9.3 Pharmacokinetics

Pharmacokinetics describes how drugs are distributed in and eliminated from the body. Three broad approaches have traditionally been used in this field. The first is **compartmental modeling**, where the body is divided into one or more discrete compartments (often aggregations of multiple tissue types) into which drug distributes. Typically, exchange occurs between the compartments according to first-order kinetics. **Physiological modeling** employs anatomical compartments (plasma, liver, kidney, etc.) and blood flow rates between the compartments govern drug distribution. **Model-independent** pharmacokinetics are used to characterize the distribution and elimination of drug in plasma without using any complex underlying model. Simple quantities such as plasma area under the curve (AUC), volume of distribution (V_D), and elimination half-life ($t_{1/2}$) are measured under this approach.

One-compartment model. It is necessary to introduce a few basic concepts, which we do so in the context of a one-compartment model with intravenous drug infusion. The dynamics of such a model are shown schematically in Figure 9.3. The first important concept is the (apparent) volume of distribution (V_D), which is the volume that the drug in plasma is "effectively" diluted in. That is, the plasma concentration, C, is related to the amount of drug in the body, A, as follows:

$$C = \frac{A}{V_D}. \tag{9.31}$$

This yields V_D:

$$V_D = \frac{A}{C}. \tag{9.32}$$

The volume of distribution gives an idea of how extensively a drug distributes from the plasma to body tissues. Typical volumes of body compartments are 3 L for the plasma (the non-cellular component of blood), 5 L for blood, 15 L for all extracellular water, and 42 L for total body water [68]. Therefore, if $V_D = 3$ L, then the drug is sequestered entirely in the plasma; a V_D of 5 L implies that it stays within the blood but freely enters erythrocytes. In general, the larger the V_D the more extensively a drug distributes into body tissues. Many drugs also bind extensively to plasma proteins, principally albumin. Such bound drug is very slow to exit the plasma, and only free drug should be considered in V_D calculations.

The first order plasma elimination rate, k_{el}, is (obviously) the rate at which drug is eliminated from plasma. It is important to note that this rate depends upon the distribution of drug between plasma and tissues, and thus depends upon the volume of distribution.

The plasma clearance Cl_p, unlike k_{el}, is independent of V_D. Plasma clearance rate has units of volume per time (e.g. L/hr), and reflects the volume of plasma from which drug is completely removed in a unit of time. Thus, it simultaneously measures both the rate at which the clearing organ excretes the substance and the rate of plasma flow through the organ [64], which is typically the kidney. Assume the kidneys are responsible for the clearance. Plasma flow through the kidneys is constant and a certain volume of this plasma will be cleared per unit time; this is precisely the clearance parameter Cl_p. If the volume of distribution is large, the plasma drug concentration will be small. Only a small amount of drug will be cleared per unit time and therefore k_{el} will be small. Mathematically, we have the following relationships:

$$\frac{dA}{dt} = -Cl_p C = -k_{el} A = -k_{el} V_D C, \tag{9.33}$$

$$Cl_p = k_{el} V_D. \tag{9.34}$$

The plasma area under the curve (AUC) is simply the integral of drug concentration $C(t)$ over time:

$$AUC = \int_0^\infty C(t)\, dt. \tag{9.35}$$

The AUC has units concentration × time (e.g., mg hr L^{-1}). Interestingly, dose, plasma clearance, and plasma AUC are all related. For bolus of size D, the initial drug concentration is $C_0 = D/V_D$, and $C(t) = C_0 \exp(-k_{el}t)$. Integrating gives:

$$AUC = \int_0^\infty \frac{D}{V_D} e^{-k_{el}t}\, dt = \frac{D}{V_D k_{el}} = \frac{D}{Cl_p}. \tag{9.36}$$

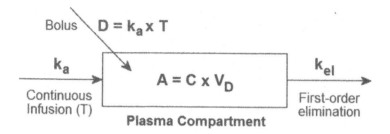

FIGURE 9.3: Schematic for one-compartment pharmacokinetic model with either continuous drug infusion at rate k_a for time T, or a bolus of total dose D.

This relationship also holds for any infusion schedule as long as the total dose delivered is D.

Finally, using these concepts we can easily frame the dynamics of the one-compartment model with an initial bolus of size D in the following way:

$$\frac{dA}{dt} = -k_{el}A = -k_{el}V_D C = -Cl_p C = -Cl_p \frac{A}{V_D} = -\frac{D}{AUC}C, \qquad (9.37)$$

$$\frac{dC}{dt} = \frac{1}{V_D}\frac{dA}{dt}, \qquad (9.38)$$

$$A(0) = D, \qquad (9.39)$$

$$C(0) = \frac{D}{V_D}. \qquad (9.40)$$

For a continuous infusion at rate k_a from time 0 to T the equations are similar, but with a "$+k_a H(T - t)$" term on the right side.

Two-compartment model. We now present the two-compartmental pharmacokinetic model, which explicitly models the distribution of drug from the central compartment (plasma) to the peripheral compartment (tissue) (Figure 9.4). The time course of plasma drug concentration for a two-compartment model is markedly different from the one-compartment model. It is characterized by two phases: the initial distribution (α) phase where plasma drug concentration rapidly drops as it is sequestered in tissue, and a much slower elimination (β) phase where plasma and tissue drug are essentially at equilibrium and drug clearance (i.e., k_{el}) is slower. This pattern is shown in Figure 9.5.

Using the so-called microscopic rate constants, the differential equations describing the amount of drug in the two compartments are the following:

$$\frac{dA_1}{dt} = k_{21}A_2 - k_{12}A_1 - k_{el}A_1, \qquad (9.41)$$

$$\frac{dA_2}{dt} = k_{12}A_1 - k_{21}A_2, \qquad (9.42)$$

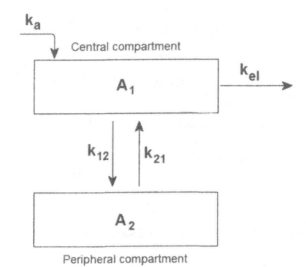

FIGURE 9.4: Schematic for a two-compartment pharmacokinetic model.

where A_1 is the amount of drug in the central compartment (plasma) and A_2 is the amount in the peripheral compartment (tissue). Typically, the drug concentration in the central compartment (plasma) is of interest and can be easily measured.

For a bolus infusion, the plasma concentration, C, is usually reported using the following model:

$$C(t) = Ae^{-\alpha t} + Be^{-\beta t}. \tag{9.43}$$

This model is popular because of its simplicity and because the so-called macroscopic rate constants A, α, B, β can be estimated graphically from a plot of plasma concentration vs. time. In general $A \gg B$ and $\alpha \gg \beta$. Heuristically, this model can be thought of as the superposition of two separate models. Because $A \gg B$, in early time $A\exp(-\alpha t) \gg B\exp(-\beta t)$, and during this time the drug is distributing from the plasma to peripheral tissues, and $C(t) = A\exp(-\alpha t)$ is the model for the distribution phase. The concentration in this model quickly goes to zero, as α is large. Following distribution, $C(t) = B\exp(-\beta t)$ is the model for plasma concentration during the elimination phase. Explicit expressions relating the macroscopic and microscopic rate constants can be derived (see [68] for details).

Extensions. The two-compartment model above can be extended to consider an arbitrary number of compartments, which may have various physical interpretations. The usual notation for a three-compartment model is:

$$C(t) = Ae^{-\alpha t} + Be^{-\beta t} + Ce^{-\gamma t}. \tag{9.44}$$

Another useful extension is to consider infusing a total dose D over the time

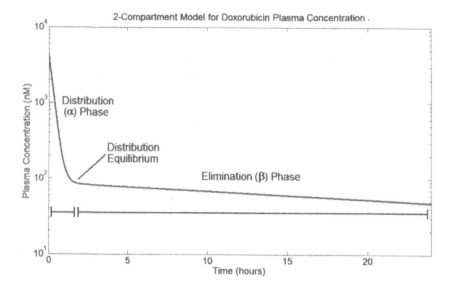

FIGURE 9.5: Two-phase plasma drug concentration dynamics for a two-compartment model. Shown is the two-compartment model for doxorubicin plasma concentration from Greene et al. [31] with $A = 4425$ nM, $\alpha = 4.159$ hr^{-1}, $B = 87$ nM, and $\beta = 0.02530$ hr^{-1}.

interval $(0, T)$, considering the two-compartment model as the explicit summation of two independent compartments:

$$\frac{\mathrm{d}C_1(t)}{\mathrm{d}t} = \frac{A}{T}H(T - t) - \alpha C_1, \qquad (9.45)$$

$$\frac{\mathrm{d}C_2(t)}{\mathrm{d}t} = \frac{B}{T}H(T - t) - \beta C_2, \qquad (9.46)$$

$$C(t) = C_1(t) + C_2(t). \qquad (9.47)$$

An analytic solution for $C(t)$ can be found by integration. (We leave this as an exercise.)

9.4 The Norton-Simon hypothesis and the Gompertz model

Having covered some basics, we devote the next block of this chapter to models focusing on the optimal scheduling of treatment regimes. The influential Norton-Simon and Goldie-Coldman models are our major focus.

We begin our discussion of modeling in vivo cancer treatment with a coupling of the log-kill model and the Gompertz model for in vivo tumor growth. The Gompertz model is a sigmoidal growth curve often used to describe tissue growth. While derived from an actuarial model unrelated to biology, it can be interpreted as a model for growth that is initially exponential, but the rate of growth decreases exponentially with either time or organ size. There are multiple mathematical forms of the model, and the reader can consult Chapter 2 for a detailed discussion of growth curves in cancer. This simple growth law has been applied to human tumors and has motivated "dose-dense" treatment schedules for breast cancer chemotherapy.

9.4.1 Gompertzian model for human breast cancer growth

In 1988, Norton fit a Gompertzian model to human breast cancer growth [53], which we use as our tumor growth model for the remainder of this section. Recall from Chapter 2 that the model can take the following form:

$$\frac{dN}{dt} = bN \ln\left(\frac{N_\infty}{N}\right), \tag{9.48}$$

which has solution,

$$N(t) = N_0 \exp(k(1 - \exp(-bt))), \tag{9.49}$$

where

$$k = \ln\left(\frac{N_\infty}{N_0}\right).$$

Using this model, Norton estimated that breast cancer was typically diagnosed at a volume of about 5 mL of packed cells, corresponding to $N_0 = 4.8 \times 10^9$ cells. Death occurred at a total cancer burden of 1 L, giving the lethal load $N_L = 10^{12}$ cells. Furthermore, $N_\infty = 3.1 \times 10^{12}$ cells (3.1 liters), and b had a log-normal distribution with mean -2.9 and standard deviation 0.71.

9.4.2 The Norton-Simon hypothesis and dose-density

Chemotherapy primarily targets rapidly dividing cells, and when multiple treatments are delivered, there is tumor regrowth between them. Under the Gompertzian model, the tumor growth rate attenuates with tumor age and size, implying that smaller tumors are relatively more susceptible to chemotherapy. Using this as theoretical motivation, Norton and Simon proposed the "Norton-Simon hypothesis":

> *Therapy results in a rate of regression in tumor volume that is proportional to the rate of growth that would be expected for an unperturbed tumor of that size.*

It follows from the Gompertz model (or any model of sigmoidal growth) that more closely spaced treatments will result in greater cell kill, as less regrowth will occur between treatments. Moreover, the tumor at the start of each treatment will be smaller, implying that the growth rate will be greater than for a larger tumor, and per the Norton-Simon hypothesis therapy will result in greater rates of regression.

9.4.3 Formal Norton-Simon model

In Norton and Simon's original 1977 work [54], they proposed that cell kill due to some chemotherapy occurs at a rate proportional to the tumor growth fraction, giving the general tumor growth law under therapy:

$$\frac{dN}{dt} = GF(N)N(t) - L(t)GF(N)N(t), \tag{9.50}$$

where $GF(N)$ is the tumor growth fraction and $L(t)$ is the current "level" of chemotherapy. Applying this concept to the Gompertz model, we have:

$$\frac{dN}{dt} = \underbrace{b\ln\left(\frac{N_\infty}{N(t)}\right)}_{\text{growth fraction}}(N(t) - L(t)N(t)). \tag{9.51}$$

Note that if $L(t) = L_0$ is constant over time, then the analytic solution is simply

$$N(t) = N_0 \exp\left(\left(\frac{N_\infty}{N_0}\right)(1 - e^{-b(1-L_0)t})\right). \tag{9.52}$$

Using this simple theoretical model, we can make several predictions concerning optimal scheduling of a total dose of chemotherapy, D_T, delivered in K treatments. Norton and Simon's primary result was that *dose-density* should be maximized. That is, the time interval between doses should be minimized. Dose-density is a somewhat nebulous concept, as there is no precise way to compare the "density" of two different regimes. Note that dose-dense is not equivalent to high-dose, as high doses widely spaced may not be very "dense" in time. Dose-density is also not equivalent to *dose-intensity*, which is the dose delivered per unit time. We can say that for a fixed total dose, a more intense schedule is also more dense.

The efficacy of a treatment regime is typically measured using two primary metrics:

1. **Cancer cell nadir**: The lowest cancer cell count achieved. If the predicted nadir is very low, perhaps less than several hundred cells, then chemotherapy may be curative. That is, there is a reasonable probability that all cancer cells are eliminated and the cancer is totally cured. If there is a reasonable chance of cure, then the goal of chemotherapy is to minimize the nadir.

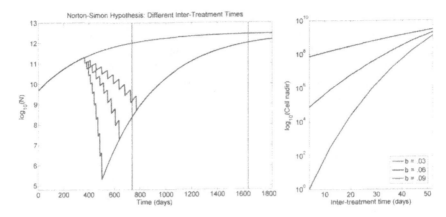

FIGURE 9.6: The left figure shows 10 treatments spaced either 15, 30, or 45 days apart under Gompertzian growth and the Norton-Simon hypothesis. The former, more dose-dense regime results in a lower cancer cell nadir, but all regimes converge to the same asymptotic growth curve and overall survival time is equivalent. The upper dashed line is the unperturbed growth curve, and patient death is expected to occur at the solid line for unperturbed growth and at the dotted line when treatment is applied. The right figure plots the cancer cell nadir vs. inter-treatment time (for 10 treatments) for tumors with different growth rate parameters (b), demonstrating that the benefit to dose density is greater for faster growing tumors.

   2. **Survival time**: The total amount of time that the patient survives when chemotherapy is not curative. If there is not a reasonable chance of chemotherapy being curative (the tumor burden is too high, the tumor is metastatic, or poorly responsive to therapy, etc.) then the goal of chemotherapy is to maximize survival time.

We examine the Norton-Simon hypothesis applied to the Norton model for Gompertzian breast cancer growth computationally, using $N_0 = 4.8 \times 10^9$ and assuming that death occurs at the lethal load $N_L = 10^{12}$ cells. For each treatment, we impose a constant level of drug, $L(t)$, over some relatively brief interval—e.g., 24 hours. Our results may be summarized as follows:

   1. For a given dose D_T divided into K treatments, the earlier that treatment is initiated the lower the cancer cell nadir. However, survival time is not affected by the time of treatment initiation.

   2. Spacing treatments more closely—i.e., using a more dose-dense treatment schedule—results in a lower cell nadir, but inter-treatment time does not affect survival time (see Figure 9.6).

   3. The number of treatments, K, into which a constant total dose is divided

does not affect survival time, except if K is so large that tumor growth is markedly positive. Cell nadir increases monotonically with K.

4. Increasing dose-density is relatively more beneficial, in terms of cell nadir, for more rapidly growing tumors.

Overall, the Norton-Simon model predicts that survival time, in the absence of a cure, is remarkably independent of the treatment schedule: the time at which therapy is initiated, the number of treatments, and the inter-treatment time are all unimportant. However, if the goal of therapy is to cure the tumor by eradicating all cancer cells—i.e., minimize the cancer cell nadir— then treatment should be initiated as soon as possible, inter-treatment time should be minimized (dose-density), and the total dose should be delivered in as few infusions as possible.

Since decreasing the cancer cell nadir transiently reduces the tumor burden without affecting overall survival time, dose-dense regimes may improve early survival and perhaps increase median survival time without affecting the long-term survival rate.

9.4.4 Intensification and maintenance regimens

Norton and Simon [54] argued that their model supports a treatment plan of initially moderately intense treatment to decrease tumor size and hence increase growth rate, followed by late intensification to take advantage of the increased growth fraction of a smaller tumor. Moreover, they argued that low-dose maintenance therapy likely has little effect on survival. From simulations, however, we can see that if the same total dose is delivered, it is irrelevant whether treatment employs a constant dose or early or late intensification. Under any model where cell-kill is proportional to drug concentration, if intensification involves an increase in total dose then it will be beneficial, but this is no profound result.

9.4.5 Clinical implications and results

In practice, drug sensitivity in tumors is heterogeneous, so combination chemotherapy with several agents is used, with agents given either sequentially, concurrently, alternating, or in some combination. *The Norton-Simon model is interpreted to predict that dose-dense schedules with agents given in sequence, rather than concurrently or in alternating schedules, are best.* The strategy is to eliminate all cells sensitive to each agent independently, which is best accomplished by a dose-dense schedule. Sequential delivery is therefore preferred because the dose level and density can be maximized when toxicity from concurrently delivered agents is absent.

When tested in clinical trials, dose-dense strategies have enjoyed modest success. The major trial supporting dose-density is the Cancer and Leukemia

Group B Trial 9741 (CALGB 9741) [14], in which patients were treated with combination therapy by doxorubicin (A), paclitaxel (T), and cyclophosphamide (C), in one of four sequences: (1) A-T-C with doses every 3 weeks, (2) A-T-C with doses every 2 weeks, (3) AC-T (i.e., A and C concurrent) every 3 weeks, or (4) AC-T every 2 weeks.

Dose-density was associated with improved disease-free and overall survival, and sequential and concurrent schedules were equally effective. This trial is important as it was purely a test of dose-density, while the interpretation of most trials purporting to test dose-density is complicated due to variations in treatment arms (e.g. total dose, number of cycles, etc. See [12] for a recent meta-analysis and review of such studies).

As an aside, we point out that any model with tumor regrowth between treatments, even simple exponential growth, would predict that a more dose-dense regime would result in a lower cancer cell nadir. Therefore, we cannot reasonably infer from modest clinical results that the Gompertz model is an accurate description of the natural history of individual breast cancers. See Chapter 7 for a discussion of associated controversies concerning the natural history of breast cancer.

9.4.6 Depletion of the growth fraction

We can see from its basic mathematical expression that the Norton-Simon model assumes that tumor growth fraction following treatment is equivalent to that of an unperturbed tumor allowed to grow to the same size:

$$\frac{dN}{dt} = GF(N)N(t) - L(t)GF(N)N(t). \tag{9.53}$$

The growth fraction is a function only of the current tumor size; i.e., $GF = GF(N)$. This directly implies the basic tenant of the Norton-Simon hypothesis:

> *The efficacy of chemotherapy is proportional to the rate of growth of an unperturbed tumor of equivalent size.*

Unfortunately, this ignores the perturbation in the underlying population age-structure caused by treatment. Proliferating cells are exclusively targeted under the Norton-Simon hypothesis, so there must be a dramatic reduction in proliferating cells following treatment. The growth fraction (at least immediately) following treatment is most likely reduced, not increased. Therefore, we have $GF \neq GF(N)$, and the growth fraction is a function of the number of actively proliferating cells, which is not a simple function of N (at least when the tumor cell population is not at a steady state distribution, which it most definitely will not be following cytotoxic treatment).

Gyllenberg and Webb showed that a system in which cells transition between quiescent (Q) and proliferating (P) compartments can exhibit Gompertzian growth kinetics [33]. (See Chapters 2 and 4 for details.) Kozusko,

Bajzer, and colleagues [42, 43] later studied treatment under this model, and concluded that following treatment the tumor "is not the same tumor" and the *Gompertz growth equation cannot be directly reapplied to determine the growth trajectory*. Nor can it, then, necessarily be used to predict response to treatment.

In [43], analytic solutions for tumor growth following treatment are derived which demonstrate that treatment *fundamentally* alters the growth trajectory by changing the asymptotic tumor size, N_∞, and if N/P is sufficiently small, tumor growth will be negative following the end of treatment. The authors also concluded, like Norton and Simon, that a dose-dense schedule is preferable and faster growing tumors respond better to dose density. However, they compared increasing, constant, and decreasing dose schemes and found that a *decreasing schedule is best*, contradicting Norton and Simon, who suggested dose intensification as a treatment strategy. Thus, a model giving gross Gompertzian kinetics but with an underlying population structure can differ in its predictions for optimizing therapy.

9.5 Modeling the development of drug resistance

While kinetic resistance plays a role in cancer treatment failure, evolution of intrinsically resistant cells is the major cause of treatment failure and ultimately death in cancer. This situation is somewhat analogous to the evolution of widespread antibacterial resistance among bacteria populations, which represents a growing threat to public health. Therefore, in this section we extend our discussion to models for antibacterial resistance as well. We begin with the enormously important work of Luria and Delbrück and later extensions by Lea and Coulson [45] which established the cause of resistance to viral attack in bacteria.

9.5.1 Luria-Delbrück fluctuation analysis

In their landmark 1943 paper [47], Luria and Delbrück attempted to determine the cause of bacterial resistance to bacteriophage infection. When bacterial cultures are challenged with a phage virus, the vast majority of bacteria are killed. However, the cultures tend to regrow, and when they do they resist infection by the same phage. Prior to Luria and Delbrück's paper, two hypotheses had been suggested to explain this phenomenon. The first supposed that mutations conferring resistance arise prior to infection. When exposed to the phage, clones carrying these resistance mutations survive and repopulate the culture, passing their resistance to their offspring. In other words, resistance evolves by natural selection, although, oddly, Luria and

Delbrück and most researchers following in their footsteps resist expressing the hypothesis in explicit Darwinian terms. The second hypothesis posits that bacteria develop a sort of immunity to the phage, analogous to mammalian immune memory. Under this hypothesis, the initial bacterial die-off is not complete; some bacteria, for whatever reason, heritable or otherwise, resist being killed. Those that survive then acquire some mechanism that blocks bacterial infection, and this acquired mechanism is somehow passed to their offspring. So, in summary, hypothesis one claims that resistance to the phage evolves, whereas hypothesis two supposed resistance is acquired.

Luria and Delbrück's experimental test of these hypotheses has led to an experimental design called *fluctuation analysis*. Here is the approach:

Step 1. From a large parental culture of bacteria, take a number of samples and plate them on phage-impregnated media, so that only resistant variants will grow. If resistance is *acquired*, then some fraction of cells will acquire resistance and, assuming the probability of becoming resistant is small, the number of resistant variants will be approximately Poisson distributed (the actual distribution is binomial). That is, the variance equals the mean. If resistance is due to spontaneously arising, pre-existing mutants, the number of resistant variants recovered simply reflects the prevalence of mutants in the parental population and will also be Poisson distributed.

Step 2. Now, take a number of small samples from the parental culture (so that the likelihood of pre-existing mutants is vanishingly small) and expand these cultures in parallel to a fixed size. Then re-plate on selective media to determine the number of resistant variants. Under the acquired resistance hypothesis, resistance is not induced until the final challenge, and the total number of resistant variants should again be Poisson distributed. However, if resistance is due to selection for resistance mutations, the number of resistant cells will vary ("fluctuate") widely as a consequence of rare mutations occurring at different points in time. In this case, variance will be much greater than the mean (rare early mutations imply a large mutant population). Analysis of these fluctuations provides a means to estimate mutation rates.

In the following sections we largely follow Luria and Delbrück's original argument. Although it is a bit old-fashioned and not up to modern standards of rigor, we stick with it to help those interested in reading the original paper interpret the result. In outline, Luria and Delbrück first generate equations for the expected number of mutation events and the expected (or average) number of resistant cells (with some hand-waving). Then, they calculate the variance for resistant cells under the two competing hypotheses. Experimental results confirmed the predictions associated with the mutation-selection hypothesis, indicating that resistance arises by selection on pre-existing mutants rather than an acquired resistance. This process is summarized schematically in Figure 9.7.

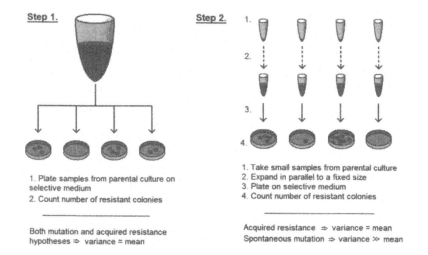

Step 1.
1. Plate samples from parental culture on selective medium
2. Count number of resistant colonies

Both mutation and acquired resistance hypotheses ⇒ variance = mean

Step 2.
1. Take small samples from parental culture
2. Expand in parallel to a fixed size
3. Plate on selective medium
4. Count number of resistant colonies

Acquired resistance ⇒ variance = mean
Spontaneous mutation ⇒ variance ≫ mean

FIGURE 9.7: Schematic for Luria-Delbrück fluctuation analysis to distinguish between hypotheses for resistance to viral attack in bacteria.

9.5.1.1 Step 1: Expected number of mutations and resistant cells

Luria and Delbrück start with the assumption that laboratory bacterial populations grow exponentially. They normalized time such that a unit of time equals the average bacterial division time divided by $\ln 2$. Therefore,

$$\frac{\mathrm{d}N}{\mathrm{d}t} = N \Rightarrow N(t) = N_0 e^t, \tag{9.54}$$

where $N(t)$ is the bacterial population size at time t. (They use the notation N_t for $N(t)$.) They also assumed that resistance mutations occur with a fixed probability with each cell division. They argue that, given their choice of time units, the probability that a single bacterium acquires a resistance mutation in a small time interval $\mathrm{d}t$ is $\alpha\,\mathrm{d}t$, and number of mutations, $\mathrm{d}m$, which occur in this interval is given by

$$\mathrm{d}m = \alpha N(t)\,\mathrm{d}t. \tag{9.55}$$

To get the average number of mutations that occur in $t \in (0, T)$, they integrate the following:

$$m = \int_0^T \mathrm{d}m = \int_0^T \alpha N(t)\,\mathrm{d}t = \int_0^T \alpha N_0 e^t\,\mathrm{d}t = \alpha(N(T) - N_0). \tag{9.56}$$

Assuming that the rate of mutations is small, Luria and Delbrück argue that the number of mutation events is Poisson-distributed. Therefore, the probability a culture of bacteria experiences no mutational events, P_0, would be

$$P_0 = e^{-m}. \tag{9.57}$$

This is all well and good, but we are really interested in the number of resistant cells, not just the number of resistance mutations, that have arisen. Since mutations are passed vertically from mother to daughter cells, the number of mutant cells expands exponentially from all original mutant "foundresses." So, let $\rho(t)$ be the number of resistant cells in a bacterial population. Assuming these cells proliferate with the same kinetics as drug-sensitive cells,

$$\frac{d\rho}{dt} = \rho + \alpha N(t). \tag{9.58}$$

Luria and Delbrück then integrate over the interval $(0, T)$ assuming that no resistant cells are initially present ($\rho(0) = 0$) to obtain the expected (average) number of resistant cells:

$$\rho = \alpha T N(T). \tag{9.59}$$

From here, Luria and Delbrück argue that most experimental bacterial populations will not conform to this average. Because mutations in early generations are highly unlikely due to the small population size, in real experiments one typically sees $\rho \ll \alpha T N(T)$. Occasionally, however, early mutations will generate a very large resistant population, so it can happen that $\rho \gg \alpha T N(T)$. Nevertheless, for experiments of limited size, the observed average will typically be lower than that predicted by equation (9.59), making the average less useful in actual experiments. So, Luria and Delbrück derive an expression for the "likely average ρ" one can expect to see in real experiments, which they denote r. To find r, they solve equation (9.58) over the interval (T_0, T) to obtain

$$r = \alpha T (T - T_0) N(T), \tag{9.60}$$

where T_0 is set to be the average time by which only a single mutation occurred in a group of C cultures. (So, T_0 is a function of C.) From this and equation (9.56) they find that

$$1 = \alpha C (N(T_0) - N_0). \tag{9.61}$$

Assuming $N_0 = 0$ and noting that $N(T_0) = N(T) \exp(T_0 - T)$, substitution and some rearrangement of equation (9.61) yields

$$T - T_0 = \ln[\alpha C N(T)]. \tag{9.62}$$

Substituting this into equation (9.60) yields a transcendental equation for r,

$$r = \alpha N(T) \ln[\alpha C N(T)]. \tag{9.63}$$

From here, two methods have been developed to estimate the mutation rate, α. The first, called the **method of means**, is to solve the transcendental equation (9.63) numerically for α. The second, or P_0 **method**, uses P_0, the fraction of cultures without resistance present. As above,

$$P_0 = e^{-m} = e^{-\alpha(N(T) - N_0)} \approx e^{-\alpha N(T)} \Rightarrow \alpha = -\frac{m}{N(T)} - \frac{\ln P_0}{N(T)}. \tag{9.64}$$

We have now generated predictions assuming the mutation hypothesis is true. The second step of Luria-Delbrück fluctuation analysis is to analyze the fluctuations to test these predictions.

9.5.1.2 Step 2: Test the mutation-selection hypothesis

To test hypothesis, Luria and Delbrück produce an estimate of the variance in number of resistant cells. They begin by estimating the partial distributions of number of resistant bacteria due to mutations arising in the time interval $(T - \tau, T - \tau + d\tau)$. We estimate that, in this interval,

$$dm = \alpha N(\tau)d\tau = \alpha N(T)e^{-\tau}d\tau. \tag{9.65}$$

At the time of observation, T, they argue that each mutant clone arising in $d\tau$ has grown to size e^{τ}. Therefore, the distribution of number of mutants from $d\tau$ has a mean e^{τ} times greater and a variance $e^{2\tau}$ times greater than the mean of the distribution of number of mutation events. Since this latter distribution is assumed to be Poisson, one has that

$$d\rho = \mu N(T)\, d\tau, \tag{9.66}$$

$$\text{var}_{d\rho} = \alpha N(T)e^{\tau}\, d\tau. \tag{9.67}$$

Integrating over either $(0, T)$ or (T_0, T) gives:

$$\text{var}_\rho = \alpha N(T)\left(e^{T} - 1\right), \tag{9.68}$$

$$\text{var}_r = \alpha N(T)\left(e^{T-T_0} - 1\right). \tag{9.69}$$

Substituting $T - T_0$ from equation (9.62) yields

$$\text{var}_r = C\alpha^2 N(T)^2. \tag{9.70}$$

Finally, the ratio of variance to mean becomes

$$\frac{\text{var}_r}{r} = \frac{\alpha C N(T)}{\ln \alpha C N(T)}. \tag{9.71}$$

It follows that $\text{var}_r \gg r$ if $\alpha C N_T \gg 1$. This latter quantity represents the total number of mutants, so the variance will be very high as long as we expect a significant number of mutants to arise. In experiments, it is indeed observed that variance $\gg$ mean, confirming the mutation-selection hypothesis.

9.5.1.3 Later work

In 1949, Lea and Coulson [45] derived a probability generating function for the distribution of mutants and proposed several new methods to estimate the number of mutants, and hence rate of mutation. For a review of theoretical work that followed over the next half century, see Zheng [70]. Foster [24] gives a fairly concise review of practical methods for estimating mutation rates.

9.5.2   The Goldie-Coldman model

Work following that of Luria and Delbrück confirmed and amplified their results so that by the time Goldie and Coldman published their original 1979 paper [28], it had become well established that microbial populations develop drug resistance through spontaneous, inheritable mutations that arise independently of the agent, and that the emergence of resistant phenotypes is the primary cause of antibiotic treatment failure. Assuming a similar process occurs during cancer treatment, Goldie and Coldman derived a simple, testable model for cancer chemotherapy resistance. The model assumes that all cancer cells have unlimited replicative potential, considers treatment with only a single drug, and assumes that cells are either sensitive or totally resistant to the drug. Let N represent the tumor size (say, mass of cells), μ be the mass of resistant cells, and let α and β be the rates per cell division at which cells mutate in and out of the resistant class, respectively. Furthermore, normal and mutant cells are assumed to follow the same growth kinetics. From these assumptions, Goldie and Coldman argued that the change in μ with a small change in N (ΔN) can be expressed as follows:

$$\mu(N + \Delta N) = \mu + \underbrace{\frac{\mu}{N}\Delta N}_{\text{growth of mutants}} + \underbrace{\alpha\left(1 - \frac{\mu}{N}\right)\Delta N}_{\text{mutation in}} - \underbrace{\beta\frac{\mu}{N}}_{\text{mutation out}}. \tag{9.72}$$

Expanding $\mu(N + \Delta N)$ using Taylor's theorem yields

$$\mu(N + \Delta N) = \mu + \frac{d\mu}{dN}\Delta N + O(\Delta N^2). \tag{9.73}$$

Equating the two expressions, dividing by ΔN, and letting $\Delta N \to 0$ gives

$$\frac{d\mu}{dN} = \frac{\mu}{N} + \alpha\left(1 - \frac{\mu}{N}\right) - \beta\frac{\mu}{N}. \tag{9.74}$$

If one assumes that resistance is a permanent condition ($\beta = 0$), then we have the following solution for μ under initial conditions $N(0) = 1$ and $\mu(0) = 0$:

$$\mu(t) = \left(1 - N(t)^{-\alpha}\right)N(t). \tag{9.75}$$

From this it easily follows that the fraction of resistant cells within the tumor at any time, F, is given by:

$$F(t) = \frac{\mu}{N} = 1 - N^{-\alpha}, \tag{9.76}$$

where time dependencies on the r.h.s. have been suppressed for brevity.

Alternatively, as mentioned in Chapter 7, one can derive the model from a differential equation giving the rate of change of μ with respect to time from first principles:

$$\frac{d\mu}{dt} = \underbrace{\left(\frac{\mu}{N}\right)\frac{dN}{dt}}_{\text{growth of mutants}} + \underbrace{\alpha\frac{dN}{dt}\left(1 - \frac{\mu}{N}\right)}_{\text{mutation in}} - \underbrace{\beta\frac{dN}{dt}\left(\frac{\mu}{N}\right)}_{\text{mutation out}}. \tag{9.77}$$

Note that model (9.77) is not particular to chemotherapy resistance; it represents a general model for the emergence of any mutant from a wild-type population.

An attractive property of this model is that the number of mutants does not depend upon any underlying growth law for the primary tumor: the instantaneous tumor size at any time gives the expected number of mutants. However, we can see that this is a consequence of the assumption that the cell division rate is precisely equal to dN/dt, which is not necessarily the case. For example, rapid proliferation nearly balanced by rapid cell death gives a very small dN/dt, but mutants will emerge far faster than predicted by the model. An alternative model can be derived that accounts for this behavior, and we suggest it as a student project.

Now let us return to the model and its predictions. As pointed out in the original paper, rare mutations early in time lead to a large mutant population later in time. Such mutations have a disproportional effect upon the mean number of mutant for a given tumor size, and therefore the number of mutant has a highly right-skewed distribution, and the expected value, μ, gives limited information. Goldie and Coldman confirmed this using a stochastic simulation of the model.

Now assume that the presence of any resistant mutants leads to treatment failure. To determine the probability that a tumor can be cured, we must determine the probability that a resistant colony has arisen at the time of diagnosis—i.e., the probability that at least one mutation has occurred. Because of its right-skewed distribution, the mean number of *mutant cells* μ is not a surrogate for the expected number of *mutation events*. Using arguments similar to those above, we can derive an expression for the probability that no resistant colonies exist in a tumor of size N, which we designate as $\hat{P}(N)$. Expanding $\hat{P}$ around N using Taylor's theorem gives:

$$\hat{P}(N + \Delta N) = \hat{P}(N) + \frac{d\hat{P}}{dN}\Delta N + O(\Delta N^2). \tag{9.78}$$

For the small population increment ΔN, the rate at which mutants arise is approximately equal to the constant value $\alpha \Delta N$. Therefore, the (random) number of mutations is approximately Poisson-distributed, and the probability that no mutation occurs during the interval ΔN is approximately

$$\hat{P}(\Delta N) = \exp(-\alpha \Delta N). \tag{9.79}$$

Finally, we have the following relationships:

$$\hat{P}(N + \Delta N) = \hat{P}(N) \times \hat{P}(\Delta N) = \hat{P}(N)\exp(-\alpha \Delta N) \tag{9.80}$$
$$= \hat{P}(N)(1 - \alpha \Delta N + O(\Delta N^2)).$$

The last step uses the Taylor-series expansion for $\exp(-\alpha \Delta N)$. Equating the right-hand sides of (9.78) and (9.80), dividing by ΔN and neglecting terms of

$O(\Delta N^2)$ gives the following:

$$\frac{d\hat{P}}{dN} = -\alpha\hat{P}(N). \tag{9.81}$$

Integrating and applying initial conditions $\hat{P}(1) = 1$ yields our final expression for $\hat{P}(N)$:

$$\hat{P} = \exp(-\alpha N + \alpha). \tag{9.82}$$

So, the probability that no cell in the population acquires a mutation decreases exponentially with the total tumor size at a rate equal to α, the mutation rate per cell cycle. This and the expected number of mutant cells, μ, as a function of the logarithm of tumor size, $\ln N$, are shown in Figure 9.8. From the figure it is apparent that the model predicts that with increasing tumor size there is a rapid transition from nearly curable to nearly incurable. This can also be seen with a few simple calculations. Suppose $\hat{P} = 0.95$; then we can solve for the tumor size with that characteristic, which we denote N_{95}. We can likewise solve for a tumor size yielding $\hat{P} = 0.05$, namely N_{05}. Then, for any value of α, we have

$$\ln\left(\frac{N_{05}}{N_{95}}\right) = 4.067 = e^{\ln(2)\times 5.87}. \tag{9.83}$$

The last point shows that under an exponential tumor growth model it takes just under six doubling times for a tumor to become nearly incurable. Therefore, the model predicts that treatment should be initiated as soon as possible, lest the window of opportunity be missed.

A reasonable range for α is between 10^{-5} and 10^{-8}. Since about 10^9 cells must be present for cancer diagnosis, the probability that no treatment-resistant mutant exists by the time of diagnosis is essentially negligible. Even under very optimistic parameter values—e.g., $\alpha = 10^{-8}$ and $N = 10^8$—the probability of cure with a single agent is small. Therefore, the basic Goldie-Coldman model makes the following broad predictions:

1. Treatment-resistant mutants arise early in a tumor's natural history and are likely present at the time of clinical diagnosis.

2. The tumor fraction that is resistant increases with tumor size.

3. Treatment must be initiated as early as possible, as tumors rapidly transition from curable by a single agent to nearly incurable.

9.5.3   Extensions of Goldie-Coldman and alternating therapy

9.5.3.1   Two-drug Goldie-Coldman model

Several years after they published their original model, Goldie and Coldman extended their basic model to consider resistance to two non-cross-resistant

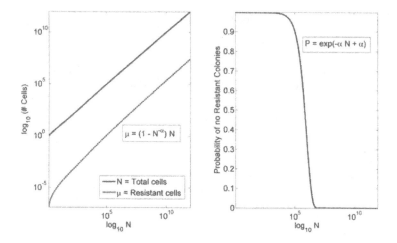

FIGURE 9.8: The expected number of mutant cells, μ, and the probability that no mutant colony exists, $\hat{P}$, as a function of total tumor size N. Here, $\alpha = 10^{-6}$. Also note that when $\hat{P} = 0.5$, we have $\mu \approx 9.32$.

drugs [30]. From this model, they concluded that *non-cross-resistant drugs should be alternated as quickly as possible to minimize the probability that multi-drug resistant mutants arise during treatment.* In the literature, this is generally regarded the major prediction of the Goldie-Coldman hypothesis.

The model considers two drugs, T_1 and T_2, and three populations of resistant cells. We represent singly-resistant mutants by R_1 and R_2, which are resistant to T_1 and T_2, respectively. Doubly resistant mutants are represented by R_{12}. Cells mutate to become resistant to T_1 and T_2 at rates α_1 and α_2 per cell division; these rates are the same whether mutating from sensitive to singly resistant or from singly to doubly resistant classes. It is assumed that resistance to both agents must be acquired in a step-wise manner. Furthermore, the tumor is assumed to grow exponentially. These assumptions give a basic model similar to the one above:

$$\frac{\mathrm{d}R_1}{\mathrm{d}N} = \alpha_1 \left(1 - \frac{R_1 + R_2}{N} \right) - \alpha_1 \frac{R_1}{N} + \frac{R_1}{N}, \tag{9.84}$$

$$\frac{\mathrm{d}R_2}{\mathrm{d}N} = \alpha_2 \left(1 - \frac{R_1 + R_2}{N} \right) - \alpha_2 \frac{R_2}{N} + \frac{R_2}{N}, \tag{9.85}$$

$$\frac{\mathrm{d}R_{12}}{\mathrm{d}N} = (\alpha_1 + \alpha_2) \frac{R_1 + R_2}{N} + \frac{R_1 + R_2}{N}. \tag{9.86}$$

This model can also be rewritten using differential equations with respect to time, which we leave as an exercise. As above, the mean number of resistant cells does not tell the whole story, as rare mutants that arise early in time expand the mutant population greatly. The distribution of mutants is right-skewed, and we must determine the probability that no doubly resistant colony has arisen at the time of diagnosis or following treatment. To this end, first assume that when there are N_0 sensitive cells, the number of cells resistant to T_i is $R_i(N_0)$. Following some interval of tumor growth after which N sensitive cells are present, $R_i(N)$ is approximated as follows:

$$R_i(N) = R_1(N_0)\frac{N}{N_0} + \alpha_i N \ln\left(\frac{N}{N_0}\right). \tag{9.87}$$

Using arguments similar to those for the single mutant model (see [30] for details), one can derive the following expression for the probability that no doubly resistant mutants exist when the tumor has size N_2, given that no such mutants were present at size N_1:

$$P(R_{12} = 0 \text{ when } N = N_2 | R_{12} = 0 \text{ when } N = N_1) = \tag{9.88}$$
$$\exp(-\alpha_2[R_1(N_2) - R_1(N_1)] - \alpha_1[R_2(N_2) - R_2(N_1)] + 2\alpha_1\alpha_2(N_2 - N_1)),$$

where $R_i(N_i)$ is the number of singly resistant mutants at tumor size N_i. For initial conditions:

$$N_1 = 1, \ R_1(1) = 0, \text{ and } R_2(1) = 0,$$

equation (9.88) reduces to:

$$P(R_{12} = 0 \text{ for a tumor of size } N) = \exp(-2\alpha_1\alpha_2[1 + (\ln(N) - 1)N]). \tag{9.89}$$

For a given initial tumor size, equation (9.89) gives the probability that the tumor is incurable—i.e., a doubly resistant colony exists. Then a treatment regimen with tumor regrowth may be imposed. Equations (9.87) and (9.88) can be used to determine the probability that resistance arises as a tumor regrows following a treatment that instantaneously reduces the tumor size. Such calculations can be done for different regimens of T_1 and T_2 and are tedious but straightforward. Goldie and Coldman performed such numerical calculations assuming that a single treatment of either T_1 or T_2 reduces the population of sensitive cells by two orders of magnitude (i.e., 2 log kill). Testing different regimes, they concluded that, if the two treatments could not be given simultaneously, then it is best to alternate them as rapidly as possible to minimize the probability that doubly resistant mutants arise.

9.5.3.2 Stochastic two-drug Goldie-Coldman model

Goldie and Coldman proposed a stochastic version of their two-drug model [29] using a simple stochastic birth-death process to model tumor growth. In this model, only a subset of cells were considered to be tumor stem cells—i.e., have unlimited replicative potential. Using this model they made an interesting prediction concerning the response of slow-growing versus fast-growing tumors.

Rapidly growing tumors typically respond much better to chemotherapy. For a number of acute leukemias, multi-agent chemotherapy is now frequently curative. However, chronic slow-growing leukemias have a much poorer response and, while survival in such cancers is long due to their indolent clinical course, treatment is very rarely curative. Such differences in response are traditionally ascribed to kinetic differences and the much lower growth fraction of chronic versus acute cancers. However, as Goldie and Coldman pointed out, many chronic leukemias are characterized by a so-called *blast crisis*—a terminal phase in which growth accelerates rapidly. If kinetics were the only mechanism for resistance, then cancers should become highly responsive in the blast crisis, *yet response to traditional chemotherapy is dismal.* The stochastic Goldie-Coldman model suggests an explanation.

Slow-growing tumors are characterized by a near balance between birth and death that is tipped only slightly in favor of birth, while rapidly growing tumors have relatively little internal death. Therefore, far more generations may pass in a chronic tumor, inducing genetic heterogeneity that is orders of magnitude greater than that seen in acute tumors, implying that chronic tumors are far more likely to harbor many multi-agent resistant clones than do acute tumors.

9.5.3.3 Relaxation of the "symmetry assumption"

Day, in 1986 [16], relaxed the "symmetry assumption" that both drugs are equally effective against the target population of the original Goldie-Coldman model. Day developed a stochastic birth-death model of resistance to two drugs in which growth for all cell classes is approximately exponential. It is essentially a stochastic version of the two-drug Goldie-Coldman model [30] in which all cells are stem cells. The model scheme is outlined in Figure 9.9. The essential feature of this model is that the cytotoxic effect of the two drugs on the target populations vary asymmetrically, and this significantly changes the model predictions for optimal scheduling. Day also considered asymmetry in other parameters, such as birth rate. Treatment follows a stochastic log-kill model, implying that the number of survivors is binomially distributed.

Birth-death process. While we have minimized our discussion of stochastic processes (with the notable exception of the Poisson process), we briefly motivate the stochastic birth-death process and develop the basic algorithm for simulating Day's model. Consider a one-dimensional birth-death process for a single population of cells, letting $N(t)$ represents the population size at time t. Let

$$\lambda = \text{per-capita growth rate, and}$$
$$\mu = \text{per-capita death rate.}$$

We assume that births occur at the exponentially distributed rate $\lambda_N = \lambda N$, and deaths occur at the exponentially distributed rate $\mu_N = \mu N$. This gives

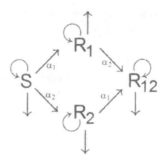

FIGURE 9.9: Schematic for Day's 1986 [16] stochastic version of the Goldie-Coldman model.

us a set of parameters, $\{\lambda_N\}_{N=0}^{N=\infty}$, $\{\mu_N\}_{N=0}^{N=\infty}$, that describe transition rates for our Markov chain. Thus, at any state N, we have:

$$P(\text{birth event}) = P_{N,N+1} = \frac{\lambda_N}{\lambda_N + \mu_N}$$

$$P(\text{death event}) = P_{N,N-1} = \frac{\mu_N}{\lambda_N + \mu_N}$$

The overall rate of transition is exponentially distributed with rate $\lambda_N + \mu_N$. This process approximates exponential growth with rate $\lambda - \mu$.

Implementation of the Day model. For the Day model, we have a birth-death process that is "distributed" between 4 populations. The simplest implementation of this model is given by the following algorithm:

```
Define T_max as time to run simulation.
N = Total number of cells.
Loop from time T = 0 until T = T_max
    1. Determine time until the next event according to random draw
       from an exponential distribution with mean 1 / (mu + lambda)
    2. Randomly determine if event is birth or death.
    3. If birth, set N = N + 1, else set N = N - 1.
    4. If birth, randomly choose progenitor cell based on frequency of
       each type.
        4A. If mother = S, generate one daughter R1 or R2 with
            probability alpha_1 and alpha_2, respectively
        4B. If mother = R1, generate one daughter R12 with probability
            alpha_2. Similar if mother = R2.
    5. If death, randomly select a population, S, R1, R2, or R12 to
       reduce by 1, based on the relative frequency of each.
    6. Advance time by time to next event.
    7. Impose cell death at appropriate treatment times.
End Loop
```

Unfortunately, this algorithm becomes prohibitively computationally expensive for more than about 10^8 cells. To handle larger cell counts, we develop a hybrid stochastic and deterministic model. The basic algorithm for the case with resistance to a single drug follows (i.e. only S and R_1 are considered):

```
Define N_threshold as the population size to switch from stochastic
to deterministic dynamics.
Iterate through each day:
    1. If S < N_threshold:
        A. Run a birth-death process for 1 day and track the number
           of mutations.
        B. Transfer mutant cells to the R population
    2. If S > N_threshold
        A. Increase S according to the exponential growth function.
        B. Calculate the number of cell divisions that occurred.
        C. From the # of divisions, calculate the expected number
           of mutations.
        D. If the number of mutations is > than M_threshold, just
           add them to R.
           If not, iterate through each division and randomly
           determine if a mutation event occurred and add individual
           mutations to R.
    3. If R < N_threshold, run a birth-death process for 1 day.
    4. If R > N_threshold, expand the R population exponentially.
    5. Impose cell death at appropriate treatment times.
```

It is straightforward but somewhat tedious to extend this approach to the two drug situation, and we leave this as an exercise.

Predictions. Day confirmed, using simulations, that alternating schedules are excellent in the symmetric case, but in many asymmetric cases, a simple alternating schedule is often far from optimal. While no universal principle applied to every case, Day found a "worst drug rule" to be broadly applicable. That is, use the weaker of two drugs either first or for a greater duration than the stronger drug; an example of such a treatment is given in Figure 9.10. This approach, somewhat counterintuitively, tends to increase survival as it better suppresses cells resistant to the stronger drug and hence minimizes the emergence of two-drug resistant clones. This prediction runs counter to common clinical practice.

9.5.3.4 Goldie-Coldman contra Norton-Simon

The Goldie-Coldman and Norton-Simon models are somewhat at odds in their predictions for optimizing therapy. Goldie-Coldman attempts to minimize the emergence of new resistance during the treatment period by alternating drugs. As also discussed in Section 9.4.5, the Norton-Simon model prescribes *sequential* dose-dense therapy, under the implicit premise that clinically relevant heterogeneity in drug sensitivity is already present when treat-

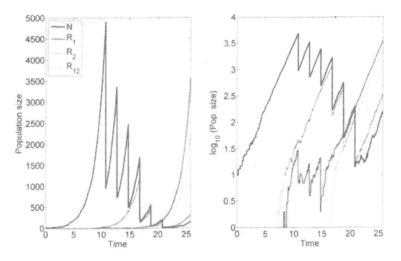

FIGURE 9.10:   Example realization of treatment under the stochastic Day model using a simple birth-death process.   Three treatments of the weak drug are followed by three of the strong; parameters are $\lambda = 1.0$, $\mu = 0.4$, $\alpha_1 = \alpha_2 = 0.001$. These are chosen so that the granularity of the stochastic process can be displayed, and resistance emerges in an unrealistically small tumor.

ment is initiated. Indeed, the Goldie-Coldman model itself predicts that drug resistance arises very early in a tumor's natural history.

Several clinical trials have tested alternating therapy versus standard chemotherapy for breast cancer [10, 17], but all have failed to demonstrate any benefit to alternating over standard schedules. The CALGB 9741 trial [14] is the major trial supporting dose-density, although it does not address sequential versus alternating therapy. Several other trials evaluated sequential versus concurrent or alternating therapy, and seem to suggest a slight benefit to sequential therapy. For example, Bonadonna et al. [11] found sequential DOX → CMF schedule superior to alternating DOX/CMF. However, CMF → DOX was no better than standard CMF, suggesting a simple "sequential is better" interpretation is not really correct, and the order in which these particular agents are administered matters.

A problematic argument based on kinetic resistance suggests that sequential therapy should be generally superior to alternating schedules. Assume that a tumor is evenly divided between two subpopulations, each of which is sensitive to only one of two agents. Then alternating the two agents might be expected to increase kinetic resistant since individual subpopulations would have more time to regrow between treatments. However, if growth rate is modulated by the total tumor population and not the size of any single subpopulation, then the overall response of each subpopulation will be similar whether drugs are alternated or given sequentially, assuming each is equally effective against the sensitive subpopulation.

From this, we argue that, while the clinical data lends more support to the Norton-Simon approach over the Goldie-Coldman, there is no fundamental conflict between the two models. Rather, the conflict arises because the dichotomy between treatment strategies that focus on dose-density vs. those that focus on alternating (or concurrent) treatment are clinically incompatible. The data suggests that maximizing dose-density is more important.

9.5.4 The Monro–Gaffney model and palliative therapy

In the previous two sections we discussed the most influential models in modern clinical oncology. In particular, the log-kill model and the Norton-Simon hypothesis have helped motivate a treatment model where therapy is given as soon as possible in a relatively brief, intense period. Under the Norton-Simon model, an intensive dose-dense regime yields the lowest cancer cell nadir and consequently the best chance for cure. However, as we shall see, when reasonable hope for a cure cannot be entertained, lower-dose, long-term therapy may better increase survival time.

9.5.4.1 The model

Recently, Monro and Gaffney [51] proposed a relatively simple dynamical model that combines the Norton-Simon and Goldie-Coldman notions—that is, the model assumes that chemotherapy efficacy and per-unit time probability of a resistance mutation are both proportional to the unperturbed Gompertzian tumor growth rate. One drug and two classes of cells—drug-sensitive (S) and resistant (R)—are considered. Let, at time t,

$N_S(t) =$ number of chemosensitive cells, and

$N_R(t) =$ number of cells with complete resistance to the drug.

The total number of cells in the tumor is therefore $N(t) = N_S(t) + N_R(t)$. Both classes proliferate via identical Gompertzian kinetics and mutate between class. Mutation from the sensitive to the resistant class occurs at rate τ_1 per cell cycle, and back mutations occur at rate τ_2 per cell cycle. These considerations give the following model:

$$\frac{dN_S}{dt} = \beta \ln\left(\frac{N_\infty}{N}\right) [N_S + \tau_2 N_R - \tau_1 N_S - \lambda C N_S], \qquad (9.90)$$

$$\frac{dN_R}{dt} = \beta \ln\left(\frac{N_\infty}{N}\right) [N_R + \tau_1 N_S - \tau_2 N_R]. \qquad (9.91)$$

The first term in both equations gives Gompertzian growth modulated by the total cell count. The per-capita growth rate is $\beta \ln(N_\infty/N)$, and this term governs the rates of mutation between the classes and the response to treatment. From this it follows that τ_1 and τ_2 are mutation rates per cell cycle, rather than simply per unit time. The term $-\lambda C N_S$ represents the death rate of sensitive cells in response to some concentration of the drug, C, which is also proportional to the proliferation rate. We can also lump λC into a single parameter, C_0.

This model represents an explicit union of the two most influential cancer treatment models, and it is simple but potentially powerful framework with which we can evaluate the Norton-Simon notion of dose-density while explicitly taking Goldie-Coldman into account. Moreover, it would be a straightforward exercise to expand this model to consider the effects of two non-cross-resistant drugs and determine if the Goldie-Coldman prediction that drugs should be rapidly alternated is preserved when Norton-Simon is taken into account. These are important and straightforward projects, but Monro and Gaffney turned their attention toward the problem of maximizing survival in the setting of palliative chemotherapy, rather than maximizing cure probability by minimizing cell nadir or the probability of a resistant mutant arising.

9.5.4.2 Predictions of the Monro-Gaffney model

Optimal dose. Monro and Gaffney hypothesized that overly aggressive treatment in a palliative setting could accelerate selection of treatment resistant mutants and therefore ultimately decrease response to therapy and survival. Numerical simulation of the model indeed supports this suggestion. This is because resistance emerges quickly in response to high-dose therapy, leading to treatment failure no matter how much chemotherapy is given. However, an overly low dose does not sufficiently inhibit sensitive cells, and survival is also decreased. The optimum dose suppresses sensitive cells, but not so much that resistant cells have a strong survival advantage. This is summarized in Figures 9.11 and 9.12.

Treatment delay. A second interesting result of the Monro-Gaffney model is its suggestion that a moderate treatment delay can improve survival time, as shown in Figure 9.12. This is corroborated by results for the clinical treatment of some slow-growing, chronic cancers, in which no survival advantage is conferred by initiating treatment earlier. For example, chronic lymphocytic leukemia (CLL) is incurable but is characterized by a long, indolent clinical course. Early treatment offers no survival benefit in this disease, and therapy is generally deferred until there is clear evidence of disease progression [38].

A large study [48] found, in accord with other results, that androgen deprivation therapy for localized, low-risk prostate cancer actually decreased cancer-specific survival (although it did not affect overall survival). The authors suggested one possible explanation:

> Suppression of moderately or well-differentiated cells not destined to harm a patient's overall survival may allow for the establishment or overgrowth of more rapidly growing malignant clones.

While this study examined hormonal treatment for prostate cancer, the results of the Monro-Gaffney model are likely generalizable to it.

The results in [48] also highlight one aspect of drug resistance that we have not yet taken into account in a model: resistance can be achieved by fairly general mechanisms such as increased proliferation rate or reduced apoptosis, which are expected to translate into a more aggressive phenotype. We also note that, under the model, the longer therapy is delayed, the greater the optimum level of chemotherapy.

Dose-density. Monro and Gaffney also studied non-continuous dosing computationally, and found that protracted dose-densification and intensification schemes generally decrease survival time in the palliative setting, which is unsurprising given the results for continuous therapy. However, they noted that brief dose-dense schemes for "marginally curable" tumors may give an increased chance of cure and minimally affect survival time, and therefore may be appropriate for potentially curable cancers.

9.5.4.3   Summary

The Norton-Simon and Goldie-Coldman models predict that, to maximize probability of a cure, therapy should be initiated as soon as possible and be as dose-dense as possible. That is, "hit 'em hard and fast." The Goldie-Coldman model and variations of it make further predictions about the alternating or sequential scheduling of agents. This is when the goal of therapy is to eliminate all cancer cells from the body. Sadly, for most advanced or metastatic solid tumors, curative therapy remains elusive, and the job of the medical oncologist is largely to provide palliative care, using chemotherapy to extend life or reduce tumor-related symptoms, but not to cure. In this case the Monro-Gaffney model suggests the "less-is-more" approach just discussed. We note that the role of chemotherapy toward the end of life is a complex emotional and sociological issue, and the reader may consult [37] for an introduction to this perplexing but important subject.

At least some data supports the notion that less aggressive therapy in the palliative setting is superior. Palliative chemotherapy has clearly been shown to extend survival in many malignancies [37], but beyond first- or second-line chemotherapy the benefit is less clear. In the case of non-small cell lung cancer (NSCLC), one study [49] found the response rate to third- or fourth-line therapy to be 2.3% and 0%, respectively. The deleterious effect of aggressive therapies on the failing physiologies of terminal cancer patients also doubtless plays a role in the influence on survival, either positive or negative, of chemotherapy given near the end of life. Von Gruenigen et al. [32] found that aggressiveness of care near the end of life in ovarian cancer patients was associated with lower survival, as was chemotherapy during the last three months of life. Hospice care, which forgoes any attempt at curative chemotherapy or other aggressive treatments, was associated with longer survival compared to non-hospice care for certain terminally ill cancer patients [15]. These results

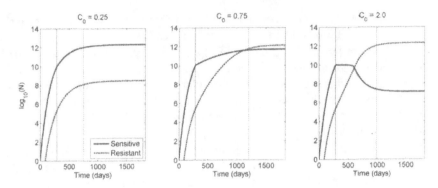

FIGURE 9.11: Response to therapy for three different levels of continuous chemotherapy under the Monro-Gaffney model. The solid vertical line indicates the start of therapy, and the dotted line is the expected time of death. Too little therapy, and the growth of sensitive cells is not sufficiently suppressed (left panel), but too much promotes the rapid overgrowth of resistant cells (right panel). The optimal level, shown in the center panel, leads to a rough balance between sensitive and resistant cells. Parameter values are $N_\infty = 2 \times 10^{12}$, $\beta = 5.928 \times 10^{-3}$ day^{-1}, $\tau_1 = \tau_2 = 10^{-6}$. Death is assumed to occur at a total tumor burden of 5×10^{11} cells, and treatment is initiated when $N = 10^{10}$ cells.

are likely more related to the effect of aggressive therapy on the patient rather than any effect on tumor biology.

9.5.5   The role of host physiology

We have exclusively focused on the effect of chemotherapy on cancer cells, but we must at least note the importance of host physiology, as toxicity to healthy tissues is dose-limiting in cancer therapy, and also can have a detrimental effect on survival. Several authors have considered bone marrow suppression, the most common toxicity. Panetta, for example, considered this in a very simple context [56], while Scholz et al. [61] have proposed a complex model for granulopoiesis, suppression by chemotherapy, and exogenous granulocyte colony-stimulating factor (G-CSF, a growth factor promoting granulopoiesis) support.

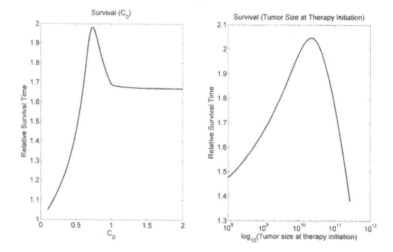

FIGURE 9.12: The left panel gives relative survival time (survival time under treatment / survival time without treatment) for continuous chemotherapy following initiation at a tumor size of 10^{10} cells under different values of C_0, representing the strength of chemotherapy. The right panel gives relative survival time when therapy is initiated at different tumor burdens with a fixed $C_0 = 0.75$.

9.6 Heterogeneous populations: The cell cycle

A number of models have examined treatment on a tumor with sensitive and resistant populations where sensitivity is determined (at least partly) by whether the cell is quiescent or passing through the cell cycle. Therefore, we turn our attention to the cell cycle and how it influences resistance to and scheduling of chemotherapy.

9.6.1 The Smith-Martin conceptual model

It is worthwhile to take this opportunity to explicitly present the cell cycle and some models of the process. The cell cycle classically consists of four phases: G_1 (Gap 1), S (Synthesis), G_2 (Gap 2), and M (Mitosis). In G_1, the cell increases in size while proteins necessary for DNA synthesis are produced. DNA replication occurs in S phase. The cell is readied for mitosis in G_2, and the DNA is divided between two daughter cells in M phase. Interphase comprises G_1, S, and G_2. Chemotherapy generally acts on actively cycling cells, and a number of agents primarily or only affect cells in particular phases. For example, mitotic inhibitors such as the taxanes and vinca alkaloids are

G_2/M phase-specific, while antimetabolites like methotrexate inhibit DNA synthesis and are therefore specific for S phase. Cells that are not actively passing through the cycle are sometimes said to be in G_0, which may be considered a precursor or extension of G_1.

Smith and Martin proposed a conceptual kinetic model for cell proliferation in a classic 1973 paper, "Do Cells Cycle?" [65]. They proposed that following mitosis cells enter some state (A) in which they remain for a variable length of time, but exit with a constant probability, P. That is, the time until exit is an exponential random variable, and in a continuous model the rate of exit is first-order with respect to A-state cells. On leaving A-state, cells enter B-phase, which lasts a fixed amount of time, T_B, and represent passage through the cell-cycle. Following mitosis, cells re-enter the A-state. Thus, the aggregate growth rate of a cell population is a function of the probability (i.e. exponentially distributed rate) at which cells exit A-state, the length of B-phase, and the cellular death rate. For the interested reader, Smith and Martin [65] present equations with which P can be explicitly determined from the gross cell growth rate and T_B. The model is illustrated schematically in Figure 9.13.

Smith and Martin also noted that, if the cell cycle were a fixed-length process and all cells were committed to cycling, then cells starting from the same point in the cell cycle should divide in synchrony. This prediction can be tested by labeling cells with a pulse of $[^3H]dT$ (radio-labeled Thymidine, a component of DNA). Only cells in S phase will pick up the label, so in a culture with cells in various stages of the cycle, a single cohort can be marked. If cells progress through the cell cycle together, then this cohort will all divide at the same time, and no non-labeled cells will divide at that time (because unlabeled cells must have been in a different portion of the cell cycle when the label was applied and would therefore enter mitosis at a different time if the cell cycle were of unvarying length for all cells). In such experiments, one can measure, at any given moment, what fraction of cells undergoing mitosis are labeled (in the marked cohort). This measure is called the fraction of labeled mitoses, or FLM. So, if the cell cycle length were invariant, then the FLM would either be 100% (the cohort is in mitosis) or 0% (the cohort is in interphase). Therefore, the FLM as a function of time would (ideally) be a Heaviside function, jumping instantly to 1 when the cohort entered mitosis and dropping instantly to 0 as the cohort leaves mitosis. This pattern is observed to some degree in actual experiments; however, over time successive peaks or jumps are dampened; i.e., they become more spread out until eventually the FLM approaches an equilibrium proportion between 0 and 1. Smith and Martin explained this dampening by suggesting that the time spent in the A-phase varies among cells, and their model predicted that the dampening in the FLM oscillations should be exponential. This result generally agrees well with experiment. The period of the oscillations are not affected by P (the probability a cell exits the A-state), but the rate of dampening is.

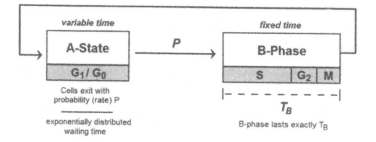

FIGURE 9.13: Schematic of the conceptual kinetic model for cell proliferation proposed by Smith and Martin (see also Figure 1 of [65]).

9.6.2 A delay differential model of the cell cycle

The Smith-Martin model translates perfectly into a delay differential equation. Suppose we have a class of cells in the A-state—i.e., they are non-cycling, quiescent, or in G_0. Let $Q(t)$ be the number of these quiescent cells. The number of actively cycling B-state cells, on the other hand, we denote $P(t)$. We assume that cells transition from quiescent to cycling states at constant rate k, taking exactly τ units of time to complete the cell cycle. Following passage through the cell cycle, each cell gives rise to two daughter cells that re-enter the quiescent compartment. Disregarding cell death for the moment, we can express this biology with the following model:

$$\frac{dQ(t)}{dt} = -kQ(t) + 2kQ(t-\tau), \tag{9.92}$$

$$\frac{dP(t)}{dt} = kQ(t) - kQ(t-\tau), \tag{9.93}$$

with appropriate initial conditions. This basic model can be extended to include details of the cell cycle. For example, one can subdivide the cycling cell class ($P(t)$) into number of cells in interphase, say $I(t)$, and number in mitosis, $M(t)$. Furthermore, define τ_0 and τ_1 to be the time it takes to pass through interphase and mitosis, respectively. Ignoring death again gives

$$\frac{dQ(t)}{dt} = -kQ(t) + 2kQ(t-\tau_0-\tau_1), \tag{9.94}$$

$$\frac{dI(t)}{dt} = kQ(t) - kQ(t-\tau_0), \tag{9.95}$$

$$\frac{dM(t)}{dt} = kQ(t-\tau_0) - kQ(t-\tau_0-\tau_1), \tag{9.96}$$

again with appropriate initial conditions. This slightly more complicated model agrees extremely well with the dynamics predicted by Smith and Martin. Starting with a population of quiescent cells at time 0 with all other

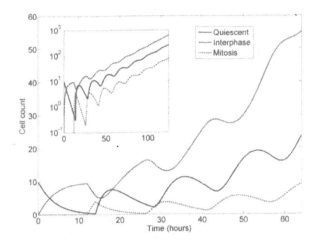

FIGURE 9.14: Basic dynamics of a delay model for the cell cycle considering quiescence, interphase, and mitosis derived from Smith and Martin's conceptual model. The insert shows the dynamics on a log-scale and demonstrates that the model converges to exponential growth.

variables and history set to 0, we observe damped oscillations in the number of cells in mitosis (as well as interphase and quiescence; Figure 9.14). Moreover, as predicted, these oscillations are characterized by a rapid rise to a peak followed by exponential decline. For smaller k (slower transition from quiescence to the cell cycle), oscillations are damped more quickly, which was also predicted by Smith and Martin. Following an early period of irregular growth with oscillations in the different compartments, there is fairly rapid convergence to a constant distribution of cells among states and exponential growth (Figure 9.14). Genrally, dynamics of this and the simpler model that considers only P and Q are similar.

Having formulated a basic model that gives realistic dynamics, we now introduce cell cycle-specific treatment to this framework. First we consider a treatment that targets only actively cycling cells with the simpler formalism. Suppose drug efficacy (cell-kill rate, δ) is proportional to concentration, $C(t)$, and suppose $C(t)$ is piecewise constant—i.e., either chemotherapy is present at a constant dose or it is not present at all. Our model then becomes the following:

$$\frac{dQ(t)}{dt} = -kQ(t) + 2kQ(t-\tau)e^{-\delta \int_{t-\tau}^{t} C(s)ds}, \tag{9.97}$$

$$\frac{dP(t)}{dt} = kQ(t) - kQ(t-\tau)e^{-\delta \int_{t-\tau}^{t} C(s)ds} - \delta C(t)P, \tag{9.98}$$

$$\tag{9.99}$$

where

$$C(t) = \begin{cases} C_0 & \text{(treatment on)}, \\ 0 & \text{(treatment off)}. \end{cases}$$

With this model one can examine how an exponentially growing tumor responds to cell cycle-specific drugs while explicitly accounting for age structure. During the period of time in which the drug is administered, $t \in [0, T]$, the drug concentration, measured as area under the curve (AUC), is

$$AUC = \int_0^T C(s)\,ds = \text{constant}. \qquad (9.100)$$

We use two metrics when evaluating drug efficacy: (1) tumor size at the end of treatment, and (2) tumor size after some large time, T_{max}. The former is similar to cell nadir but accounts for positive growth during treatment time, and the latter is a surrogate for survival time. Note that tumor size after T is not a reliable marker for long-term response because of perturbations in age-structure. The basic dynamics of response to treatment for several simple infusion schedules are shown in Figure 9.15.

According to this model, if the same cumulative dose is delivered in a single constant infusion, it is universally better to prolong the infusion time (Figure 9.16). Also, increasing the concentration of chemotherapy for a single infusion of fixed time gives diminishing returns with increasing dose, comparable to the plateau in cytotoxicity discussed in the section of dose-response curves.

Numerical investigations reveal that dividing a single treatment into multiple treatments delivered in the same net amount of infusion time gives results that vary with cell proliferation rate, k, and with the drug's efficacy, δ. In general, the cell nadir is lowest when a single treatment is given, an unsurprising result as this scenario represents the most dose-dense option. (See the discussion of the Norton-Simon model above.) However, dividing the dose into many treatments, while typically increasing cell nadir, can also increase survival time, unlike the basic Norton-Simon model. However, the picture is not that simple. In certain regions of parameter space, a smaller number of longer treatments is inferior to more, shorter treatments, especially in the case of a rapidly growing tumor (large k) or a strong drug (large δ). Thus, the model suggests that for highly responsive tumors either very few treatments or very many treatments should be given. The former should be done if the curative therapy is the goal, the latter if one is attempting to maximize survival. We leave confirmation of these results as an exercise.

9.6.3 Age-structured models for the cell cycle

Now we return our attention to age-structured models (see Chapter 4). The basic model for quiescent and cycling cells we studied above is in fact equivalent to an age structured partial differential equation (PDE) model, as we

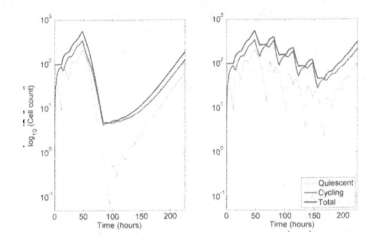

FIGURE 9.15: Time-series of tumor response to either one bolus infusion of chemotherapy or the same dose divided between four episodes.

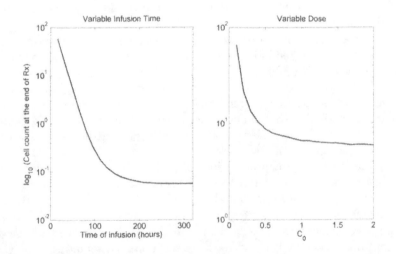

FIGURE 9.16: This figure demonstrates that the same dose delivered over a long time dramatically improves response, but the efficacy eventually plateaus. Increasing the dose delivered in a fixed time also results in increasing cytotoxicity that plateaus with increasing dose due to depletion of the sensitive cell population.

now demonstrate. Recall that $P(t)$ is the number of cells in the cycling compartment. Cells enter this compartment at rate $kQ(t)$ and complete division in exactly τ units of time. We let

$p(t, a) =$ The population density of proliferating cells of age a, $a \in [0, \tau]$.

This is the density of cells of age a, the integral of which gives us number of cells. Now, we have that cycling cells suffer mortality at rate $\delta C(t)$ at time t. This gives us an advection PDE describing the passage of cells through time and age:

$$\frac{\partial p}{\partial t} = -\frac{\partial p}{\partial a} - \delta C(t) p(t, a). \tag{9.101}$$

We impose boundary conditions at age 0 and age τ as:

$$p(t, 0) = kQ(t) \text{ and } p(t, \tau) = kQ(t - \tau) \exp\left(-\delta \int_{t-\tau}^{t} C(s)\, ds\right). \tag{9.102}$$

To complete the model, we include the governing equation for $Q(t)$,

$$\frac{dQ(t)}{dt} = -kQ(t) + 2kQ(t - \tau). \tag{9.103}$$

To see that the age-structure model is equivalent to the delay differential model, we integrate over age to get the number of cycling cells at time t:

$$P(t) = \int_{0}^{\tau} p(t, s)\, ds. \tag{9.104}$$

Integrating (9.101) from 0 to τ with respect to age a yields

$$\frac{dP(t)}{dt} + p(t, \tau) - p(t, 0) = -\delta C(t) P(t). \tag{9.105}$$

Applying the boundary conditions defined in (9.102) returns us to the expression for $dP(t)/dt$ in (9.99):

$$\frac{dP(t)}{dt} = kQ(t) - kQ(t - \tau) \exp\left(-\delta \int_{t-\tau}^{t} C(s)\, ds\right) - \delta C(t) P(t). \tag{9.106}$$

A number of authors have proposed various age-structured models of the cell cycle for tumor cells and their response to cell cycle-specific chemotherapy. Spinelli et al. [66] proposed a model for proliferating and quiescent cells somewhat similar to that above. Basse et al. proposed several models for the cell cycle in tumor cells which explicitly consider multiple stages of the cell cycle [2, 3, 4]. However, in these models cells cycle continuously with no quiescent state (A-state). The most general formulation is given in [4].

Gaffney [25, 26] also proposed several age-structured models that considered mutation to resistance, as in the Goldie-Coldman model, and used these models to study the scheduling of S-phase specific chemotherapies.

A number of models have also been proposed for the response of cycling cells to continuous low-dose irradiation, including several age-structured models [13, 35, 36, 69]. For example, Chen et al. [13] proposed a similar model to that studied above. Letting $n(t, a)$ represent the population density of cells with age a, $a \in [0, \infty)$, we have the model,

$$\frac{\partial n}{\partial t} = -\frac{\partial n}{\partial a} - \dot{D}\alpha(a)n - g(a)n, \qquad (9.107)$$

where $\dot{D}$ is the radiation dose-rate, $\alpha(a)$ gives first order killing that depends upon cell age, and hence position in the cell cycle, and $g(a)$ represents the rate at which cells of age a produce two daughter cells of age 0. The daughters immediately re-enter the cell cycle with no quiescent phase. The domain here is semi-infinite, as cells exit the cycle according to the age-dependent probability distribution $g(a)$ and no right boundary condition is imposed. On the left boundary, however,

$$n(0, t) = 2 \int_0^\infty g(s)n(s, t) \, ds. \qquad (9.108)$$

While this model was meant to represent continuous irradiation, it is equivalent to the case of continuous chemotherapy we consider above. Note that this model ignores any cell death that is quadratic with the dose (as in the linear-quadratic model) or repair of sublethal damage; such models are discussed more extensively in Chapter 11. Similar to our conclusion using the delay model, the model of Chen et al. [13] and later extensions by Hahnfeldt and Hlatkey [35, 36] suggest that protracted irradiation schedules may kill cells better than more intense irradiation. However, intracellular DNA repair/misrepair dynamics of radiation-induced lesions imply that shorter irradiation times lead to greater cell death (again, discussed in Chapter 11). Thus, cell cycle redistribution and DNA repair are competing intracellular dynamics that determine optimal treatment time.

This last point highlights the very important observation that applies to chemotherapy. The pharmacokinetics of the drug in plasma, delivery to the target tumor site, cellular uptake, detoxification, and repair of intracellular lesions will affect optimal treatment in a way that cannot be simply accounted for with first-order cell killing.

9.6.4   More general sensitivity and resistance

Many other models designed to study sensitive and resistant tumor cell populations use simpler ODE formulations for the cell cycle or for transfer between quiescent and proliferating compartments. One of the earliest was that of Birkhead et al. [9], who considered proliferating and quiescent cells which were either genetically susceptible or resistant to some chemotherapy. Let $S_P(t)$ and $S_Q(t)$ represent the number of proliferating and quiescent cells that are sensitive to the drug, respectively. Similarly, let $R_P(t)$ and $R_Q(t)$ be the amount of proliferating and quiescent resistant cells, respectively. Cells transition between quiescent and cycling states at constant per-capita rates, and sensitive cells mutate to become resistant at rate ϵ per cell division. Only genetically susceptible cells that are actively proliferating are vulnerable to chemotherapy. This leads to Birkhead et al.'s formal model:

$$\frac{dS_P}{dt} = \alpha S_P - \eta S_P \mu S_P - \mu S_P - \epsilon S_P + \lambda S_Q, \qquad (9.109)$$

$$\frac{dS_Q}{dt} = \mu S_P - \lambda S_Q, \qquad (9.110)$$

$$\frac{dR_P}{dt} = \epsilon S_P + \alpha R_P - \eta R_P - \mu R_P + \lambda R_Q, \qquad (9.111)$$

$$\frac{dS_Q}{dt} = \mu S_P - \lambda S_Q. \qquad (9.112)$$

This is a basic ODE framework in which the notions of cell cycle specifity considered in the previous section may be combined with mutation to resistance as in the Goldie-Coldman models.

Panetta and Adam [55] later considered a model for cell cycle specific chemotherapy essentially identical to that of Birkhead et al., but without a class of resistant cells. Hahnfeldt et al. [34] also used a similar framework to study two types of proliferating cells, with transitions between compartments, that are differentially sensitive to treatment.

These models tend to agree that intense treatment yields diminishing returns as treatment-sensitive cells become depleted. Hahnfeldt et al. concluded that long term, low dose, "metronomic" therapy may be a better strategy to suppress tumor growth in the long term rather than trying to destroy all cells in the short term.

9.7 Drug transport and the spatial tumor environment

The deranged tumor microenvironment represents a significant barrier to drug transport into the tumor, which can become another mechanism generating treatment resistance. In this section we present some basic transport principles, and then mention how these concepts relate to the tumor spheroid, and tumor cord, geometries (see Chapter 3).

9.7.1 Solute transport across tumor capillaries

Solute exchange between the vascular and extravascular spaces is governed by diffusion and convection. Specifically, exchange is determined by transcapillary concentration gradient, the hydrostatic and osmotic pressure gradients, vascular surface area for exchange, A (mm^2), and three key transport parameters: the hydraulic permeability (or hydraulic conductivity) L_p (mm / hr / mmHg), the diffusional permeability P (mm / hr), and the osmotic reflection coefficient σ (unit-less) [39]. The reader may consult the 1987 review by Jain [39] for an extensive discussion of transcapillary transport in tumors.

Consider the concentration of some solute, S, in a small section of capillary. According to Starling's hypothesis, fluid flow across the capillary wall, J_F, is

$$J_F = L_p A[P_V - P_E - \sigma(\Pi_V - \Pi_E)]. \tag{9.113}$$

Here, $P_V - P_E$ (mmHg) is the hydrostatic pressure difference, and $\Pi_V - \Pi_E$ (mmHg) is the osmotic pressure difference.

The total flow of solute, J_S, in units of mass per time (mg / hr), from plasma into the tumor extracellular space occurs by passive diffusion and convective transport across the capillary wall according to the Staverman-Kedem-Katchalsky equation [39]:

$$J_S = \underbrace{PA(S_V - S_E)}_{\text{diffusion}} + \underbrace{J_F(1 - \sigma_F)\Delta S_{lm}}_{\text{convection}}, \tag{9.114}$$

$$\Delta S_{lm} = \frac{S_V - S_E}{\ln(S_V/S_E)}, \tag{9.115}$$

where S_V is the solute concentration (mg / mm^3) on the vascular side of the capillary, and S_E is the concentration on the extracellular side. Therefore, ΔS_{lm} is the solute log-mean concentration difference. The first term in equation (9.114) gives transport by diffusion, and the second is transport by convection. Constants P and A are defined above.

Another key transport parameter, the hydraulic permeability L_p, is defined as the flow of water through a unit area of capillary wall per unit difference in hydrostatic pressure across the wall [50]. The diffusional permeability P is the mass transport of the solute per unit area of capillary wall per unit of solute concentration difference when the fluid flow is zero [50].

The osmotic pressure is determined by the molarity of dissolved solutes in solution; i.e., $\Pi \propto MRT$, where M is molarity, R is the universal gas constant, and T is temperature. The osmotic pressure gradient across the capillary wall induces fluid flow. The osmotic reflection coefficient σ can be thought of as the fraction of solute that is "reflected" at the capillary wall during ultrafiltration, as opposed to that which flows freely across [50]. Only the fraction reflected can exert osmotic pressure. The value of σ depends, in part, on size of the solute; it is close to 1 for macromolecules such as albumin, and closer to 0 for micromolecules.

The solvent-drag reflection coefficient, σ_F, measures the coupling between fluid flow and solute flow [39]. If σ_F is small, then the solute is readily transported by fluid flow. For a given substance the solvent-drag reflection coefficient, σ_F, tends to be similar in value to the osmotic reflection coefficient, σ [19].

Another important ratio is the Peclet number (Pe):

$$Pe = \frac{J_F(1 - \sigma_F)}{PA} = \frac{L_p(1 - \sigma_F)}{P}[(P_V - P_E) - \sigma(\Pi_V - \Pi_E)]. \qquad (9.116)$$

The Peclet number indicates the relative importance of diffusive versus convective transport [39]. If Pe is small, which tends to be the case for micromolecules, then diffusive transport is dominant. In this case, J_S can be approximated using an effective permeability coefficient P_{Eff}, as follows:

$$J_S = P_{\text{Eff}} A(S_V - S_E). \qquad (9.117)$$

In tumors, the intercapillary distance is typically much greater than normal tissue, hampering drug delivery. Moreover, blood flow is heterogenous and vessels are prone to collapse, causing heterogeneity in delivery. The vasculature is malformed, immature, and quite leaky. This fact allows a great deal of fluid extravasation into the tumor and leakage of macromolecules, giving high hydrostatic and osmotic pressure in the tumor interstitium. Thus, the hyrdostatic pressure and osmotic pressure differences are minimal. On the other hand, the diffusional permeability, P, and hydraulic conductivity, L_p, are much higher in leaky tumor vessels compared to normal, well-formed capillary beds.

9.7.2 Fluid flow in tumors

Flow through the tumor interstitium may be modeled by *Darcy's law* for flow in a porous medium, which gives volumetric flow rate, Q (mm^3 / hr), as:

$$Q = -AK\nabla P, \qquad (9.118)$$

where A (mm^2) is the cross-sectional area of flow, K (mm^2 / mmHg / hr) is the hydraulic conductivity of the medium, and ∇P is the interstitial pressure gradient. Dividing by A gives the flux (or "Darcy flux"), $\mathbf{u}$:

$$\mathbf{u} = -K\nabla\mathbf{P}. \qquad (9.119)$$

We also have the continuity equation for fluid flow:

$$\frac{d\rho}{dt} + \nabla \cdot (\rho\mathbf{u}) = \phi(\mathbf{x}), \qquad (9.120)$$

where ρ is density and ϕ is the source/sink term (in units of volumetric flow per unit volume of tissue). In the case of an incompressible fluid, this reduces to:

$$\nabla \cdot u = \phi(\mathbf{x}). \qquad (9.121)$$

Combining equations (9.119) and (9.121) gives:

$$\nabla \cdot (-K\nabla\mathbf{P}) = \phi(\mathbf{x}). \qquad (9.122)$$

Baxter and Jain applied these basic equations to describe fluid flow in a tumor spheroid [5], and also describe interstitial solute flow in the spheroid by a convection-diffusion equation.

Volumetric flow through a capillary is proportional to the product of the hydrostatic pressure drop from the arterial to the venous ends of the capillary and the resistance to flow [40]; that is,

$$Q = \frac{\Delta P}{FR}. \tag{9.123}$$

In the case of laminar fluid flow in a cylinder of length L and radius R the *Hagen-Poiseuille law* gives flow resistance as:

$$FR = \frac{8\eta L}{\pi R^4}. \tag{9.124}$$

Pozrikidis and Farrow [58] further describe fluid flow in a tumor at the microvessel level.

9.7.3 Tumor spheroid

The tumor spheroid geometry has been widely used to model delivery of cytotoxic drugs. A series of papers by Baxter and Jain [5, 6, 7] modeled fluid flow and macromolecule (e.g. antitumor monoclonal antibodies) delivery to tumors using the spheroid as the base geometry. One of the most important results is that tumors have high interstitial fluid pressure in the core, but this pressure drops off in the periphery. This causes radially outward fluid flow and convective transport of macromolecules out of the tumor. Also, continuous or divided drug infusions can better maintain a transcapillary solute gradient.

9.7.4 Tumor cord framework

The other major geometry that has been used to model in vivo tumors is the tumor cord. Tumor cords are one of the fundamental microarchitectures of solid tumors, consisting of a microvessel nourishing nearby tumor cells [8]. This simple architecture has been used by several authors to represent the in vivo tumor microenvironment [8, 19, 67], and a whole solid tumor can be considered an aggregation of a number of tumor cords.

Doxorubicin (DOX), an important and potent antitumor agent, is characterized by poor penetration into the tumor mass. Eikenberry [19] extended a previous model by El Kareh and Secomb [21], which modeled DOX plasma pharmacokinetics (using a three-compartment model by Robert et al. [59]) and diffusional transport into a non-spatial tumor, to consider transport into a tumor cord geometry using the solute transport laws discussed above.

9.8 Exercises

Exercise 9.1: Consider two exponentially expanding populations of cells, one treated, N_T, and one a control, N_C. Assume the treatment population is exposed to a constant concentration of drug, C, and exposure induces death at the per capita rate β per unit of drug. Show that for a total exposure time T, we arrive at the log-linear model:

$$S = \frac{N_T}{N_C} = e^{-\beta C \times T} \qquad (9.125)$$

Exercise 9.2: We do not arrive at a simple log-kill model if the underlying cellular growth dynamic is not exponential. Consider the case of logistic growth:

$$\frac{dN_C}{dt} = \alpha N_C \left(1 - \frac{N_C}{K} \right) \qquad (9.126)$$

$$\frac{dN_T}{dt} = \alpha N_T \left(1 - \frac{N_T}{K} \right) - \beta C N_T. \qquad (9.127)$$

1. Find the analytic solution to the system and confirm that for constant C the surviving fraction can be expressed as:

$$S = \frac{N_T}{N_C} = \frac{e^{-\alpha T}(K - N_0)(\beta_0 + \alpha) + N_0(\beta_0 + \alpha)}{e^{T(\beta_0 - \alpha)}(K\beta_0 + K\alpha - N_0\beta_0) + N_0\beta_0} \qquad (9.128)$$

   where we let $\beta_0 = \beta C$.

2. Find conditions on the parameters for a log-kill model to approximate survival, i.e. when:

$$S \approx e^{-\beta C \times T} = e^{-\beta_0 T} \qquad (9.129)$$

   Why are these conditions unsurprising?

3. Reproduce Figure 9.2, which plots S versus either β_0 or T on a semi-logarithmic scale. These plots demonstrate a shoulder region when concentration is varied and a roughly bi-exponential curve when exposure time is varied. This suggests non-exponential cell growth as a possible explanation for departures from the log-kill model.

Exercise 9.3: Norton and Simon assume that the tumor growth fraction is equivalent to the per-capita growth rate, $b \ln \left(\frac{N_\infty}{N(t)} \right)$. Show that this is in error (consider the cell cycle time). Does this error affect the model's predictions?

Exercise 9.4: Apply the general Norton-Simon model, $dN/dt = GF(N)N(t) - L(t)GF(N)N(t)$, to different basic growth laws for tumor-growth, e.g. exponential, logistic, or von Bertalanffy. Determine which model predictions are preserved for sigmoidal growth versus the simple exponential growth.

Exercise 9.5: So far, we have only considered a constant level of chemotherapy that is either on or off. Implement 1-, 2-, and 3-compartment pharmacokinetics models for chemotherapy into the model. Parameterized 2- and 3-compartment models for the agent doxorubicin can be found in [20, 31, 59]. Do more realistic pharmacokinetics change the qualitative predictions?

References

[1] Adams DJ: In vitro pharmacodynamic assay for cancer drug development: application to crisnatol, a new DNA intercalator. *Cancer Res* 1989, 49:6615–20.

[2] Basse B, Baguley BC, Marshall ES, Joseph WR, van Brunt B, Wake G, Wall DJ: A mathematical model for analysis of the cell cycle in cell lines derived from human tumors. *J Math Biol* 2003, 47:295–312.

[3] Basse B, Baguley BC, Marshall ES, Joseph WR, van Brunt B, Wake G, Wall DJ: Modelling cell death in human tumour cell lines exposed to the anticancer drug paclitaxel. *J Math Biol* 2004, 49:329–57.

[4] Basse B, Ubezio P: A generalised age- and phase-structured model of human tumour cell populations both unperturbed and exposed to a range of cancer therapies. *Bull Math Biol* 2007, 69:1673–90.

[5] Baxter LT, Jain RK: Transport of fluid and macromolecules in tumors. I. Role of interstitial pressure and convection. *Microvasc Res* 1989, 37:77–104.

[6] Baxter LT, Jain RK: Transport of fluid and macromolecules in tumors. II. Role of heterogeneous perfusion and lymphatics. *Microvasc Res* 1990, 40:246–63.

[7] Baxter LT, Jain RK: Transport of fluid and macromolecules in tumors. III. Role of binding and metabolism. *Microvasc Res* 1991, 41:5–23.

[8] Bertuzzi A, D'Onofrio A, Fasano A, Gandolfi A: Regression and regrowth of tumour cords following single-dose anticancer treatment. *Bull Math Biol* 2003, 65:903–931.

[9] Birkhead BG, Rankin EM, Gallivan S, Dones L, Rubens RD: A mathematical model of the development of drug resistance to cancer chemotherapy. *Eur J Cancer Clin Oncol* 1987, 23:1421–1427.

[10] Boccardo F, Rubagotti A, Amoroso D, Perrotta A, Sismondi P, Farris A, Mesiti M, Gallo L, Pacini P, Villa E, Agostara B: Lack of effectiveness of

adjuvant alternating chemotherapy in node-positive, estrogen-receptor-negative premenopausal breast cancer patients: Results of a multicentric Italian study. The Breast Cancer Adjuvant Chemo-Hormone Therapy Cooperative Group (GROCTA). *Cancer Invest* 1997, 15:505–512.

[11] Bonadonna G, Zambetti M, Moliterni A, Gianni L, Valagussa P: Clinical relevance of different sequencing of doxorubicin and cyclophosphamide, methotrexate, and Fluorouracil in operable breast cancer. *J Clin Oncol* 2004, 22:1614–1620.

[12] Bonilla L, Ben-Aharon I, Vidal L, Gafter-Gvili A, Leibovici L, Stemmer SM: Dose-dense chemotherapy in nonmetastatic breast cancer: A systematic review and meta-analysis of randomized controlled trials. *J Natl Cancer Inst* 2010, 102:1845–1854.

[13] Chen PL, Brenner DJ, Sachs RK: Ionizing radiation damage to cells: Effects of cell cycle redistribution. *Math Biosci* 1995, 126:147–170.

[14] Citron ML, Berry DA, Cirrincione C, Hudis C, Winer EP, Gradishar WJ, Davidson NE, Martino S, Livingston R, Ingle JN, Perez EA, Carpenter J, Hurd D, Holland JF, Smith BL, Sartor CI, Leung EH, Abrams J, Schilsky RL, Muss HB, Norton L: Randomized trial of dose-dense versus conventionally scheduled and sequential versus concurrent combination chemotherapy as postoperative adjuvant treatment of node-positive primary breast cancer: First report of Intergroup Trial C9741/Cancer and Leukemia Group B Trial 9741. *J Clin Oncol* 2003, 21:1431–439.

[15] Connor SR, Pyenson B, Fitch K, Spence C, Iwasaki K: Comparing hospice and nonhospice patient survival among patients who die within a three-year window. *J Pain Symptom Manage* 2007, 33:238–426.

[16] Day RS: Treatment sequencing, asymmetry, and uncertainty: Protocol strategies for combination chemotherapy. *Cancer Res* 1986, 46:3876–3885.

[17] De Placido S, Perrone F, Carlomagno C, Morabito A, Pagliarulo C, Lauria R, Marinelli A, De Laurentiis M, Varriale E, Petrella G, et al: CMF vs alternating CMF/EV in the adjuvant treatment of operable breast cancer. A single centre randomised clinical trial (Naples GUN-3 study). *Br J Cancer* 1995, 71:1283–1287.

[18] Drewinko B, Patchen M, Yang LY, Barlogie B: Differential killing efficacy of twenty antitumor drugs on proliferating and nonproliferating human tumor cells. *Cancer Res* 1981, 41:2328–2333.

[19] Eikenberry S: A tumor cord model for doxorubicin delivery and dose optimization in solid tumors. *Theor Biol Med Model* 2009, 6:16.

[20] Eksborg S, Strandler HS, Edsmyr F, Näslund I, Tahvanainen P: Pharmacokinetic study of i.v. infusions of adriamycin. *Eur J Clin Pharmacol* 1985 28:205–212.

[21] El-Kareh AW, Secomb TW: A mathematical model for comparison of bolus injection, continuous infusion, and liposomal delivery of Doxorubicin to tumor cells. *Neoplasia* 2000, 2:325–338.

[22] El-Kareh AW, Secomb TW: A mathematical model for cisplatin cellular pharmacodynamics. *Neoplasia* 2003, 5:161–169.

[23] El-Kareh AW, Secomb TW: Two-mechanism peak concentration model for cellular pharmacodynamics of Doxorubicin. *Neoplasia* 2005, 7:705–713.

[24] Foster PL: Methods for determining spontaneous mutation rates. *Methods Enzymol* 2006, 409:195–213.

[25] Gaffney EA: The application of mathematical modelling to aspects of adjuvant chemotherapy scheduling. *J Math Biol* 2004, 48:375–422.

[26] Gaffney EA: The mathematical modelling of adjuvant chemotherapy scheduling: Incorporating the effects of protocol rest phases and pharmacokinetics. *Bull Math Biol* 2005, 67:563–611.

[27] Gilman A: The initial clinical trial of nitrogen mustard. *Am J Surg* 1963, 105:574–578.

[28] Goldie JH, Coldman AJ: A mathematic model for relating the drug sensitivity of tumors to their spontaneous mutation rate. *Cancer Treat Rep* 1979, 63:1727–1733.

[29] Goldie JH, Coldman AJ: Quantitative model for multiple levels of drug resistance in clinical tumors. *Cancer Treat Rep* 1983, 67:923–931.

[30] Goldie JH, Coldman AJ, Gudauskas GA: Rationale for the use of alternating non-cross-resistant chemotherapy. *Cancer Treat Rep* 1982, 66:439–449.

[31] Greene RF, Collins JM, Jenkins JF, Speyer JL, Myers CE: Plasma pharmacokinetics of adriamycin and adriamycinol: Implications for the design of in vitro experiments and treatment protocols. *Cancer Res* 1983, 43:3417–3421.

[32] von Gruenigen V, Daly B, Gibbons H, Hutchins J, Green A: Indicators of survival duration in ovarian cancer and implications for aggressiveness of care. *Cancer* 2008, 112:2221–2227.

[33] Gyllenberg M, Webb GF: Quiescence as an explanation of Gompertzian tumor growth. *Growth Dev Aging* 1989, 53:25–33.

[34] Hahnfeldt P, Folkman J, Hlatky L: Minimizing long-term tumor burden: The logic for metronomic chemotherapeutic dosing and its antiangiogenic basis. *J Theor Biol* 2003, 220:545–554.

[35] Hahnfeldt P, Hlatky L: Resensitization due to redistribution of cells in the phases of the cell cycle during arbitrary radiation protocols. *Radiat Res* 1996, 145:134–143.

[36] Hahnfeldt P, Hlatky L: Cell resensitization during protracted dosing of heterogeneous cell populations. *Radiat Res* 1998, 150:681–687.

[37] Harrington SE, Smith TJ: The role of chemotherapy at the end of life: "when is enough, enough?" *JAMA* 2008, 299:2667–2678.

[38] Hallek M, Cheson BD, Catovsky D, Caligaris-Cappio F, Dighiero G, Dhner H, Hillmen P, Keating MJ, Montserrat E, Rai KR, Kipps TJ; International workshop on chronic lymphocytic leukemia: Guidelines for the diagnosis and treatment of chronic lymphocytic leukemia. A report from the International Workshop on Chronic Lymphocytic Leukemia updating the National Cancer Institute Working Group 1996 guidelines. *Blood* 2008, 111:5446–5456.

[39] Jain RK: Transport of molecules across tumor vasculature. *Cancer Metastasis Rev* 1987, 6:559–593.

[40] Jain RK: Determinants of tumor blood flow: A review. *Cancer Res* 1988, 48:2641–2658.

[41] Jumbe N, Louie A, Leary R, Liu W, Deziel MR, Tam VH, Bachhawat R, Freeman C, Kahn JB, Bush K, Dudley MN, Miller MH, Drusano GL: Application of a mathematical model to prevent in vivo amplification of antibiotic-resistant bacterial populations during therapy. *J Clin Invest* 2003, 112:275–285.

[42] Kozusko F, Bajzer Z: Combining Gompertzian growth and cell population dynamics. *Math Biosci* 2003, 185:153–167.

[43] Kozusko F, Bourdeau M, Bajzer Z, Dingli D: A microenvironment based model of antimitotic therapy of Gompertzian tumor growth. *Bull Math Biol* 2007, 69:1691–1708.

[44] Kuh HJ, Jang SH, Wientjes MG, Au JL: Computational model of intracellular pharmacokinetics of paclitaxel. *J Pharmacol Exp Ther* 2000, 293:761–770.

[45] Lea DE, Coulson CA: The distribution of the numbers of mutants in bacterial populations. *J Genet* 1949, 49:264–285.

[46] Levasseur LM, Slocum HK, Rustum YM, Greco WR: Modeling of the time-dependency of in vitro drug cytotoxicity and resistance. *Cancer Res* 1998, 58:5749–5761.

[47] Luria SE, Delbrück M: Mutations of bacteria from virus sensitivity to virus resistance. *Genetics* 1943, 28:491–511.

[48] Lu-Yao GL, Albertsen PC, Moore DF, Shih W, Lin Y, DiPaola RS, Yao SL: Survival following primary androgen deprivation therapy among men with localized prostate cancer. *JAMA* 2008, 300:173–181.

[49] Massarelli E, Andre F, Liu DD, Lee JJ, Wolf M, Fandi A, Ochs J, Le Chevalier T, Fossella F, Herbst RS: A retrospective analysis of the outcome of patients who have received two prior chemotherapy regimens including platinum and docetaxel for recurrent non-small-cell lung cancer. *Lung Cancer* 2003, 39:55–61.

[50] Michel CC: Capillary permeability and how it may change. *J Physiol* 1988, 404:1–29.

[51] Monro HC, Gaffney EA: Modelling chemotherapy resistance in palliation and failed cure. *J Theor Biol* 2009, 257:292–302.

[52] Mueller M, de la Peña A, Derendorf H: Issues in pharmacokinetics and pharmacodynamics of anti-infective agents: Kill curves versus MIC. *Antimicrob Agents Chemother* 2004, 48:369–377.

[53] Norton L: A Gompertzian model of human breast cancer growth. *Cancer Res* 1988, 48:7067–7071.

[54] Norton L, Simon R: Tumor size, sensitivity to therapy, and design of treatment schedules. *Cancer Treat Rep* 1977, 61:1307–1317.

[55] Panetta JC, Adam J: A mathematical model of cycle-specific chemotherapy. *Mathl Comput Modelling* 1995, 22:67–82.

[56] Panetta JC: A mathematical model of breast and ovarian cancer treated with paclitaxel. *Math Biosci* 1997, 146:89–113.

[57] Papac RJ: Origins of cancer therapy. *Yale J Biol Med* 2001, 74:391–398.

[58] Pozrikidis C, Farrow DA: A model of fluid flow in solid tumors. *Ann Biomed Eng* 2003, 31:181–194.

[59] Robert J, Illiadis A, Hoerni B, Cano JP, Durand M, Lagarde C: Pharmacokinetics of adriamycin in patients with breast cancer: Correlation between pharmacokinetic parameters and clinical short-term response. *Eur J Cancer Clin Oncol* 1982, 18:739–745.

[60] Sadowitz PD, Hubbard BA, Dabrowiak JC, Goodisman J, Tacka KA, Aktas MK, Cunningham MJ, Dubowy RL, Souid AK: Kinetics of cisplatin binding to cellular DNA and modulations by thiol-blocking agents and thiol drugs. *Drug Metab Dispos* 2002, 30:183–190.

[61] Scholz M, Engel C, Loeffler M: Modelling human granulopoiesis under poly-chemotherapy with G-CSF support. *J Math Biol* 2005, 50:397–439.

[62] Skipper HE: Kinetics of mammary tumor cell growth and implications for therapy. *Cancer* 1971, 28:1479–1499.

[63] Skipper HE, Schabel FM, Wilcox WS: Experimental evaluation of potential anticancer agents. XIII. On the criteria and kinetics associated with "curability" of experimental leukemia. *Cancer Chemother Rep* 1964, 35:1–111.

[64] Seldin DW: The development of the clearance concept. *J Nephrol* 2004, 17:166–171.

[65] Smith JA, Martin L: Do cells cycle? *Proc Natl Acad Sci USA* 1973, 70:1263–1267.

[66] Spinelli L, Torricelli A, Ubezio P, Basse B: Modelling the balance between quiescence and cell death in normal and tumour cell populations. *Math Biosci* 2006, 202:349–370.

[67] Thalhauser CJ, Sankar T, Preul MC, Kuang Y: Explicit separation of growth and motility in a new tumor cord model. *Bull Math Biol* 2009, 71:585–601.

[68] Welling PG: *Pharmacokinetics: Processes, mathematics, and applications.* Washington, DC: American Chemical Society, 1997.

[69] Zaider M, Minerbo GN: A mathematical model for cell cycle progression under continuous low-dose-rate irradiation. *Radiat Res* 1993, 133:20–26.

[70] Zheng Q: Progress of a half century in the study of the Luria-Delbrück distribution. *Math Biosci* 1999, 162:1–32.

[71] Zhi JG, Nightingale CH, Quintiliani R: Impact of dosage regimens on the efficacy of piperacillin against *Pseudomonas aeruginosa* in neutropenic mice. *J Pharm Sci* 1988, 77:991–992.

Chapter 10

Major Anticancer Chemotherapies

10.1 Introduction

Diseases desperate grown
By desperate appliance are relieved,
Or not at all

William Shakespeare, *Hamlet*, Act IV, Scene 3

Cancer has been treated surgically since antiquity. Following the discovery of X-rays in 1896, physicians and scientists were quick to adapt radiation as an anticancer therapy. With the development of modern systemic chemotherapy in the latter half of the 20th century, the treatment of solid tumors came to follow a "cut-burn-poison" model. In the previous chapter we covered the "poison" of chemotherapy, and we discuss the "burn" of radiation in the next. Here, we follow the previous chapter with additional history and background on some of the particular poisons in common use. The name is apt—white arsenic, a favorite poison of the Middle Ages, was used effectively against chronic leukemia in the late 1800s. Although arsenic subsequently fell out of favor as an anticancer drug, arsenic trioxide was recently revived as a highly effective therapy for promyelocytic leukemia (PML) [39, 53]. Indeed, arsenicals enjoy a colorful history as a therapeutic for a variety conditions that dates to antiquity [53]. Just as appropriately, modern cancer chemotherapy arose from poison gas research that followed the First World War and blossomed during the second.

In early 1942, several investigators at Yale University, in coordination with the wartime Office of Scientific Research and Development, began studying the nitrogen mustards [19], a class of chemical derived from the sulfur mustards, the infamous "mustard gas" that had been used to terrible effect on the battlefields of World War I. By this time, it was well established that both mustard gas and its less reactive cousin, nitrogen mustard, were potently cytotoxic to rapidly proliferating cells, particularly those of the hematopoietic (blood-forming) system and the gastrointestinal tract. Moreover, the effect of nitrogen mustard on cells appeared similar to that of X-rays (and indeed, both induce similar breaks in the DNA, although by different mechanisms). Motivated by such data, Gilman and colleagues took, at first, a single mouse with a transplanted lymphoid tumor and treated it with nitrogen mustard until the tumor completely disappeared. After some time, the tumor recurred and was treated again, only this time regression was less complete. When it recurred a third time, no further treatment affected it.

Despite this and other studies with similar results, the compound eventually was tested in humans.[1] Initially, a patient with terminal lymphosarcoma, moribund and riddled with tumors, was selected and treated to extraordinary effect, his tumors shrinking and disappearing in days. However, mirroring the mouse, the cancer recurred, and a second course gave a diminished response. When the disease began to progress again, further treatment had little effect. The clinical trajectories of these early trials in both mouse and human foreshadowed the clinical experience of chemotherapy in the decades to come, and it remains a common trajectory to this day: the development of treatment resistance has always been the hobgoblin of chemotherapy (see Chapter 9 for more details.)

Because of the secrecy of the chemical gas program, these and other trials were conducted behind closed doors, with patient charts even reporting treatment by "0.1 mg. per kg. compound X given intravenously" [19]. In 1946 wartime secrecy restrictions were lifted, and a series of publications reporting these results ushered in the modern era of cancer chemotherapy. In the next two decades, a host of similar agents with antitumor activity would be developed. Our goal in this chapter is to introduce the biological and medical aspects of some particular chemotherapies in widespread use, as well as some particular mathematical treatments, contra the more general treatment given in the prior chapter.

10.2 Alkylating and alkalating-like agents

The alkylating agents make up the largest and oldest class of chemotherapeutics. They target the DNA, RNA, and proteins, although their interaction with DNA is generally considered to be the dominant mechanism for cytotoxicity. Many alkylating agents are small organic molecules with two reactive functional groups. The first forms a covalent link to a nitrogenous base in the DNA, forming what is called a monofunctional adduct. The second functional group can then react with another base, covalently cross-linking them to form a bifunctional adduct. These cross-links may be either intra-strand

---

[1]A ubiquitous myth, appearing widely in the literature and even in textbooks, holds that a 1943 wartime incident in Bari, Italy, was the inspiration for chemotherapy. The story goes that sailors and civilians exposed to mustard gas were observed to experience bone marrow suppression, giving physicians the idea of using the gas as chemotherapy. This tale is certainly false, as the first human trials had begun a year *before* the Bari incident.

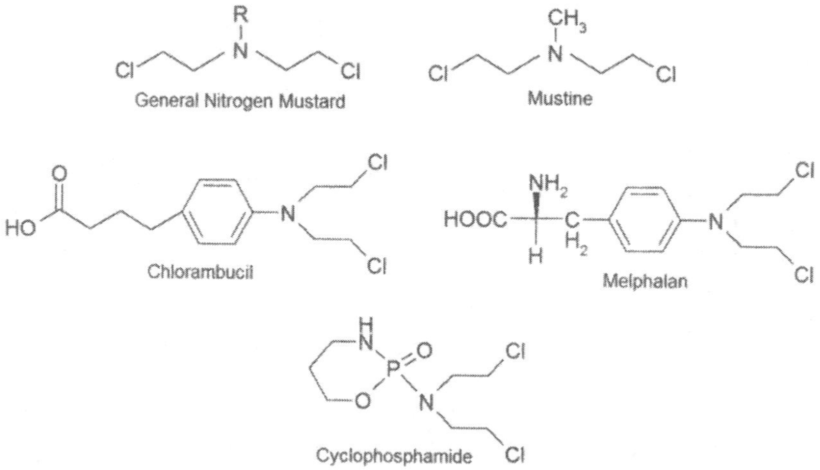

FIGURE 10.1: The basic structure of the nitrogen mustards, along with some of the specific drugs in this class.

or cross-strand, and they interfere with DNA processing machinery (e.g., for replication and (or) transcription) and often distort the helical structure of the DNA.

Alkylating agents are widely used to treat various leukemias, but they have limited selectively for tumor cells, often causing "collateral" damage to non-target cells. What selective cyotoxicity they possess is believed to be primarily due to (1) high proliferation rates of tumor cells, and (2) deficiencies in DNA repair mechanisms in tumor cells. The latter is particularly important in testicular cancer, as defective DNA repair is linked to hypersensitivity to cisplatin in this disease [33].

10.2.1 Nitrogen mustards

The nitrogen mustards were the first anticancer chemotherapies, and are derivatives of the sulfur mustards, which, as mentioned above, are chemical warfare agents. Their antitumor activity was discovered by a group studying such agents during World War II, and they were secretly tested on the first human subjects in 1942. It was not until secrecy restrictions were lifted in 1946 that these investigations were published publicly [19]. They are classified as alkylating agents and produce DNA cross-links by the mechanism outlined above. The first agents studied were tris(β-chloroethyl)amine and methyl-bis(β-chloroethyl)amine (mustine). All other drugs in the nitrogen mustard class are based on mustine, and the basic structure of such drugs is shown in Figure 10.1. Nitrogen mustards covalently bind to guanine residues in the DNA to form adducts and form cross-links between opposite strands of DNA.

In the 1950s and 60s two major classes of nitrogen mustard derivatives were developed specifically to target tumor cells. The reactivity of the nitrogen mustard base is also reduced in these drugs, which allows more of the drug to reach the DNA target intact. The aromatic nitrogen mustards replace the methyl group of nitrogen mustard with an (R-Phe) group, stabilizing the nitrogen and reducing drug reactivity. Chlorambucil and melphalan are two such drugs. Melphalan (L-phenylalanine mustard) incorporates the amino acid residue L-phenylalanine, allowing it to be taken up by the L-phenylalanine active transport mechanism. Therefore, it may selectively target cells undergoing rapid protein synthesis, such as rapidly proliferating tumor cells. An advantage of the aromatic nitrogen mustards is that they can be given orally.

The oxazaphosphorines are nitrogen mustard pro-drugs—inactive agents that become activated by metabolic mechanisms after introduction to the body. The major drug in this class is cyclophosphamide, an inactive transport form of the drug with a much higher therapeutic range than do agents with direct alkylating ability. Ifosfamide is a derivative of cyclophosphamide that is metabolized more slowly and has a somewhat different therapeutic profile. Moreover, fractionation of ifosfamide dose schedules and delivery by continuous infusions of between 1 and 5 days decreases off-target toxicity, increases the maximum allowable dose, and improves therapeutic efficacy.

The uptake and efflux mechanisms for many of the nitrogen mustards have been well quantified, and many enter the cell by carrier-mediated transport. Mustine enters cells primarily by active transport mediated by the choline carrier; choline is a natural substance similar in structure to mustine [20].

Uptake of melphalan appears to be mediated by two separate active transport mechanisms involved in amino acid transport [4, 21], while efflux likely occurs only by simple diffusion. Goldenberg et al. [21] measured cell-to-interstitium concentration ratio for melphalan to be 11.0 ± 3.1 in physiological conditions, and Begleiter et al. [2] showed that about 80% of intracellular drug was exchangeable with the extracellular compartment. The rate of change in intracellular concentration of the drug can be described as the sum of two Michaelis-Menten terms for uptake and a first-order term for efflux:

$$\frac{dC}{dt} = V_{max1} \frac{[S]}{K_{m1} + [S]} + V_{max2} \frac{[S]}{K_{m2} + [S]} - kC, \qquad (10.1)$$

where $C(t)$ and $[S]$ are intra- and extracellular drug concentrations, respectively, $V_{max1} = 1.1 \pm 0.4 \times 10^{-16}$ mol/cell/min, $K_{m1} = 8.0 \pm 3.0 \times 10^{-5}$ M, $V_{max2} = 2.2 \pm 2.4 \times 10^{-17}$ mol/cell/min, $K_{m2} = 1.0 \pm 0.7 \times 10^{-5}$ M [4], and $k = 0.13 \pm 0.05$ [2].

Cyclophosphamide is similarly transported into the cell by two saturable processes [22], and the kinetics of DNA damage by cyclophosphamide have been studied both in vitro [7] and in vivo [9, 50].

Chlorambucil is more water-soluble than other nitrogen mustards [49], and transport across the membrane occurs by passive diffusion, with intracellular

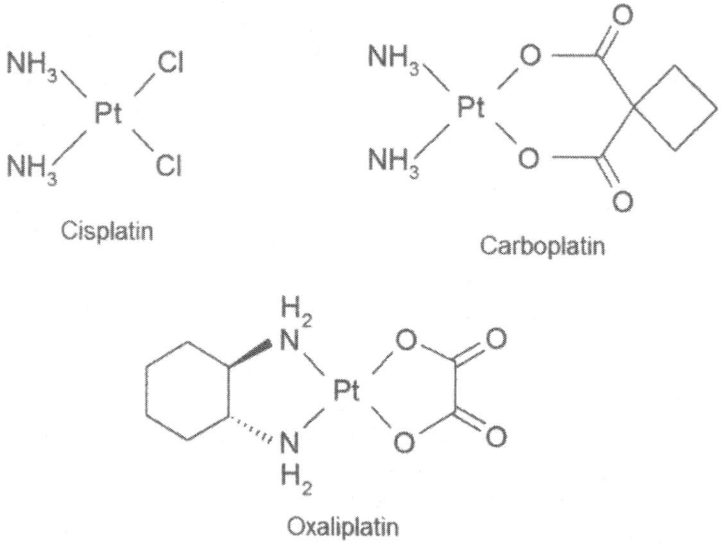

FIGURE 10.2: The three FDA approved platinum-based drugs. All are based on cisplatin.

concentration reaching equilibrium with the extracellular concentration very rapidly [1]. The kinetics of chlorambucil metabolism and binding to DNA, RNA, and proteins have been studied quantitatively by several groups [1, 27].

For a comprehensive quantitative review of membrane transport of the nitrogen mustards the reader may consult the 1980 review by Goldenberg and Begleiter [20]. To our knowledge, the nitrogen mustards have not been studied using more complicated dynamical models.

10.2.2   Platinum-based drugs

The cytotoxic action of cisplatin (cis-diamminedichloroplatinum(II)) was accidently discovered in 1965. It was first used in humans as an anticancer agent in 1971 and approved by the FDA in 1978 [33]. It remains a frontline treatment for ovarian and testicular cancer and has been the focus of several mathematical models. Metastatic testicular cancer in particular is often hypersentitive to cisplatin, and cisplatin-based chemotherapy regimes have increased 5-year survival rates from 5% to over 80% [33] making it one of the few solid metastatic tumors for which systemic chemotherapy is curative. All other platinum-based drugs are based on cisplatin, which has the chemical structure cis-[Pt Cl$_2$ (NH$_3$)$_2$]; the structures of this and several other platinum drugs are shown in Figure 10.2.

Cisplatin's major mechanism of action as a cytotoxic agent requires intracellular activation of the molecule by aquation (displacement by water) of one of the two Cl groups. This allows activated cisplatin to bind covalently to purine DNA bases to form a cisplatin-DNA adduct. Aquation of the other Cl group leads it to bind to a second DNA base, thus cross-linking the DNA strand [33]. The formation of such mono- and bifunctional adducts interferes with replication and transcription and ultimately leads to cellular apoptosis by a sequence of events that is not yet fully understood.

Platinum-based drugs are often classified as alkylating agents, and while their mechanism of action is similar to alkylating agents they contain no alkyl group and therefore such a classification is technically incorrect.

Cisplatin can cause severe damage to the kidneys (nephrotoxicity) and gastrointestinal tract. Carboplatin (cis-diammine(1,1,-cyclobutanedicarboxylato)-platinum(II)), a second-generation compound, replaces the chlorine groups with a more stable leaving group, and has vastly reduced nephrotoxicity compared to cisplatin [33, 35]. Myelosuppression (inhibition of bone marrow) is the dose-limiting toxicity for carboplatin. Cisplatin and carboplatin have similar efficacy in the treatment of ovarian cancer [33].

Cisplatin is highly polar, which limits its diffusion across the non-polar cell membrane. Copper transport proteins appear to determine cisplatin uptake and efflux. Copper transporter-1 (CTR1) plays a major role in uptake, and copper efflux proteins remove cisplatin from the cell. Intracellular thiols detoxify cisplatin, and DNA-repair mechanisms can repair cisplatin lesions [33].

There is a great deal of quantitative information on the kinetics of cisplatin-DNA binding and cross-linking as well as binding kinetics between various thiols and cisplatin. Several models of in vitro uptake and cytotoxicity have been developed. The relatively simple mechanism of action and the relative wealth of quantitative data for cisplatin and previous work make it an excellent choice for modeling.

There are several studies on the pharmacokinetics of plasma cisplatin, which can be reasonably described by a one-compartment model. The kinetics of DNA binding and cross-linking by both cisplatin and carboplatin were quantified by Knox et al. [35], and later studies have quantified the kinetics of cisplatin binding to thiols such as glutathione [8, 25] and metallothionein [24]. Based on their experimental data, Sadowitz et al. proposed a simple mathematical model accounting for cisplatin transport across the cell membrane, detoxification by intracellular thiols, and binding to the DNA [44]. El-Kareh and Secomb have developed several dynamical models of the uptake kinetics and cytotoxicity of cisplatin [15, 16].

10.2.3   Nitrosoureas

The nitrosoureas were developed in the 1960s, following the finding by the Cancer Chemotherapy National Service Center screening program that nitrosoguanidines had weak activity against intraperitoneal leukemias in mice.

The nitrosoureas are structurally similar, are very lipophilic, have a low extent of ionization, and do not readily bind plasma proteins. Therefore, they can cross the blood-brain barrier and enter the central nervous system (CNS). Interest in these agents began with the discovery that the drug was partially effective against leukemia cells implanted intracerebrally [45]. At that time, no other agents had activity against CNS tumors, so this finding spurred development of this class of drugs.

The two most successful and commonly used nitrosoureas are 1,3-bis(2-chloroethyl)-1-nitrosourea (BCNU, carmustine) and 1-(2-chloroethyl)-3-cyclohexyl-1-nitrosourea (CCNU, lomustine). Because of their ability to cross the blood-brain barrier, they are frequently used to treat CNS malignancies, including malignant gliomas (primary brain tumors). Delayed bone-marrow suppression is the dose-limiting toxicity. While there is conflicting data, the evidence overall suggests that the nitrosoureas are equally cytotoxic to non-proliferating and proliferating cells [11, 26].

In the body, BCNU and CCNU break down fairly rapidly into 2-chloroethyldiazene hydroxide and isocyanate [3, 49]. The former undergoes further reactions to form a chloroethyl carbonium ion which in turn forms mono- and bifunctional DNA adducts [49]. In solution, the half-lives for BCNU and CCNU have been variously measured as 46 and 53 minutes [52], and 57 and 64 minutes [3], respectively. Both agents appear to cross the cell membrane by passive diffusion, and the isocyanate decomposition product readily enters cells by diffusion. However, the carbonium ion that alkylates the DNA is not taken up [3], and therefore breakdown of the nitrosoureas before they have reached their cellular target may represent a barrier to therapy. This is one aspect of these drugs that could be studied with a theoretical model.

Weikam and Deen [52] developed a quantitative kinetics scheme that described DNA alkylation by the nitrosureas that successfully explained in vitro dose-response curves in 9L rat glioma cells. Kohn [34] reports some quantitative data on the kinetics of DNA crosslinking by nitrosoureas.

10.2.4   Methylating agents

Methylating agents form monofunctional DNA adducts and typically add a methyl group to guanine residues. Important methylating agents are dacarbazine, procarbazine, and temozolomide.

10.3   Antitumor antibiotics

In general, the term antibiotic can refer to any compound derived from a microbial source—like bacteria of the genus *Streptomyces*, for example—that kills cells. A number of such compounds effectively kill cancer cells and are hence called antitumor or antineoplastic antiobiotics to distinguish them from their more famous siblings, the antimicrobial antibiotics. A number of such antineoplastic antibiotics are in common use in chemotherapy. Their primary modes of action are to inhibit nucleic acid function in some way.

10.3.1 Anthracyclines

The anthracyclines are a class of antitumor antibiotics derived from *Steptomyces* bacteria with a variety of mechanisms of action. They are currently one of the most useful and widely used chemotherapies, with activity against a very broad range of neoplasms. The most commonly used are daunorubicin, doxorubicin, idarubicin, and epirubicin. All are very similar to doxorubicin, which remains the standard and most commonly used anthracycline.

Doxorubicin has received a great deal of attention from theorists, who have proposed many competing mathematical models for treatment using this agent [12, 14, 17, 28, 41, 46]. (For details, see Chapter 9.) Its plasma pharmacokinetics have been quantified using two- and three-compartment models [13, 23, 42].

10.3.2 Mitomycin-C

One agent of particular interest is mitomycin-C, which produces intrastrand DNA cross-links by a complex mechanism that requires enzymatic reduction for activation [49]. The requirement for bioreduction suggests that mitomycin-C, unlike most other chemotherapeutics, may have significant activity in hypoxic tumors since such environments are reducing. Indeed, Teicher et al. have consistently found mitomycin to be toxic to hypoxic cells [48, 47], and it is equally effective against proliferating and non-proliferating cells [11].

10.3.3 Bleomycins

The bleomycins are a family of glycopeptides first isolated from *Streptomyces verticullus* in 1966. They are used in combination chemotherapy against several lymphomas, head and neck cancers, and germ-cell tumors [6]. Bleomycin is most successfully used in cisplatin-etoposide-bleomycin combination therapy, which is 80% curative for testicular cancer [6, 33]. The bleomycins are fairly large molecules (MW $\sim$ 1500 Da) with a common core structure and a variable sugar moiety and positively charged tail [6]. The commercial formula is called Blenoxane and is a mixture primarily of bleomycins A_2 and B_2. The term bleomycin therefore refers to a variety of similar molecules, although "bleomycin" is typically regarded as a single agent in the literature.

Bleomycin targets the DNA by several related pathways that induce single and double stranded DNA breaks. Such damage is very similar to that induced by ionizing radiation (see Chapter 11). Bleomycins are hydrophilic and do not diffuse across the cell membrane. They bind to a surface receptor and enter the cell by receptor-mediation endocytosis [6, 37]. Although membrane transport of bleomycins is inefficient, internalization of only a few hundred molecules is needed to cause cell death [37].

Once inside the cell, bleomycin must be activated by an oxygen species and a reduced transition metal co-factor (nearly always Fe^{2+}). The activation cascade and subsequent mechanism for DNA damage is complex and variable. For a thorough characterization of the process that includes a partial quantification of the kinetics the reader may consult the review by Burger [5]. In brief, reduced iron (Fe^{2+}) and O_2 form a complex with bleomycin. The O_2-Fe^{2+}-BLM complex equilibrates with $O_2^{\bullet -}$-Fe^{3+}-BLM. The latter complex is converted to activated bleomycin, Fe^{3+}-BLM-OOH [5]. Several other pathways involving superoxide and hydrogen peroxide can also generate activated bleomycin [5].

In the absence of a DNA target, activated bleomycin is unstable and undergoes irreversible molecular suicide. Otherwise, it associates with DNA in a sequence specific manner. That is, it targets GpC or GpT sequences with a specificity that is also affected by nearby nucleotides [5]. Bleomycin also tends to target DNA linker regions (between nucleosomes) of genes being actively transcribed [6], and "hot spots" for attack appear to exist [5]. Once associated with the DNA, bleomycin always attacks the C4′ position of the doexyribose sugar, removing the hydrogen and forming a DNA radical. A series of radical reactions then occur following one of two pathways. If O_2 is available a break in the DNA backbone is created. If not, an abasic site in the DNA (monomer location lacking either a purine or pyrimidine) is produced [6, 5].

Following formation of the initial DNA lesion, bleomycin is rapidly reactivated. Since reactivation apparently occurs faster than dissociation of bleomycin from DNA [37], a second lesion is often formed near the first, accounting for the large number of double-stranded DNA breaks induced by bleomycin. It has been estimated that a single bleomycin molecule can induce 8–10 DNA strand breaks [37], and 10–20% of DNA cleavage events are double-stranded breaks [5].

The in vitro dose-response curve for bleomycin toxicity is always concave upward (it plateaus) when plotted on a semilogarithmic plot.

10.4   Antimetabolites

The principle antimetabolites in cancer chemotherapy are the antifolates, which primarily interfere with nitrogenous base synthesis, thereby inhibiting DNA and RNA synthesis. The antifolates are generally considered to be S-phase specific.

Folic acids are part of the biosynthetic pathway that produces thymidylate, or deoxythymidine monophosphate (dTMP), one of the nucleosides required for DNA synthesis. The pathway converts dUMP (deoxyuridine monophosphate) to dTMP. The key enzyme is thymidylate synthase (TS), which oxidizes N^5,N^{10}-methylenetetrahydrofolate (5,10-CH_2FH_4) to 7,8-dihydrofolate (7,8-FH_2), after which tetrahydrofolate must be regenerated. This is accomplished in two steps: the first is catalyzed by dihydrofolate reductase (DHFR) and the second by serine hydroxymethyl transferase.

Antifolates disrupt dTMP synthesis either by directly targeting TS or by inhibiting DHFR and thus depleting the 5,10-CH_2FH_4 necessary for TS activity. Tetrahydrofolates are necessary for de novo purine synthesis, which is also blocked by DHFR inhibition.

DHFR inhibitors have been in use for over 60 years, beginning with aminopterin in 1948. It was replaced by methotrexate (MXT) in 1956, which is still commonly used. Another compound, 5-fluorouracil, is a pyrimidine analog that binds to and permanently deactivates TS. As a uracil analogue, it also has antitumor activity through incorporation into RNA.

10.5   Mitotic inhibitors

The mitotic inhibitors primarily interfere with microtubule dynamics, generating a dysfunctional mitotic spindle causing G2/M phase block or death during M phase. Therefore, although they do affect cells during interphase, they are considered M-phase specific drugs. The two major classes of mitotic inhibitors are the vinca alkaloids and taxanes, although other mitotic inhibitors exist. Such drugs have been the (at least indirect) focus of many mathematical models studying the efficacy of cell-cycle specific chemotherapies (see Chapter 9).

10.5.1   Taxanes

The taxanes are a relatively recent development in cytotoxic chemotherapy, only coming into widespread use in the 1990s, although their development spanned several decades. In the 1960s, National Cancer Institute (NCI) screening programs found that the crude extract of Pacific yew tree (*Taxus brevifolia*) bark had anticancer activity. In 1971, the active ingredient paclitaxel was identified, and in 1983 clinical trials began. Trials were hampered by hypersentivity reactions, which were eventually tempered with pre-treatment steroids and a switch to long (e.g., 24 hours) infusion times. Finally, in 1992 the FDA approved paclitaxel for advanced ovarian cancer [40].

The two major taxanes in clinical use are paclitaxel (Taxol) and docetaxel (Taxotere). Like the vinca alkaloids (see below), they are considered mitotic spindle inhibitors, although they act by stabilizing, rather than inhibiting, microtubule formation. Both pacli- and docetaxel comprise a complex taxane ring system and an ester side chain. The drugs differ in this side chain [40].

Paclitaxel acts on tubulin, a major building-block of the cytoskeleton. Tubulin is a dimer, with subunits α and β, that polymerize to form microtubules. Microtubules are normally in dynamic equilibrium with tubulin dimers, and paclitaxel binds reversibly to tubulin (preferably the β subunit) shifting the equilibrium to favor microtubule formation [40].

Paclitaxel binding greatly increases microtubule stability, but microtubules are inherently dynamic polymers that undergo constant and rapid remodeling. The two remodeling processes are dynamic instability, where microtubules transition between lengthening and shortening phases at their ends, and treadmilling, where tubulin is added to one end of the microtubule at the same rate at which it is removed from the other [31]. At low concentrations, paclitaxel inhibits these dynamic processes, and the microtubules that form are dysfunctional. Microtubules are one of the polymers that make up the cellular cytoskeleton, and microtubule dysfunction can affect cell motility, intracellular transport, and response to growth factors. However, their most dramatic role is in forming the mitotic spindle during mitosis, which has a deranged morphology in taxane-treated cells [31]. The taxanes are therefore considered to be M-phase specific, inhibiting cell proliferation primarily by blocking the cell cycle at the G2-M phase transition [43]. Cytotoxicity increases three-fold from G1 to G2 [10]. Taxanes also may retard the cell's response to growth factors during interphase [43].

Cells treated with paclitaxel develop abnormal asters and bundles of microtubules. The asters appear in G2/M phase and have not been associated with toxicity. On the other hand, microtubule bundles form throughout the cell cycle and are associated with toxicity, while decline of these structures correlates with paclitaxel resistance [43]. However, the jury is still out regarding the precise mechanism of cytotoxicity, and more subtle effects than these on microtubule dynamics may be clinically more important [31].

The taxanes are characterized by biphasic plasma pharmacokinetics and are highly protein-bound [40]. Transport across the membrane occurs by passive diffusion, but P-glycoprotein expression associated with multi-drug resistance (MDR) causes active efflux of the drug [40]. At equilibrium, the intracellular drug concentration is several hundred times greater than extracellular concentration, as nearly all intracellular drug binds to cellular components [36]. Kuh et al. [36] developed a model of intracellular pharmacokinetics for paclitaxel. Panetta [38] proposed a mathematical model of paclitaxel treatment of an in vivo tumor consisting of proliferating and quiescent cells (see Chapter 9).

10.5.2   Vinca alkaloids

The vinca alkaloids were isolated from the Madagascar periwinkle plant (*Catharanthus roseus*, formerly *Vinca rosea*) in the 1960s. Periwinkle had been used extensively in folk medicine to treat various disorders, including diabetes. This prompted laboratory investigation, and although the anti-diabetic activity failed to be substantiated, the vinca alkaloids were found to suppress the bone marrow and have antileukemic activity [29]. The earliest, best studied, and most widely used vinca alkaloids are vincristine and vinblastine.

At high concentrations, the vinca alkaloids cause concentration-dependent disintegration of microtubule polymers, and this was the first mechanism for cytotoxicity identified. However, at much lower concentrations the vinca alkaloids kinetically stabilize microtubules, causing subtle disorganization of the mitotic spindle [30]. This inhibition of mitotic spindle function causes cells to become arrested at metaphase; this in turn correlates strongly with inhibition of cell proliferation.

Therefore, while the taxanes induce microtubule polymerization and the vinca alkaloids inhibit polymerization, both classes of drugs inhibit microtubule dynamics at low concentrations, which may represent a common mechanism for their clinical anticancer activity [31, 32]. Interestingly, the differential effects at different concentrations may be explained (for vinca alkaloids) by the fact that vinblastine has two binding sites. At low concentrations, it binds with high affinity to sites at the microtubule ends ($K_d = 1.9$ μm, 16-17 binding sites / microtubule), thus inhibiting microtubule remodeling without significantly affecting overall mass. At higher concentrations it binds to low affinity sites on the microtubule surface to cause depolymerization [30].

The vinca alkaloids also appear to have a profound ability to destroy tumoral vasculature. The cytoskeleton of endothelial cells can be very rapidly destabilized by vinca alkaloids, leading to cell death and loss of blood flow to the tumor within minutes. Moreover, endothelial cells of the developing and immature tumor vasculature appear to be selectively targeted [32]. Jordan and Wilson [32] published a nice review of agents that target microtubules.

Like the taxanes, the vinca alkaloids bind extensively to cellular components, and at equilibrium the intracellular concentration is much greater than the extracellular concentration [30].

10.6   Non-cytotoxic and targeted therapies

A number of treatment modalities other than cytotoxic chemotherapies exist, as we briefly discuss here. Prostate and breast cancers are treated hormonally: androgen deprivation by castration or anti-androgens has long been the cornerstone of prostate cancer treatment (see Chapter 5 for a thorough discussion), and anti-estrogens are used to prevent breast cancer recurrence. More recently, monoclonal antibodies targeting either cancer-specific, mutated proteins, or cytokines involved in cancer progression, have also been used. In [18], the authors discussed at length another targeted therapy, the treatment of chronic myelogenous leukemia with the drug, imatinib.

References

[1] Bank BB, Kanganis D, Liebes LF, Silber R: Chlorambucil pharmacokinetics and DNA binding in chronic lymphocytic leukemia lymphocytes. *Cancer Res* 1989, 49:554–559.

[2] Begleiter A, Grover J, Goldenberg GJ: Mechanism of efflux of melphalan from L5178Y lymphoblasts in vitro. *Cancer Res* 1982, 42:987–991.

[3] Begleiter A, Lam HP, Goldenberg GJ: Mechanism of uptake of nitrosoureas by L5178Y lymphoblasts in vitro. *Cancer Res* 1977, 37:1022–1027.

[4] Begleiter A, Lam HY, Grover J, Froese E, Goldenberg GJ: Evidence for active transport of melphalan by two amino acid carriers in L5178Y lymphoblasts in vitro. *Cancer Res* 1979, 39:353–359.

[5] Burger RM: Cleavage of nucleic acids by bleomycin. *Chem Rev* 1998, 98:1153–1170.

[6] Chen J, Stubbe J: Bleomycins: Towards better therapeutics. *Nat Rev Cancer* 2005, 5:102–112.

[7] Crook TR, Souhami RL, McLean AE: Cytotoxicity, DNA cross-linking, and single strand breaks induced by activated cyclophosphamide and acrolein in human leukemia cells. *Cancer Res* 1986, 46:5029–5034.

[8] Dabrowiak JC, Goodisman J, Souid AK: Kinetic study of the reaction of cisplatin with thiols. *Drug Metab Dispos* 2002, 30:1378–1384.

[9] DeNeve W, Valeriote F, Edelstein M, Everett C, Bischoff M: In vivo DNA cross-linking by cyclophosphamide: Comparison of human chronic lymphatic leukemia cells with mouse L1210 leukemia and normal bone marrow cells. *Cancer Res* 1989, 49:3452–3456.

[10] Donaldson KL, Goolsby GL, Wahl AF: Cytotoxicity of the anticancer agents cisplatin and taxol during cell proliferation and the cell cycle. *Int J Cancer* 1994, 57:847–855.

[11] Drewinko B, Patchen M, Yang LY, Barlogie B: Differential killing efficacy of twenty antitumor drugs on proliferating and nonproliferating human tumor cells. *Cancer Res* 1981, 41:2328–2333.

[12] Eikenberry S: A tumor cord model for doxorubicin delivery and dose optimization in solid tumors. *Theor Biol Med Model* 2009, 6:16.

[13] Eksborg S, Strandler HS, Edsmyr F, Näslund I, Tahvanainen P: Pharmacokinetic study of i.v. infusions of adriamycin. *Eur J Clin Pharmacol* 1985 28:205–212.

[14] El-Kareh AW, Secomb TW: A mathematical model for comparison of bolus injection, continuous infusion, and liposomal delivery of doxorubicin to tumor cells. *Neoplasia* 2000, 2:325–338.

[15] El-Kareh AW, Secomb TW: A mathematical model for cisplatin cellular pharmacodynamics. *Neoplasia* 2003, 5:161–169.

[16] El-Kareh AW, Secomb TW: A theoretical model for intraperitoneal delivery of cisplatin and the effect of hyperthermia on drug penetration distance. *Neoplasia* 2004, 6:117–127.

[17] El-Kareh AW, Secomb TW: Two-mechanism peak concentration model for cellular pharmacodynamics of doxorubicin. *Neoplasia* 2005, 7:705–713.

[18] Everett RA, Zhao Y, Flores KB, Kuang Y: Data and implication based comparison of two chronic myeloid leukemia models. *Math Biosc Eng* 2013, 10:1501–1518.

[19] Gilman A: The initial clinical trial of nitrogen mustard. *Am J Surg* 1963, 105:574–578.

[20] Goldenberg GJ, Begleiter A: Membrane transport of alkylating agents. *Pharmacol Ther* 1980, 8:237–274.

[21] Goldenberg GJ, Lam HY, Begleiter A: Active carrier-mediated transport of melphalan by two separate amino acid transport systems in LPC-1 plasmacytoma cells in vitro. *J Biol Chem* 1979, 254:1057–1064.

[22] Goldenberg GJ, Land HB, Cormack DV: Mechanism of cyclophosphamide transport by L5178Y lymphoblasts in vitro. *Cancer Res* 1974, 34:3274–3282.

[23] Greene RF, Collins JM, Jenkins JF, Speyer JL, Myers CE: Plasma pharmacokinetics of adriamycin and adriamycinol: Implications for the design of in vitro experiments and treatment protocols. *Cancer Res* 1983, 43:3417–3421.

[24] Hagrman D, Goodisman J, Dabrowiak JC, Souid AK: Kinetic study on the reaction of cisplatin with metallothionein. *Drug Metab Dispos* 2003, 31:916–923.

[25] Hagrman D, Goodisman J, Souid AK: Kinetic study on the reactions of platinum drugs with glutathione. *J Pharmacol Exp Ther* 2004, 308:658–666.

[26] Hahn GM, Gordon LF, Kurkjian SD: Responses of cycling and non-cycling cells to 1,3-bis(2-chloroethyl)-1-nitrosourea and to bleomycin. *Cancer Res* 1974, 34:2373–2377.

[27] Hill BT: Studies on the transport and cellular distribution of chlorambucil in the Yoshida ascites sarcoma. *Biochem Pharmacol* 1972, 21:495–502.

[28] Jackson TL: Intracellular accumulation and mechanism of action of doxorubicin in a spatio-temporal tumor model. *J Theor Biol* 2003, 220:201–213.

[29] Johnson IS, Armstrong JG, Gorman M, Burnett JP Jr.: The vinca alkaloids: A new class of oncolytic agents. *Cancer Res* 1963, 23:1390–1427.

[30] Jordan MA, Thrower D, Wilson L: Mechanism of inhibition of cell proliferation by vinca alkaloids. *Cancer Res* 1991, 51:2212–2222.

[31] Jordan MA, Toso RJ, Thrower D, Wilson L: Mechanism of mitotic block and inhibition of cell proliferation by taxol at low concentrations. *Proc Natl Acad Sci USA* 1993, 90:9552–9556.

[32] Jordan MA, Wilson L: Microtubules as a target for anticancer drugs. *Nat Rev Cancer* 2004, 4:253–265.

[33] Kelland L: The resurgence of platinum-based cancer chemotherapy. *Nat Rev Cancer* 2007, 7:573–584.

[34] Kohn KW: Interstrand cross-linking of DNA by 1,3-bis(2-chloroethyl)-1-nitrosourea and other 1-(2-haloethyl)-1-nitrosoureas. *Cancer Res* 1977, 37:1450–1454.

[35] Knox RJ, Friedlos F, Lydall DA, Roberts JJ: Mechanism of cytotoxicity of anticancer platinum drugs: Evidence that cis-diamminedichloroplatinum(II) and cis-diammine-(1,1-cyclobutanedicarboxylato)platinum(II) differ only in the kinetics of their interaction with DNA. *Cancer Res* 1986, 46:1972–1979.

[36] Kuh HJ, Jang SH, Wientjes MG, Au JL: Computational model of intracellular pharmacokinetics of paclitaxel. *J Pharmacol Exp Ther* 2000, 293:761–770.

[37] Mir LM, Tounekti O, Orlowski S: Bleomycin: Revival of an old drug. *Gen Pharmac* 1996, 27:745–748.

[38] Panetta JC: A mathematical model of breast and ovarian cancer treated with paclitaxel. *Math Biosci* 1997, 146:89–113.

[39] Papac RJ: Origins of cancer therapy. *Yale J Biol Med* 2001, 74:391–398.

[40] Pazdur R, Kudelka AP, Kavanagh JJ, Cohen PR, Raber MN: The taxoids: Paclitaxel (Taxol) and docetaxel (Taxotere). *Cancer Treat Rev* 1993, 19:351–386.

[41] Ribba B, Marron K, Agur Z, Alarcón T, Maini PK: A mathematical model of doxorubicin treatment efficacy for non-Hodgkin's lymphoma: Investigation of the current protocol through theoretical modelling results. *Bull Math Biol* 2005, 67:79–99.

[42] Robert J, Illiadis A, Hoerni B, Cano JP, Durand M, Lagarde C: Pharmacokinetics of adriamycin in patients with breast cancer: Correlation between pharmacokinetic parameters and clinical short-term response. *Eur J Cancer Clin Oncol* 1982, 18:739–745.

[43] Rowinsky EK, Donehower RC, Jones RJ, Tucker RW: Microtubule changes and cytotoxicity in leukemic cell lines treated with taxol. *Cancer Res* 1988, 48:4093–4100.

[44] Sadowitz PD, Hubbard BA, Dabrowiak JC, Goodisman J, Tacka KA, Aktas MK, Cunningham MJ, Dubowy RL, Souid AK: Kinetics of cisplatin binding to cellular DNA and modulations by thiol-blocking agents and thiol drugs. *Drug Metab Dispos* 2002, 30:183–190.

[45] Schabel FM Jr, Johnston TP, McCaleb GS, Montgomery JA, Laster WR, Skipper HE: Experimental evaluation of potential anticancer agents VIII. Effects of certain nitrosoureas on intracerebral L1210 leukemia. *Cancer Res* 1963, 23:725–733.

[46] Sinek JP, Sanga S, Zheng X, Frieboes HB, Ferrari M, Cristini V: Predicting drug pharmacokinetics and effect in vascularized tumors using computer simulation. *J Math Biol* 2009, 58:485–510.

[47] Teicher BA, Holden SA, al-Achi A, Herman TS: Classification of antineoplastic treatments by their differential toxicity toward putative oxygenated and hypoxic tumor subpopulations in vivo in the FSaIIC murine fibrosarcoma. *Cancer Res* 1990, 50:3339–3344.

[48] Teicher BA, Lazo JS, Sartorelli AC: Classification of antineoplastic agents by their selective toxicities toward oxygenated and hypoxic tumor cells. *Cancer Res* 1981, 41:73–81.

[49] Thurston DE: *Chemistry and Pharmacology of Anticancer Drugs.* Boca Raton, FL: CRC Press, 2006.

[50] Wang JY, Prorok G, Vaughan WP: Cytotoxicity, DNA cross-linking, and DNA single-strand breaks induced by cyclophosphamide in a rat leukemia in vivo. *Cancer Chemother Pharmacol* 1993, 31:381–386.

[51] Urano M, Fukuda N, Koike S: The effect of bleomycin on survival and tumor growth in a C3H mouse mammary carcinoma. *Cancer Res* 1973, 33:2849–2855.

[52] Weinkam RJ, Deen DF: Quantitative dose-response relations for the cytotoxic activity of chloroethylnitrosoureas in cell culture. *Cancer Res* 1982, 42:1008–1014.

[53] Zhu J, Chen Z, Lallemand-Breitenbach V, de Thé H: How acute promyelocytic leukaemia revived arsenic. *Nat Rev Cancer* 2002, 2:705–713.

Chapter 11

Radiation Therapy

11.1 Introduction

On December 28, 1895, Wilhelm Röntgen (often spelled "Roentgen" in English) presented to the Physical Society of Würzburg a report detailing his discovery of a type of extraordinary new rays. These rays were capable of passing through black cardboard to light a paper, coated in barium platinocyanide, with a "brilliant fluorescence." These unknown "X-rays" (so named since x is often the unknown variable) passed through all materials, but with varying permeabilities that related loosely to the materials' density, biological materials included; in Röntgen's original report:[1]

> If the hand be held before the fluorescent screen, the shadow shows the bones clearly with only faint outlines of the surrounding tissues.

Photographic plates were also affected by X-rays, as demonstrated famously by the skeletal image of a hand with a visible ring shown in Figure 11.1, and their diagnostic potential was immediately recognized. Within a few short months high quality medical radiographs had already been produced, and, before the end of 1896, the Edison Roentgen Ray Apparatus was being manufactured by General Electric. In one illustration of how quickly X-rays were adopted for medical use, Williams wrote in 1903, in the preface to what was already the third edition of his textbook (the first had been published in 1901), *The Roentgen Rays in Surgery and Medicine,*

> I planned to include as complete a list as possible of the publications on the medical and surgical uses of the Roentgen rays, but when it was found that the list would add nearly one hundred pages, it was omitted...

The exact origins and credit for using X-rays therapeutically, as opposed to diagnostically, are shrouded in some mystery and controversy. A fair num-

---

[1] On a new kind of rays. See [32] for an 1896 English translation of the original paper. The interested reader may also have easier access to a more recent reprint [33] and accompanying commentary [34].

ber of sporadic attempts to treat various cancers and other maladies by X-rays were carried out in the first few years following Röntgen's discovery, but these were generally unscientific, poorly documented interventions of uncertain benefit.[2] Leopold Freund, now regarded as the father of radiotherapy as a scientific specialty, performed the first rigorous investigations of radiotherapy. Beginning on November 24, 1886, he performed three experimental X-ray treatments of a giant hairy nevus (a congenital pigmented skin lesion) [23]. Freund also founded the method of fractionated radiotherapy, where a large cumulative radiation dose is divided into small daily doses. Fractionated radiotherapy endures today, and we devote a significant portion of this chapter to models geared toward optimizing fractionation schedules. By the very early 1900s, a number of reports of X-ray treatments for various cancers had appeared in the literature (e.g., by Pusey in 1902 [29]), and an early consensus began to emerge that radiotherapy affected biological tissue, and it could at least slow the course of disease in many patients.

It is of note that Röntgen never applied for any patents related to X-rays, nor did he ever benefit financially from his monumental discovery. When, in 1901, he received the first Nobel Prize in Physics, Röntgen donated the prize money to the University of Würzburg.[3]

X-rays represent the first and still most important type of medically used ionizing radiation. By far the most important non-diagnostic use of radiation is cancer treatment, and this is the sole topic of the remainder of this chapter. Ionizing radiation exerts its effects by generating ions in biological material, which in turn induce damaging radical chain reactions that damage many cellular components. The most important target is the DNA, and radiation-induced DNA lesions can cause mutations, loss of genetic material, genetic and genomic instability, and cell death. This cytotoxic effect is the basis for its efficacy against a wide variety of cancers (as well as other neoplastic and hyperproliferative conditions).

Quantitative models have been fairly widely used in radiation therapy, and at least three major areas have received significant theoretical attention: (1) the underlying mechanisms for lethality in relation to dose-response curves, (2) the kinetics of radiation-induced damage repair, and (3) optimal fractionation of radiation regimes for various cancers. DNA repair is the central dynamic determining the cellular response to radiation, and all models focus, in essence, on the interaction between DNA lesions and associated repair enzymes.

---

[2]Emil Grubbé, a medical student at the time X-rays were discovered and a later radiotherapy pioneer, may have attempted the first therapeutic X-ray treatment for a recurrent breast carcinoma on January 29, 1896 [23].

[3]Such a tradition of eschewing monetary gain was continued with the later work of the Curies. Notably, Edison disagreed with this view and sought to profit from the X-rays. However, after his assistant Clarence Dally died from radiation exposure, Edison abandoned X-ray work, saying in 1903, "Don't talk to me about X-rays, I am afraid of them."

FIGURE 11.1: Photographic plate of a living human hand exposed to X-rays from Röntgen's original 1895 report. Figure adapted from [33], a reprint of the translation in [32].

While dose-response curves in chemotherapy have received relatively modest theoretical attention, dose-response curves in radiotherapy have been studied extensively with mechanistic models. Explaining the details of such curves has been the primary motivation for models of the mechanisms of radiation-induced damage. These models also form the foundation for quantitative predictions of dose/fractionation dependence in modern radiotherapy. Such models also must take into account the kinetics of damage repair, and therefore are the most fundamental to our study of radiotherapy.

All dose-response curves are exponential for sufficiently high doses, but most are characterized by a shoulder region at low doses [27], as shown in Figure 11.2. A great deal of work has focused on understanding the underlying mechanism producing the shoulder, and at least three non-mutually exclusive hypotheses exist.

The **classical "target-hit" hypothesis** posits that there are discrete "targets," presumably double-stranded DNA, which can be "hit" by ionizing radiation. Such hits are assumed to be independently and randomly distributed throughout the cell population and occur in direct proportion to the dose of radiation delivered. Cell death is assumed to follow from absorption of a sufficient number of hits. This hypothesis accounts for no repair processes or further interactions between lesions [15]. Classical theory predicts that the shoulder region arises as consequence of multiple hits being necessary for lethality. Indeed, a cell almost certainly must suffer many hits for death to ultimately occur, but as we shall see, this is more a consequence of various repair processes rather than an intrinsic property of the cellular damage response. A single double-stranded break is sufficient to cause death [28]. While in many ways insightful, classical theory is fatally flawed in its disregard of repair.

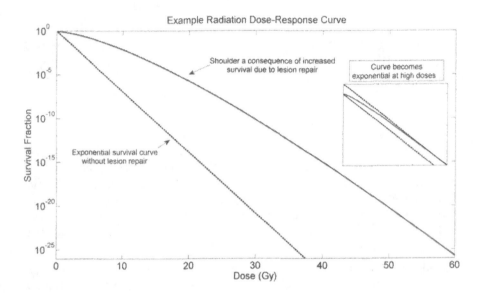

FIGURE 11.2: Example dose-response curve for irradiated cells on a log-linear scale. The shoulder region, as discussed in the text, is predicted to be a consequence of DNA lesion repair by most models. As shown in the insert, all dose-response curves become exponential for sufficiently large radiation doses. Note that this curve is only for demonstrative purposes, and it is not based on any particular data.

Two more modern hypotheses for radiation damage explain the shoulder region in a more biologically plausible way. The first, which leads to **lesion-interaction models**, posit that two (or more) sublethal or potentially lethal lesions initially formed by independent radiation tracks interact to become lethal lesions. The second, expressed as **saturable repair models**, suggests that saturable enzymatic repair processes can repair some otherwise lethal damage induced by radiation. Both hypotheses assume a time-window in which repair reactions occur. After this time lesions become permanent, or as we say somewhat awkwardly, "fixed." (Here, the word "fix" is used to mean "made stable" or unchanging, as in fixation. It does not mean to mend.) Models of these hypotheses tend to assume that the fraction of the cell population with unrepaired lesions dies.

A quasi-mechanistic model of the lesion-interaction type—the linear-quadratic (LQ) formalism—successfully describes the shoulder region of radiation dose-response curves and is now used nearly universally to predict response under different delivery schemes for radiation therapy. This simple formalism encapsulates a great deal of underlying biology and is predictive and powerful. It has supplanted older empirical relationships such as nominal standard dose (NSD), time, dose, and fractionation (TDF), and cumulative radiation effect (CRE), formulas that all have obvious problems and have lead a leading practitioner to declare that, "Perhaps the main [problem with these empirical methods] is the availability of easy-to-use tables that have discouraged clinicians from thinking about the real biological factors involved" [11].

Many of the more mechanistic lesion-interaction and even saturable repair models of radiation-induced cell death approximate the LQ model within the clinically relevant range of radiation doses. However, for high-dose radiation schemes, predictions of the LQ model can diverge significantly from those of other models. In this section, we first discuss the molecular mechanisms by which radiation causes cell death and various models of these mechanisms. We also discuss models of DNA break repair kinetics, as this is intimately related to cytotoxicity. Finally, we discuss the LQ model and its most common mechanistic interpretation, and the application of the LQ model to radiation fractionation scheduling.

11.2 Molecular mechanisms

Ionizing radiation exerts its biological effects by imparting energy to the target material. This energy induces ionizations which leads to the rapid production of free radicals. Such radicals diffuse and react with cellular components, the most important being DNA, and damages them. This damage is detected by the cell, which activates a cellular response that can lead either to successful damage repair or cell death. Improper or incomplete repair of DNA damage can cause cell death or a variety of mutations, chromosomal aberrations, and genetic instability [28].

Radiation damage occurs almost instantaneously. Following an acute dose of radiation, ionizations are complete within 10^{-15} s, and DNA lesions have been induced within a microsecond. However, delayed radical reactions may occur over a time-scale of minutes, and cellular responses to radiation can occur over hours or days [28].

In the past, radiation was measured using units of radiation exposure. Now radiation is measured using units of absorbed dose. The SI unit for absorbed dose is the gray (Gy), which is defined as the absorption of 1 joule of energy per kilogram of material (1 J/kg). The Gy has replaced the older rad (radiation absorbed dose) unit, and 1 Gy = 100 rad. To simplify historical comparisons and avoid conversion, the centigray (cGy) is sometimes used in the literature [27].

While the Gy unit reflects the total amount of energy imparted to target material, the notion of linear energy transfer (LET) quantifies the spatial distribution of energy deposition along a radiation track. For a given radiation type, LET is roughly defined as the energy lost per unit distance through the target material, or as the ratio of total energy lost to the material over the path length through the material. LET is typically measured in units of KeV / μm. Low-LET radiation includes X-rays, β particles (electrons) and secondary effects from γ-rays, while high-LET radiation includes neutrons, protons, and α-particles (helium nuclei). Low-LET radiation is sparsely ionizing, causing ionizations over a large target volume [28]. High-LET radiation, on the other hand, is densely ionizing.

A related concept is that of relative biological effectiveness (RBE). For a given test radiation type, its RBE is defined as the ratio of a standard X-ray dose to the dose of the test type required to produce the same biological effect (e.g. mutation, cell kill, etc.). That is,

$$RBE = \frac{\text{Dose of standard X-rays (250 KeV)}}{\text{Dose of test radiation type}}.$$

All low-LET radiation has an RBE of 1, while generally RBE > 1 for high-LET radiation. While dose-response curves for low-LET radiation display a shoulder region, the curves for high-LET radiation are purely exponential [27, 28]. However, all curves become exponential for sufficiently large radiation doses [27]. Most medical radiation is low-LET.

11.2.1 Ions and radical reactions

Acute doses of ionizing radiation rapidly induce double-stranded breaks (DSBs) through free-radical reactions in the nuclear DNA (chromatin), and such breaks are widely believed to be the principle cause of radiation-induced cell death. Low-LET radiation interacts with electrons in the target material and displaces high-energy electrons from their atomic orbitals. Displaced electrons (δ-electrons) with sufficient energy can further react to displace other electrons. Such electron chain reactions are produced at the track ends of low-LET radiation [28]. These ionizing events form a positive ion and a free electron. The abnormal ions are charged and have an unpaired electron, making them extremely unstable, and they quickly react to form uncharged free radicals (elements containing an unpaired electron) [27]. Oxygen radicals in particular are extremely reactive and can set off damaging chain reactions. Massive radical-induced damage can cause cell death either by apoptosis or necrosis.

Radiation may directly ionize various biological molecules, but of primary importance is the ionization of water, which sets off a series of reactions that produce potent oxygen radicals. Water ionization rapidly produces free electrons, e^-, hydrogen cations (free protons), H^+, hydrogen radicals, $H^\bullet$, and hydroxyl radicals, $HO^\bullet$, by the following reactions [27]:

$$H_2O \longrightarrow H_2O^{\bullet+} + e^-,$$
$$H_2O^{\bullet+} \longrightarrow H^+ + HO^\bullet,$$
$$H^+ + e^- \longrightarrow H^\bullet.$$

The hydroxyl radical is potently reactive and can damage a variety of organic molecules. It is the principle species responsible for DNA damage, being able to react with both the base and sugar moieties of DNA [28]. Oxygen can react with the other products of water ionization to form the superoxide anion, $O_2^{\bullet-}$, and hydrogen peroxide, H_2O_2. Superoxide and hydrogen peroxide, which along with the hydroxyl radical (among others) are referred to as reactive oxygen species or ROS, can participate in many subsequent chain reactions to generate hydroxyl radicals, radical halogens, and other damaging species. Therefore, tumor oxygenation is an important determinant of the effectiveness of radiation therapy [27]. Moreover, iron can react with $O_2^{\bullet-}$ and H_2O_2 to produce $HO^\bullet$ in the Haber-Weiss cycle.

The radical products of ionization are not generated homogenously throughout the cell, but are localized to "clusters of ionization" generated at radiation track ends in low-LET radiation [28]. DNA lesions therefore tend to occur in clusters [27]. The cascade of ionizations induced by irradiation is depicted in Figure 11.3.

Radiation-induced radicals can damage an array of cellular components. Their most important effects are single- and double-stranded DNA breaks. Single-strand breaks (SSBs) can be repaired relatively easily, and most DSBs are also repaired [27]. Unrepaired DSBs can cause the cell to undergo apoptosis when it attempts to pass through mitosis.

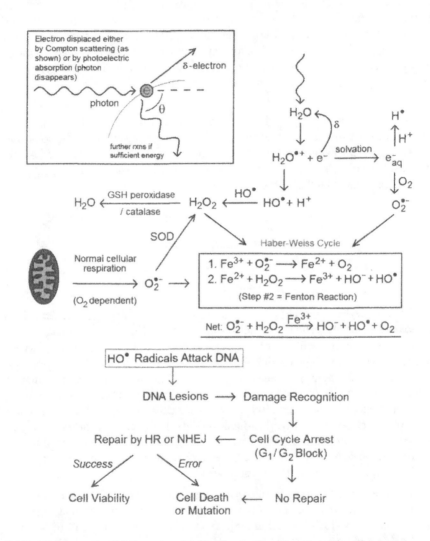

FIGURE 11.3:   Schematic of radical reactions leading to cell injury.

11.2.1.1 DSB repair mechanisms

Following formation of DSBs, cell-cycle arrest occurs and DNA repair processes are activated. Cell-cycle arrest is necessary for cells to have sufficient time to repair damage, and several signalling kinases are known to induce G_1 and G_2 block in response to DSB formation [44]. DSBs in mammalian cells are repaired by two known enzymatic mechanisms: *homologous recombination* (HR) and *non-homologous DNA end-joining* (NHEJ). In HR, a homologous section of DNA in either the sister chromatid or (rarely) the homologous chromosome "invades" the broken DNA strand and serves as a template for repair. In NHEJ, two free DNA ends are ligated together in a process that requires little or no homology between the strands.

HR is a highly accurate repair process, while NHEJ is error-prone. While NHEJ is believed to be the dominant repair process in mammalian cells, the relative contributions of the two processes is controversial. Between 10% and 50% of DSBs may be repaired by HR in mammalian cells. Because HR generally requires the presence of a sister chromatid, it is likely only active in the S and G_2 stages of the cell cycle. It is generally accepted that NHEJ is by far the dominant process in the G_1/G_0 phases [43].

11.2.1.2 Radical detoxification mechanisms

Cells are constitutively exposed to endogenous oxidative stress resulting from cellular respiration and inflammatory processes. Therefore, they are equipped with ROS detoxifying molecules and damage surveillance systems. ROS-detoxifying systems can prevent radiation damage to some degree. Radical scavengers with a thiol (sulphydryl, SH) group can detoxify free radicals by donating a hydrogen atom to the radical. Superoxide dismutase (SOD) reduces superoxide ($O_2^{\bullet-}$) to H_2O_2, which can then be further reduced to water by catalase or glutathione (GSH) peroxidases.

11.2.2 Oxygen status

As mentioned above, tumor oxygenation status helps determine response to therapy, and tumor hypoxia is an important cause of radioresistance. Within most tumors there exists a radioresistant and severely "hypoxic fraction," which ranges from 0–50% of the tumor mass before treatment. Cells are maximally radioresistant at or below an oxygen partial pressure (P_{O_2}) of about 0.5 mmHg and are fully radiosensitive above 20 mmHg [47]. A very important parameter is the *oxygen enhancement ratio* (OER) which is defined as the ratio of the dose under hypoxic conditions over the dose under aerobic conditions necessary to produce the same cell kill. Perhaps more intuitively, one can view it as a "dose multiplier:"

$$\text{OER} \times \text{Dose under normoxia} = \text{Dose under hypoxia}$$

to achieve the same kill. Experiments using high radiation doses yield an OER of 2.5–3.0 for severely hypoxic cells [8]. Although written and thought of as a constant coefficient, the OER appears to be a function of the kill efficacy, which in turn depends on dose under normoxia. It turns out that the OER is somewhat smaller for lower "normal" doses, suggesting that tumor hypoxia is slightly less of a barrier to therapeutic success for lower dose medical radiation than it is in experimental settings.

11.2.3　The four R's

Finally, we mention the "four R's" of radiotherapy: *repair, regrowth, reoxygenation,* and *redistribution* (of cells within the cell cycle). The first is ubiquitous to our discussion; the latter three are most relevant in the context of fractionation schedules, because during prolonged courses of radiation, tumors have opportunity for regrowth, reoxygenation (as the tumor shrinks so does the hypoxic fraction), and cell cycle redistribution. We address these with the LQ model toward the end of this chapter.

11.3　Classical target-hit theory

The formal multi-target, multi-hit models of classical radiation theory only consider immediate effects of radiation on DNA. As outlined above, the classical theory posits that if a sufficient number of DNA "targets" are "hit" or damaged by a volley of ionizing radiation, then a cell dies. No repair processes or further interactions between lesions are accounted for [15].

The *multi-target single-hit model* is founded on the assumptions that multiple targets exist, some number of which must be hit at least once to cause cell death. It is essentially a stochastic model that assumes "hits" occur as described by a Poisson process. The main result is the following expression of the survival fraction, S, as a function of radiation dose, D:

$$S = 1 - \left(1 - e^{-D/D_0}\right)^N. \tag{11.1}$$

This expression can be derived from the following argument. Suppose a large number, n, of DNA targets exist in a single cell. For a fixed dose of sparsely ionizing radiation, we assume that the target "hit" probabilities for all n targets are independent and identically distributed with constant probability p. Let X be the number of targets hit in a single course of radiation treatment. Then X is a binomially-distributed random variable with parameters n and p, i.e., $X \sim \mathrm{BIN}(n, p)$. For n sufficiently large and p sufficiently small, the number of hits is approximated by a Poisson random variable with parameter $\lambda = np$. If we assume that death follows from only a single hit, then

$$\text{Pr(lethal damage)} = \text{Pr}(X \geq 1) = 1 - \text{Pr}(X = 0) = 1 - e^{-\lambda}. \qquad (11.2)$$

Similarly, if death requires N damaged targets, then the probability of lethal damage is

$$P(X_1 \geq 1, X_2 \geq 1, ..., X_N \geq 1) = \prod_{i=1}^{N} P(X_i \geq 1)$$
$$= \left(1 - e^{-\lambda}\right)^N. \qquad (11.3)$$

Since the expected value of a Poisson random variable is λ and it is assumed that the number of hits is directly proportional to the dose D, we have $\lambda = D/D_0$, where D_0 is the dose to give a mean of 1 hit per target [24]. The parameter D_0 is also frequently given as the dose at which $S = \exp(-1) \approx .37$ and this parameter is used to quantify the intrinsic radiosensitivity of a cell lineage. Note that the two definitions for D_0, i.e., the dose for a mean of 1 hit per target and the dose for a surviving fraction of $\exp(-1)$, are only equivalent if a single hit is lethal. The latter is the definition more commonly encountered in the literature. Finally, the surviving fraction,

$$S = 1 - P(\text{lethal damage}) = 1 - \left(1 - e^{-D/D_0}\right)^N. \qquad (11.4)$$

According to this derivation N, also called the extrapolation number, is the number of targets that must be hit to cause cell death. However, on parameter estimation many dose-response curves fail to give integer values for N [27].

If death is caused by a single hit, then

$$S = \exp(-D/D_0) \qquad (11.5)$$

Such a "log-linear" relationship motivates most later models for radiation damage. Fowler [10] discusses a more general multi-target multi-hit model and the multi-hit single-target model. Dienes [9] proposed a simple kinetics model where radiation damage occurs in a series of kinetic steps. He showed that equations from hit theory can be derived from this kinetics model.

Classical target theory does not take into account the cell's ability to deal with damage, and incorrectly assumes that the cell's fate is sealed immediately following irradiation. Damage repair and its consequences are discussed in the next section.

11.4 Lethal DNA misrepair

While most DSBs are successfully repaired, the NHEJ repair mechanism is error-prone and can ligate free DNA ends from two different chromosomes. Misrepair of DSBs can result in catastrophic chromosomal aberrations that prevent the cell from undergoing successful mitosis. The most important of these appears to be the dicentric chromosomal aberration [3, 27], which is produced by the improper repair of two broken chromosomes (Figure 11.4).

Models that consider lethal misrepair of DNA lesions as a primary mechanism for radiation cytotoxicity are known as lesion-interaction models. Broadly, the lesion-interaction models adopt either one or both of the following assumptions: (1) there exist both lethal and sub-lethal lesions—sublethal lesions may interact to yield lethal lesions—and (2) there exist potentially lethal lesions which may be repaired but can also interact by a second order process to produce irreparable lethal lesions. In this section we discuss several historically related and very similar models: the repair-misrepair model, the lethal-potentially lethal model, and briefly, the two-lesion kinetic model. The linear-quadratic formalism, which is discussed later in this chapter, is a quasi-mechanistic model also of the lesion-interaction type.

11.4.1 Repair-misrepair model

The classical repair-misrepair (RMR) model first proposed by Tobias et al. [42] is an instructive introduction to this formalism. Here, we present a modified version similar to that discussed by Sachs et al. [37].

The RMR model considers the number of DSBs in the DNA, $U(t)$, and the number of lethal lesions, $L(t)$, that ultimately result from these breaks. The RMR model assumes that most breaks are successfully repaired, and the number of DSBs produced increases linearly with the total dose of radiation, D. The latter assumption is supported by empirical data [37, 42], which presumably implies that the rate of DSB production is linearly proportional to the rate of radiation delivery to tissue, $\dot{D}$.

The rate of repair of DNA breaks is a function of the total number of breaks; in particular, it is a first-order process. The misrepair of two broken DNA strands to form a chromosomal aberration, however, requires the interaction of two DNA strands and is therefore (at least in a well-mixed setting) a second-order process. Since broken DNA strands can be treated as equivalent, the rate of misrepair is proportional to U^2, assuming mass action. These

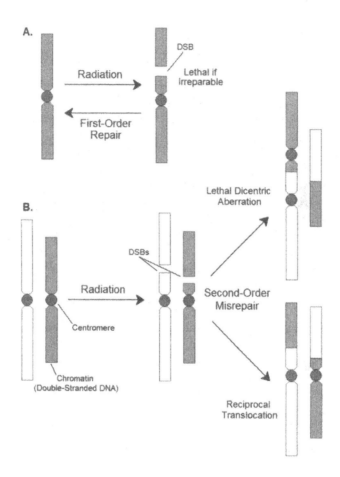

FIGURE 11.4: Induction of structural aberrations in chromosomes by radiation. DSBs are induced in direct proportion to the absorbed radiation dose. Two broken nonhomologous DNA strands can interact by second-order kinetics to form aberrant structures, like dicentric chromosomes, which possess two centromeres and tend to be lethal to the cell. Another possibility is reciprocal translocation, which is not necessarily lethal. The two structures are formed with equal probability. Note that a dicentric aberration is considered a *single* lethal lesion.

assumptions yield the following model:

$$\frac{dU}{dt} = \underbrace{\delta\dot{D}}_{\text{damage}} - \underbrace{\lambda U}_{\text{repair}} - \underbrace{\kappa U^2}_{\text{misrepair}} , \tag{11.6}$$

$$\frac{dL}{dt} = \underbrace{(1-\phi)\lambda U}_{\text{unsuccessful repair}} + \underbrace{\sigma\kappa U^2}_{\text{lethal misrepair}} , \tag{11.7}$$

where δ is the number of DSBs induced per Gy of radiation ($\delta \approx 40$ DSBs / Gy [37]), λ is the rate at which DSBs are repaired ($\lambda = 0.35$ to 1.39 per hr [1]), κ is the rate constant for second-order DSB interaction, and ϕ is the fraction of simple repairs that are successful. The fraction of misrepairs that result in a lethal lesion is σ, which was taken to be 1 in Tobias' original model. However, Sachs has argued that the rate at which lethal lesions are produced by misrepair is only $1/4$ the rate at which misrepairs occur, since it takes two DSBs to form a single lesion, and on average only half of all misrepairs result in a lethal dicentric aberration. Therefore, σ should be $1/4$ (see Figure 11.4).

Assuming that lethal lesions are Poisson-distributed among the cell population, which appears to be valid for lower dose low-LET radiation [2], the (time-dependent) surviving fraction of cells,

$$S(t) = \exp(-L(t)). \tag{11.8}$$

Lethal chromosomal lesions do not necessarily induce immediate cell death. Rather, death occurs when the cell attempts to divide. This explains the phenomenon of delayed toxicity, where tissues with cells that turnover slowly can experience toxicity months following therapy. Moreover, cells may be "sterilized," i.e., they lose reproductive capacity without being killed [27]. From the perspective of tumor control, sterilization and death are equivalent.

A simplified version of the model, presented in [42], can be used to gain deeper analytical insight. Assume an acute dose of radiation, D, is delivered, giving the initial condition $U_0 = \delta D$. Moreover, assume that fraction ϕ of first-order repairs are successful and that all misrepairs are lethal. In this case the model reduces to the following:

$$\frac{dU}{dt} = -\lambda U - \kappa U^2, \tag{11.9}$$

$$U(0) = U_0 = \delta D. \tag{11.10}$$

The solution is

$$U(t) = \frac{U_0 e^{-\lambda t}}{1 + (U_0 T)/(\epsilon)}, \tag{11.11}$$

where

$$T = 1 - \exp(-\lambda t), \tag{11.12}$$

$$\epsilon = \frac{\lambda}{\kappa}. \tag{11.13}$$

We may now integrate to calculate the total number of lesions repaired by first-order repair, $R_L(t)$, and the number of quadratic misrepairs, $R_Q(t)$:

$$R_L(t) = \int_0^t \lambda U(s)\,\mathrm{d}s = \epsilon \ln\left(1 + \frac{U_0 T}{\epsilon}\right), \tag{11.14}$$

$$R_Q(t) = \int_0^t \kappa U(s)^2\,\mathrm{d}s. \tag{11.15}$$

Also,

$$U(t) = U_0 - R_L(t) - R_Q(t). \tag{11.16}$$

Since all lesions will ultimately be repaired either by first- or second-order processes, $\lim_{t\to\infty} U(t) = 0$, which yields an expression for the number of quadratically misrepaired lesions as $t \to \infty$:

$$\lim_{t\to\infty} R_Q(t) = U_0 - \lim_{t\to\infty} R_L(t). \tag{11.17}$$

The number of lethal lesions, $L(t)$, is the number of quadratic misrepairs plus those first-order repairs which were unsuccessful, so

$$\lim_{t\to\infty} L(t) = U_0 - \phi \lim_{t\to\infty} R_L(t). \tag{11.18}$$

Combining this final expression with the relation $S = \exp(-L(t))$ yields the expected survival fraction when sufficient time is allowed for all repair process to reach completion:

$$S = \exp(-\bar{L}(t)) = e^{-U_0}\left(1 + \frac{U_0}{\epsilon}\right)^{\phi\epsilon} = e^{-\delta D}\left(1 + \frac{\delta D}{\epsilon}\right)^{\phi\epsilon}, \tag{11.19}$$

where $\bar{L}(t) = \lim_{t\to\infty} L(t)$. Note that the first term, $\exp(-\delta D)$, represents the expected log-kill curve that would result if no repair processes were active. This curve is modified by the second term, which represents the effects of the first and second-order repair processes. If the amount of time for repair is finite (say, of length t_r), and lesions at the end of that time interval, $U(t_r)$, become lethal, then

$$L(t_r) = U(t_r) + R_Q(t_r) + (1 - \phi)R_L(t_r) = U_0 - \phi R_L(t_r). \tag{11.20}$$

Thus, we arrive at a similar expression for the time-dependent survival of irradiated cells:

$$S(t) = \exp(L(t_r)) = e^{-\delta D}\left(1 + \frac{\delta D T_r}{\epsilon}\right)^{\phi\epsilon}, \tag{11.21}$$

where $T_r = 1 - \exp(-\lambda t_r)$. With this expression, we can examine how the survival curve responds to changes in parameter values. The second term is always ≥ 1, implying that while second-order misrepair contributes to the lethality of radiation, overall the repair process always increases cell survival. Moreover, increasing the time available for repair always increases survival, so there is no point at which the cost of second-order misrepair "outweighs" the benefits of first-order repair. This is an important point, as it is easy to incorrectly conclude, as some authors have, that the RMR model (and other lesion-interaction models) suggests that the shoulder region represents an increase in the number of lethal lesions that would have otherwise occurred in the absence of the repair process. Also, when there is sufficient time for all lesions to be resolved and all first-order repair is successful—i.e., $\phi = 1$—then $S \to 1$ as $\kappa \to 0$. In general,

$$\lim_{\kappa \to 0} S = e^{-\delta D(1-\phi)}, \tag{11.22}$$

and any cell death is due to first-order misrepair (we leave the proof of this result as an exercise). Therefore, the RMR model implies that repair following irradiation always increases cell survival, but misrepair is the main cause of cell death in the long term.

Increasing the time available for repair, t_r, and the first-order repair rate, λ, both increase survival, whereas a larger rate of second-order misrepair, κ, reduces survival. At the other extreme, if no time is allowed for repair then the survival curve decreases exponentially with dose, namely $\exp(-\delta D)$. Figure 11.5 shows some representative dose-response curves under the RMR model.

We can also see from equation (11.21) that as the dose D becomes large, the survival curve approaches that of a simple exponential with slope $-\delta$ on a $\log_e$-linear plot. To see this, first note that

$$\ln(S(t)) = -\delta D + \phi\epsilon \ln\left(1 + \frac{\delta DT}{\epsilon}\right). \tag{11.23}$$

Differentiating with respect to dose D gives

$$\frac{\mathrm{d}}{\mathrm{d}D}\ln(S(t)) = -\delta + \phi\delta \left(1 + \frac{\delta DT}{\epsilon}\right)^{-1}. \tag{11.24}$$

At extreme doses,

$$\lim_{D \to \infty}\left(\frac{\mathrm{d}}{\mathrm{d}D}\ln(S(t))\right) = -\delta. \tag{11.25}$$

Thus, we see that repair essentially "pushes" the unmodified log-linear survival curve to the right, and the shoulder region represents convergence to this ultimate exponential behavior. This is an important model prediction, as all empirical survival curves become exponential with sufficiently large radiation doses. As we shall see, while the RMR, lethal-potentially lethal, and saturable repair models all share this prediction, the ubiquitous linear-quadratic formalism fails to preserve this fundamental result.

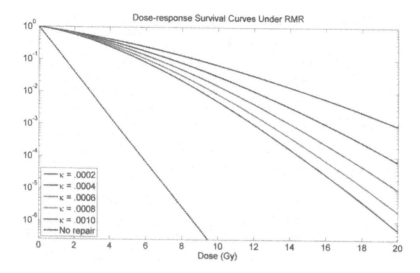

FIGURE 11.5: Dose-response curves for the RMR model as $t \to \infty$ under different values of κ, the second-order misrepair constant. Using survival data, Tobias [42] estimated parameter values of $\delta = 1.6$ Gy^{-1}, $\lambda = 0.022$ min^{-1}, $\phi = 0.89$, and $\kappa = 0.00078$ lesion^{-1} min^{-1}. These values are used for all parameters other than κ in the figure.

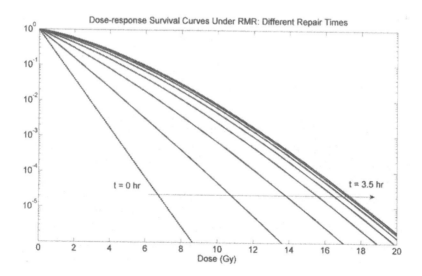

FIGURE 11.6: Dose-response curves for the RMR model for different repair time intervals, t. Note that the survival curve approaches an asymptotic curve within several hours. Other parameter values are the same as in Figure 11.5.

11.4.2 Lethal-potentially lethal model

Like the RMR model, the lethal-potentially lethal (LPL) model includes cytotoxicity from incorrectly repaired and initially irreparable, lethal DSBs. However, it also includes the notion that some reparable lesions become lethal if not repaired ("potentially lethal" lesions). Here we present the model proposed by Curtis [6] with some changes in notation. The model arises from the following assumptions:

1. Potentially lethal and lethal lesions are each produced at a rate linearly proportional to the radiation dose-rate.

2. Lethal lesions cannot be repaired and always induce cell death.

3. Potentially lethal lesions are repaired according to first-order kinetics and are misrepaired to form irreversibly lethal lesions according to second-order kinetics (binary misrepair as in the RMR model).

4. Following some prescribed amount of time, the remaining potentially lethal lesions become "fixed" and irreversibly lethal. Such lesion fixation may be due to cells entering the cell cycle.

Letting $P(t)$ and $L(t)$ represent the number of potentially lethal (reparable) and lethal (irreparable) lesions, respectively, we have the following model:

$$\frac{\mathrm{d}P}{\mathrm{d}t} = \underbrace{\delta\eta\dot{D}}_{\text{reparable damage}} - \underbrace{\lambda P}_{\text{repair}} - \underbrace{\kappa P^2}_{\text{misrepair}} , \qquad (11.26)$$

$$\frac{\mathrm{d}L}{\mathrm{d}t} = \underbrace{\delta(1-\eta)\dot{D}}_{\text{irreparable damage}} + \underbrace{\kappa P^2}_{\text{lethal misrepair}} , \qquad (11.27)$$

where η represents the fraction of radiation induced DSBs that are reparable, and all other parameters are the same as for the RMR model. We also note that, as in the RMR model, the term for the addition of lethal misrepair is more properly $(1/4)\kappa P^2$, but we have left it as is so that parameter values reflect those of the original model.

Explicit solutions to the model equations are given in [6], and they are somewhat similar in form to those for the RMR model. One interesting prediction of the LPL model, which is corroborated by data, is that the dose-rate (i.e., rate at which radiation is delivered to tissue), has a strong effect on the cell survival curve. Curtis [6] showed that for a sufficiently low dose-rate, cell survival is approximated by the expression,

$$S = \exp(-\delta(1-\eta)D). \qquad (11.28)$$

This result implies that for very low dose-rates the interaction between potentially lethal (or sublethal) lesions becomes negligible, and cytotoxicity is determined solely by the ability of single-track radiation to form irreparable DNA breaks. We will encounter this notion again when discussing the LQ model (see Section 11.7). An important potential application of this result and the model generating it is brachytherapy, where implanted radioactive seeds slowly deliver radiation to the target tissue. The effect of dose-rate upon survival under the LPL model is shown in Figure 11.7.

We have chosen to highlight the LPL model because it highlights the concept of "sublethal," potentially reparable lesions that can still induce cell death if they become fixed. Thus, we can be clear that DSBs are lethal if unrepaired, can be misrepaired to form irreversibly lethal lesions, or can be successfully repaired. There is no need to invoke a separate "sublethal" class of DNA damage. This is an important interpretation, since the precise meaning of "sublethal lesion" has generated chronic debate.

11.4.3 Parametrization

Both the RMR and LPL model can be parameterized to compare favorable with dose-response data, and we briefly discuss the implications of the derived parameter values. The first-order repair rate, λ, has been estimated to be between 0.6 and 1.32 hr^{-1} ($t_{1/2}$ between 0.53 and 1.16 hr) for the RMR model [42] and 0.5 hr^{-1} ($t_{1/2} = 1.39$ hr) for the LPL model [6]. These values are well within the empirically expected range. Interestingly, Tobias estimated that 11% of lesions are improperly repaired ($\phi = 0.89$) under the RMR model, and Curtis estimated that 18% of single-track lesions are irreparable ($\eta = 0.82$). However, under curve-fitting, both models predict much lower rates of DSB formation than indicated by empirical data. The RMR model predicts δ between 1.6 and 2.058 Gy^{-1}, and the LPL predicts $\delta = 0.7366$ Gy^{-1}. However, direct measurements estimate that about 40 DSBs are formed per Gy of radiation [27, 37]. It is possible that this disconnect is due to all misrepaired lesions being equated to a lethal lesion in both models, when in reality four misrepaired DSBs are needed on average to form single a dicentric chromosome. In any case, this suggests a problem with both models that could be resolved in future work. This is yet another demonstration of the importance of constraining parameter values by empirical data.

Finally, we briefly mention the two-lesion kinetic (TLK) model of Stewart [40], which was proposed as an extension and refinement of the RMR and LPL models. The TLK model is, in essence, the RMR model but with two classes of sublethal lesions with distinct kinetic parameters that can interact by binary misrepair. Stewart calibrated the TLK model to survival data and showed that it, unlike the LPL model, yielded reasonable predictions for both number of DSBs Gy^{-1} and rate of DSB rejoining. Interestingly, results further implied that the two classes of DSB must be repaired at vastly different rates. This model is also discussed in the section on the kinetics of damage repair.

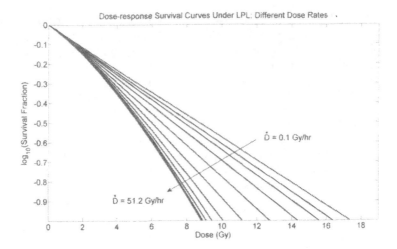

FIGURE 11.7: Dose-survival curves under the LPL model for different dose rates. The dotted line gives the analytic approximation for a vanishingly small dose-rate. Note that there is an asymptotic survival curve for both small and large dose-rates (compare to Figure 3 in [6]). Parameter values are $\delta = 0.7366$ Gy^{-1}, $\eta = 0.82$, $\lambda = 0.5$ hr^{-1}, and $\kappa = 0.055$ $lesion^{-1}$ hr^{-1}.

11.5   Saturable and enzymatic repair

While chromosomal aberrations caused by misrepair of DSBs are significant causes of lethal damage, unrepaired double-stranded breaks also cause cell death. Repair of such breaks is an enzymatic, and therefore fundamentally saturable, process. Saturability of DNA repair lies at the heart of the main hypothesis competing with lesion-interaction to explain the shoulder region of dose-response curves. In many experimental models, the half-time for DNA repair (time it takes to repair half of the DNA lesions) following irradiation increases linearly with radiation dose [31, 45]. Such studies characterize the repair process as either unsaturated (implying first-order kinetics of repair), partially saturated, or fully saturated (implying zero-order kinetics for repair). For example, the half-time for repair in rat 9L glioma cells did not depend on dose below about 6 Gy, but increased roughly linearly with doses > 12 Gy [45], suggesting saturation. Moreover, graphical comparisons of DNA repair half-time to the cell-survival curve suggest that repair saturation at least partially explains the shoulder region. Such experimental evidence has motivated the saturable-repair models we now review.

11.5.1 Haynes model

In 1966 Haynes [15] demonstrated, among other things, that cell survival in yeast cells following radiation exposure increased dramatically if the survival assay was delayed by 48 hours. This observation led him to propose a simple mathematical model for lethality as a function of the expected number of unrepaired lethal hits per cell for a given time for repair. A lethal hit here is defined as a DNA defect which irreversibly blocks DNA synthesis. Formally, the survival fraction, S, for a dose of size D is given by

$$-\ln S = F(D) - R(D), \tag{11.29}$$

where $F(D)$ is the number of lethal DNA defects per cell and $R(D)$ is the number of defects that are repaired in the available time. The exponential dependence of survival on the total number of lesions is a consequence of the assumption that hits are Poisson-distributed throughout the cell population. If $R(D) = 0$, then the model is equivalent to a classical 1-target 1-hit model. Under Haynes' model, repair can presumably continue until the cell attempts to enter S phase and replicate its DNA. As in both classical and lesion-interaction theory, the number of lethal hits is assumed to be directly proportional to dose, and $F(D) = kD$.

Haynes' method for deriving $R(D)$ is a good example of a heuristic derivation of a functional form. Haynes claimed a reasonable form of $R(D)$ should meet three requirements:

1. It must pass through the origin; i.e., no repair occurs if no damage occurs.

2. The number of defects repaired should be proportional to the number of defects at low doses.

3. $R(D)$ should plateau at high doses.

To satisfy these requirements, Haynes chose a simple saturation function,

$$R(D) = \alpha(1 - e^{-\beta D}), \tag{11.30}$$

where α is the maximum number of defects that can be repaired in the available time, and β is defined such that $\ln 2/\beta$ equals the radiation dose at which half of the repairable defects are repaired. Since what is being modeled is the saturation of repair enzymes, a Michaelis-Menten term would also be appropriate here.

From these choices of $F(D)$ and $R(D)$ we arrive at:

$$-\ln S = kD - \alpha\left(1 - e^{-\beta D}\right). \tag{11.31}$$

Much later, Sanchez-Reyes derived an equivalent model for saturable repair [38]. Because the parameter α represents the absolute number of lesions that can be repaired, it must depend upon the number of lesions actually produced and hence the dose D. It will be unique for all data-sets.

While very simple, this model makes some interesting predictions concerning dose-response curves. First, note that the derivative of equation (11.31) with respect to dose (the slope of the dose-response curve) is

$$\frac{\mathrm{d}(-\ln S)}{\mathrm{d}D} = k - \alpha\beta e^{-\beta D}, \tag{11.32}$$

which is equal to $k - \alpha\beta$ when $D = 0$. Now, Haynes defined the efficiency of repair as the fraction of lesions that are repaired:

$$\text{Efficiency} = \frac{R(D)}{F(D)} = \frac{\alpha\beta}{k}\left(1 - \frac{\beta D}{2} + \frac{\beta^2 D^2}{6} + \dots\right). \tag{11.33}$$

This can be seen from the Taylor series expansion for the exponential function. Clearly, efficiency $\approx (\alpha\beta)/k$ when βD is small. If $k = \alpha\beta$, then efficiency = 1 (100%) when $D = 0$, and therefore the initial slope of the survival curve, $k - \alpha\beta$, is 0. From this we can conclude that if the initial slope is greater than 0, then $k > \alpha\beta$, and the initial efficiency < 1. But in that case there must exist irreparable lesions (or the repair process must be fundamentally flawed). Thus, the model predicts that a non-zero initial slope indicates the induction of irreparable lesions.

Moreover, studies of survival of *E. coli* exposed to UV light suggest that the efficiency of repair is high even when significant population-wide mortality occurs [15]. When survival was reduced to a tenth ($S = 0.1$), efficiency of repair was reduced to 95%. Surprisingly, efficiency was still 88% for $S = 10^{-6}$, suggesting that even a relatively modest reduction in repair efficiency is associated with an increase in mortality by many orders of magnitude.

11.5.2 Goodhead model

An early saturable-repair model by Goodhead [13] posited that DSB repair is carried out by a pool of "suicide enzymes," leading to a simple model for the DSB repair rate:

$$\frac{\mathrm{d}n}{\mathrm{d}t} = -kcn, \tag{11.34}$$

where n is the number of DSBs, c is the number of repair enzymes, k is a mass-action constant. Goodhead assumed that repair enzymes are not recycled, so $\mathrm{d}c/\mathrm{d}t = \mathrm{d}n/\mathrm{d}t$. This model compared favorably to data and demonstrated that various dose-response phenomena could be interpreted differently under the saturable-repair framework vs. the lesion-interaction framework. However, the notion of enzymatic repair as a suicide process does not fare well against modern data.

11.5.3 General saturable-repair model

While the models above make interesting predictions, it may be more par-simonious to apply simple Michaelis-Menten kinetics and a kind of lethal-potentially lethal model to study saturable DNA repair's implications for dose-response. To that end, let $P(t)$ represent the number of potentially lethal lesions, $L(t)$ represent the number of irreversibly lethal lesions, V_{max} be the maximum rate at which lesions are repaired, and θ be the number of lesions where repair is half-maximal. We assume that fraction ϕ of initial DSBs are reparable. Incorporating the idea of DSB misrepair, we can further assume that fraction η of enzymatic-mediated repairs erroneously create a lethal aberration, yielding the following model:

$$\frac{dP}{dt} = \phi\delta\dot{D} - V_{max}\left(\frac{P}{P+\theta}\right), \qquad (11.35)$$

$$\frac{dL}{dt} = (1-\phi)\delta\dot{D} + \eta V_{max}\left(\frac{P}{P+\theta}\right). \qquad (11.36)$$

As usual, we assume that $S(t) = \exp(P(t) + L(t))$ and allow time t_r for repair. Consider the simplest case, where $\phi = 1$ and $\eta = 0$; i.e., all lesions are reparable, and none are misrepaired. Such a model yields dose-response curves with an initial slope that may be non-zero and a shoulder region that converges to a simple exponential. In this model, as in that of Goodhead [13], a non-zero initial slope may merely be the consequence of limited time available for repair, and not irreparable lesions as predicted by Haynes [15]. We leave investigation of the effects of modifying the rate of misrepair, η, and other parameters as exercises.

Significantly, if no misrepair or irreparable lesions are allowed, then this model predicts that cell survival approaches 1 as $t \to \infty$, in contrast to data. For example, Reddy et al. [31] found that repair in Chinese hamster V79 cells saturates for minuscule radiation doses, yet cell survival plateaus at a value < 1 after some time interval for repair. Such behavior is predicted by the RMR and LPL lesion-interaction models (see Figure 11.6), but not by any of the saturable repair models that we have examined so far. In fact, on the basis of this data Reddy et al. proposed that repair of damage is a saturable process, but that reparable lesions interact to yield irreparable lesions.

11.6 Kinetics of damage repair

As the models reviewed above suggest, misrepair of sublethal lesions has been traditionally and widely assumed to be a second-order process while nor-mal repair has generally been considered a first-order process. Many groups have quantified the time-course of SSB and DSB repair following irradiation in mammalian cells. Let n_0 represent the number of DNA lesions immediately following irradiation and n_t be the number remaining after time t. If damage repair is a first-order process, then

$$\frac{dn}{dt} = -\lambda n \quad \Rightarrow \quad \frac{n_t}{n_0} = e^{-\lambda t}. \tag{11.37}$$

Also this simple "mono-exponential" model performs well against some datasets, it is also contradicted by many others. It has been widely observed that the rate of DSB repair slows with time. In one study, 30-35% of DSBs were repaired within 5 minutes [14], yet the half-time for further, slow repair is generally on the order of an hour or more [26, 37]. Such observations have led to the proposal of bi- or multi-exponential models to describe the rate of damage repair. A bi-exponential model, for example, would have the form,

$$\frac{n_t}{n_0} = C_1 e^{-\lambda_1 t} + C_2 e^{-\lambda_2 t}. \tag{11.38}$$

Many authors use bi-exponential models with fast and slow components to describe DNA repair (e.g., [26]). More complex multi-exponential models typically represent hypotheses of multiple lesion types with different repair rates. Complex lesions might consist of close clusters of DSBs, a DSB in conjunction with other types of damage, or a dense cluster of non-DSB damage. Presumably, complex DSBs take longer to repair. Chromatin is heterogeneous, so some regions may also be less accessible to repair enzymes, leading to different rates of repair.

Multi-exponential models are not derived directly from any underlying dynamical model, but are rather empirical descriptions of the data. However, the hypothesis can also easily be incorporated into any of the dynamical models we have discussed. For example, the RMR model may be modified to include two types of sublethal/potentially lethal lesion [42]:

$$\frac{dU_1}{dt} = \delta_1 \dot{D} - \lambda_1 U_1 - \kappa_1 U_1^2 - \hat{\kappa} U_1 U_2, \tag{11.39}$$

$$\frac{dU_2}{dt} = \delta_2 \dot{D} - \lambda_2 U_2 - \kappa_2 U_2^2 - \hat{\kappa} U_1 U_2, \tag{11.40}$$

$$\frac{dL}{dt} = (1 - \phi_1)\lambda_1 U_1 + (1 - \phi_2)\lambda_2 U_2 + \sigma_1 \kappa_1 U_1^2$$
$$+ \sigma_2 \kappa_2 U_2^2 + \hat{\sigma} \hat{\kappa} U_1 U_2. \tag{11.41}$$

Radivoyevitch et al. [30] examined data on the disappearance of DSBs and appearance of misrepaired aberrations at high radiation doses using simple binary misrepair models. They found that the standard RMR model was inconsistent with the data, greatly overestimating the number of chromosomal aberrations. However, if two lesion types were considered, only one of which interacted with other lesions of the same type to form aberrations, the model compared favorably with data. Letting M be the number of chromosomal exchange aberrations and U_1 and U_2 be the two lesion types, the model is simply:

$$\frac{dU_1}{dt} = -\lambda_1 U_1, \tag{11.42}$$

$$\frac{dU_2}{dt} = \lambda_2 U_2 - \kappa U_2^2, \tag{11.43}$$

$$\frac{dM}{dt} = \kappa U_2^2. \tag{11.44}$$

The two-lesion kinetic (TLK) model of Stewart [40], mentioned in the section on lesion-interaction models, was proposed as a general extension of the RMR and LPL models and considers interaction between simple and complex DSBs. Lesions of either type freely interact with others of either type by second-order misrepair. DSBs are repaired by first-order kinetics, which, as in the RMR model, may be successful or unsuccessful. Unlike the RMR and LPL models, sublethal/potentially lethal lesions become fixed by some first-order physiologic process—e.g., binding of a broken DNA end to a histone protein. While formulated in more general terms, the simplified model considered by Stewart for calibration to data is identical to the modified RMR system above (equations (11.39)–(11.41)). The TLK model predicts that the half-times for simple and complex lesions vary significantly (1 hr vs. 12–15 hrs) and gives roughly bi-exponential kinetics for DSB rejoining.

More recent experimental results suggest that over 70% of complex DNA damage induced by radiation consists of clusters of either abasic, oxidized purine, or oxidized pyrimidine sites, not double-strand breaks [41]. These lesions are difficult to repair, and cellular processing of such complex lesions can lead to the production of new DSBs following initial irradiation [14].

Fowler [12] has suggested that second-order repair may be a simpler explanation for the observed lesion repair kinetics than the hypothesis of multiple classes of damage. It is somewhat curious why misrepair models have generally treated misrepair as a second-order process yet assumed that successful repair is first-order, especially when the same enzymes mediate both types of repair. A model for repair as a second-order process naturally implies that the rate of repair should slow with time, making multi-exponential models potentially unnecessary. Let $n(t)$ represent the number of free DNA breaks, with initial condition $n(0) = n_0$. The change in n via a single second-order repair process is then,

$$\frac{dn}{dt} = -Cn^2, \tag{11.45}$$

where C is some constant. Solving gives,

$$\frac{1}{n(t)} = \frac{1}{n_0} + Ct \quad \Rightarrow \quad \frac{n_0}{n(t)} = 1 + n_0 Ct. \tag{11.46}$$

We define τ to be the first half-time of repair. That is, $\tau = t$ when $n_0/n(t) = 2$, giving $\tau = 1/(n_0 C)$. Plugging τ into the equation above and rearranging gives

$$\frac{n(t)}{n_0} = \frac{\tau}{\tau + t}. \tag{11.47}$$

Solving for intervals of τ, we see that the fraction of unrepaired damage progresses as $\frac{1}{2}, \frac{1}{3}, \frac{1}{4}, \frac{1}{5}, \frac{1}{6}, \ldots$ with each interval of τ, the first half-time of repair. Comparing this to the progression for simple exponential repair, $\frac{1}{2}, \frac{1}{4}, \frac{1}{8}, \ldots$, we can see that second-order repair predicts a dramatic slowing in repair later in time, but there is reasonable agreement between the models up to two half-times of repair [12]. This pattern of damage decrease is referred to as a *reciprocal-time* pattern, and it compares favorably to several datasets. See [7] or [12] for a more thorough discussion.

Fundamentally, whether (correct) repair of DSBs is a kinetically first-order or second-order process likely depends on whether the DNA break is seen as a single lesion (single break) or two lesions (two free DNA ends). The former implies first-order kinetics for damage repair while the latter implies second-order kinetics. Since misrepair requires significant diffusion of free ends that join in a stochastic manner, it is likely that this is a second-order process at all times. However, immediately following irradiation the free ends of a DSB may be close enough that significant diffusion has not occurred and, even though repair requires the joining of the two ends, repair may resemble a first-order process.

More recently, Cucinotta and colleagues [4, 5] have proposed more explicit kinetic models for the interaction between repair enzymes and DSBs. In [4], they demonstrate that enzyme kinetics can give rise to apparently biphasic repair kinetics.

11.7 The LQ model and dose fractionation

The linear-quadratic (LQ) formalism is a semi-mechanistic model along the same lines as the RMR formalism and considered to be among the lesion-interaction class of model. However, the mechanistic justification is, to some degree, post hoc. It is approximated at low doses by many more complex models that posit second-order lesion interaction as the mechanism for cell death [37], but its success can be attributed to the fact that it describes dose-response data well, and this is a key point to keep in mind when it comes to this and other semi-mechanistic models, such as the Gompertzian model for cancer growth. The basic LQ formula states that the effect, E, of an acute radiation dose, D, is described as

$$E = \alpha D + \beta D^2. \tag{11.48}$$

The parameters α (Gy^{-1}) and β (Gy^{-2}) are constants. The surviving cell fraction for such an effect is

$$S = e^{-E} = e^{-D(\alpha + \beta D)}. \tag{11.49}$$

The mechanistic interpretation of the LQ model varies slightly from author to author, but the binary misrepair concept (second-order misrepair) is central to all. Brenner [3] uses the following assumptions to justify the LQ model (see also [1] for a variation on this theme):

1. Radiation produces double-stranded DNA breaks with a yield linearly proportional to absorbed dose, D.

2. DSBs are repaired by first-order kinetics at rate λ. The repair half-time is approximately between 0.5 and 2 hours.

3. Binary misrepair of DSBs produced independently by different radiation tracks produces lethal lesions with a yield linearly proportional to the square of the dose, D^2. This gives the βD^2 term in the LQ formalism.

4. Single radiation tracks produce irreparable lethal lesions in direct proportional to the dose, giving the αD term.

The most important and consistent interpretation is that α represents the contribution of single-track radiation damage (i.e., single lethal lesions) to cytotoxicity, while β represents the second-order interaction of sublethal or potentially lethal lesions produced by different radiation tracks to form lethal lesions. Interestingly, this interpretation is supported by the LQ approximation for the LPL model. That is, it can be shown that when the time for repair is large and the dose, D, is low, the LPL model approximates the LQ formalism with the following coefficients:

$$\alpha = \delta(1 - \eta), \tag{11.50}$$

$$\beta = \frac{(\delta\eta)^2}{2\epsilon} = \frac{(\delta\eta)^2\kappa}{2\lambda}. \tag{11.51}$$

Recall that $\delta(1 - \eta)$ is simply the number of immediately irreparable DSBs formed per Gy. The number of potentially lethal lesions per Gy is $\delta\eta$. From the approximation for β, we can also see that first-order repair and second-order misrepair are competing processes—small $\epsilon = \lambda/\kappa$ implies large β.

Similarly, the RMR model approximates the LQ formalism at low doses [42] with coefficients

$$\alpha = \delta(1 - \phi), \tag{11.52}$$

$$\beta = \frac{\delta^2\phi^2}{2\epsilon} = \frac{\delta^2\phi^2\kappa}{2\lambda}. \tag{11.53}$$

In this case, α represents unsuccessful first-order repairs, and β is similar to the LPL model.

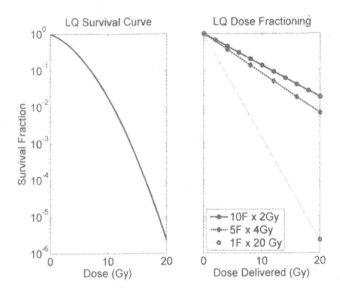

FIGURE 11.8: Dose-response curve for the LQ model (left panel) and cell survival for a total dose of 20 Gy divided into 10 fractions, 5 fractions, and a single fraction (right panel). Parameter values are $\alpha = 0.15$ Gy^{-1} and $\beta = 0.025$ Gy^{-2}.

The LQ model describes the shoulder region for dose-response curves well. It has also become extremely useful when comparing efficacy of different dose fractionation schedules. Because the effect of radiation increases nonlinearly with dose size, cytotoxicity is maximized by dividing the total dose into as few treatments as possible (Figure 11.8). However, since healthy tissues are also affected by radiation, the goal of schedule optimization is to maximize tumor kill while minimizing damage to healthy tissues.

For reference, a standard radiotherapy schedule is 35F × 2 Gy = 70 Gy in 7 weeks [11], generally given 5 times per week with weekends off. In designing a fractionation schedule, three factors must be considered: acute (early) reactions, late complications, and tumor control. The division of complications into acute and late reactions is required because irradiated cells do not die until they attempt to divide. Acute reactions occur in tissues with a high regenerative capacity and therefore high cell turnover rates. Cells quickly die from radiation, and compensatory proliferation repairs the damaged tissue. Therefore, sufficient time between treatments can ameliorate acute reactions by allowing sufficient tissue regeneration. However, late reactions become significant in tissues with limited regenerative capacity. Such tissues are less affected by inter-treatment layoff time, as cell death and compensatory proliferation may not occur until several months following treatment [11].

Most tumors respond like early-reacting tissues with rapid cell proliferation, although a few slowly growing tumors, such as prostate cancer, may actually act more like late-reacting tissues. Late reactions are generally dose-limiting. Therefore, the focus of dose fractionation is to minimize late reactions while maximizing tumor control. To this end, the LQ model and the notion of the biologically effective dose (BED) have allowed a rigorous comparison of fractionated schedules.

The BED was derived by Barendsen in 1982 [1] and has been widely used since. We divide a total dose of size D into n dose fractions of size D_n each ($D = nD_n$). Assuming that all sublethal lesions are repaired between each treatment, which is reasonable for most inter-treatment times, and that all doses are equally effective, the effect E becomes

$$E = n(\alpha D_n + \beta D_n^2) = nD_n(\alpha + \beta D_n) = D\alpha \left(1 + \frac{D_n}{\alpha/\beta}\right). \qquad (11.54)$$

A useful notion called *effectiveness per unit dose* (units Gy^{-1}) is defined, for a total dose D, as

$$E_D = \frac{E}{D}. \qquad (11.55)$$

In the case of fractionated therapy, for which E is given in equation (11.54),

$$E_D = \alpha + \beta D_n = \alpha \left(1 + \frac{D_n}{\alpha/\beta}\right). \qquad (11.56)$$

The key point here is that the effectiveness per unit dose does not depend directly on total dose, but on the dose per fraction. Dividing through by α gives the unit-less *relative effectiveness* (RE):

$$RE = \frac{E_D}{\alpha} = 1 + \frac{\beta D_n}{\alpha}. \qquad (11.57)$$

It easy to see that as D_n approaches 0, $RE \to 1$ and $E_D \to \alpha$; that is, the contribution to treatment effectiveness made by binary misrepair (second-order interaction of sublethal lesions) goes to 0, and the effectiveness per unit dose depends only upon α. Therefore, α can be understood as the efficacy of radiation that is independent of dose fractionation and size.

An example by Barendsen [1] will help clarify the utility of these observations. Suppose $\alpha/\beta = 5$ Gy, and that doses are delivered in fractions $D_n = 5$ Gy. Then $E_D = 2\alpha$ and $RE = 2$. Thus, the effectiveness per unit dose has doubled compared to the case when only the linear term contributed to effectiveness ($D_n = 0$, $E_D = \alpha$). It follows that in this case the quadratic term contributes equally to effectiveness. This leads to the interpretation that the ratio α/β is equal to the dose at which the linear and quadratic terms contribute equally to the production of lethal lesions.

Finally, dividing the effectiveness E by α gives what Barendsen termed the "extrapolated tolerance dose," although it was later renamed the *biologically effective dose* (BED):

$$BED = \frac{E}{\alpha} = D\left(1 + \frac{\beta D_n}{\alpha}\right) = D \times RE. \qquad (11.58)$$

Different tissues are characterized by different α/β ratios in a predictable way, and this fact allows the BED to be calculated for late-reacting and tumor tissues. From this, the comparative efficacy of fractionation schedules can be determined. For early-reacting tissues such as skin and mucosa, $\alpha/\beta \approx 6 - 14$ Gy, while for late-reacting tissues α/β is typically between 1.5 and 5 Gy [46]. Barendsen [1] reported $\alpha/\beta \approx 5$ Gy for connective tissue damage and CNS demyelination, and $\alpha/\beta \approx 2.5$ Gy for the lung, kidney, white matter, and CNS vasculature. For fast-growing tumors, α/β is similar to that of early-reacting tissues [11].

Heuristically, it makes sense that α/β should be large for early-reacting and small for late-reacting tissues. In the LPL model, potentially lethal lesions are repaired by first-order kinetics, interact to form lethal lesions by second-order kinetics, and are (presumably) fixed upon entry into the cell cycle. For low-LET radiation most lesions will be reparable. For late-reacting tissues, the time delay between radiation damage and proliferation is large, leaving a long time for repair before the damage becomes lethal. Hence, binary misrepair is the dominant mechanism for cytotoxicity, and α/β will be small. For early reacting tissues, there is less time for rapidly dividing cells to repair potentially lethal damage, so the effect of binary misrepair will be discounted. Thus, single-track lethal lesions are more important, and α/β will be large.

Altered fractionation regimes have received a great deal of attention in the context of head and neck squamous cell carcinomas, which are rapidly growing with high α/β ratios. Therefore, they are an ideal test model for the BED. We demonstrate the utility of the BED by comparing its predictions to the results obtained in a hyperfraction trial by the EORTC group [16], similar to an approach taken by Fowler [11]. In this trial, oropharyngeal carcinomas were treated according to standard schedules: 35F $\times$ 2 Gy = 70 Gy in 7 weeks, or the hyperfractionated schedule of 70F $\times$ 1.15 Gy = 80.5 Gy in 7 weeks. Using $\alpha/\beta = 4.5$ Gy for late-reacting tissues and $\alpha/\beta = 15$ Gy for the tumor [16], one obtains the following BEDs:

Standard (35F$\times$2Gy) Hyperfractionated (70F$\times$1.15Gy)

Tumor $BED_{Std} = 79.33$ Tumor $BED_{Hyper} = 86.67$
Late $BED_{Std} = 101.11$ Late $BED_{Hyper} = 101.07$

Thus, the BED predicts that tumor control is superior under the hyperfractionated regime, while late complications are nearly identical. Indeed, if we use the LQ model to directly predict survival, under one set of biologically

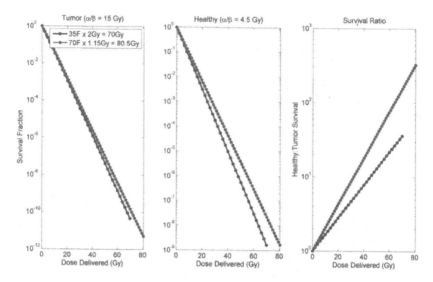

FIGURE 11.9: Cell survival in tumor vs. late-reacting healthy tissue under standard and hyperfractionated radiotherapy regimes. Parameter values for the tumor are $\alpha = 0.3$ Gy^{-1} and $\beta = 0.02$ Gy^{-2} ($\alpha/\beta = 15$ Gy); for healthy tissue $\alpha = 0.2$ Gy^{-1} and $\beta = 0.0444$ Gy^{-2} ($\alpha/\beta = 4.5$ Gy).

reasonable α and β parameters tumor cell kill is nearly an order of magnitude greater under hyperfractionation, while late-reacting tissue cytotoxicity differs by a negligible amount (Figure 11.9). Figure 11.9 also helps demonstrate that the absolute values of the BED cannot be meaningfully compared. The BED values are based on the ratio α/β, but actual survival depends on the actual values of α and β. For example, while the ratio Late BED$_{Std}$: Tumor BED$_{Std}$ is always 1.24, the actual survival ratios differ greatly depending on the choices for α and β and may be either more or less than unity for reasonable choices.

11.8 Applications

The concepts and models presented above have a wide range of clinically important applications. We finish this chapter with an outline of some of the most important.

11.8.1   Tumor cure probability

One widely used concept informed by the models above is that of *tumor cure probability* (TCP), which is defined as the probability that there are no surviving clonogenic tumor cells after completion of therapy. By clonogenic, we mean cells capable of sustaining tumor growth, so sterilized or terminally differentiated tumor cells would not be considered clonogenic. Suppose there are initially C_0 clonogenic cells. The number remaining after a treatment period of time T is simply

$$C_T = C_0 e^{-E_T}, \qquad (11.59)$$

where E_T is the total effect obtained from the standard LQ model or a modified form of it. Here, we interpret C_T as the expected (or average) number of tumor cells surviving treatment. The actual number of surviving cells in a given patient will be best represented by a random variable with mean C_T. If we assume that the probability distribution of this random variable is Poisson, the Poisson parameter λ equals C_T. Since TCP is the probability that the number of survivors is zero,

$$TCP = e^{-C_T} = e^{-C_0 e^{-E_T}}. \qquad (11.60)$$

Note that the TCP depends only upon C_T, which we can estimate using a variety of methods, including variations of the LQ model.

11.8.2   Regrowth

The major weakness of the unmodified BED is that it does not account for tumor regrowth over the course of treatment. For example, in fast growing squamous cell carcinomas the total treatment time strongly effects efficacy of radiotherapy [25], but significant tumor regrowth can occur over the course of longer schedules. The conundrum we face is to determine the optimal treatment length given these contrasting costs and benefits—increased efficacy on one hand and increase regrowth on the other. These contrasts can be quantified within the LQ and BED framework [11]. Consider a treatment schedule of total time duration T comprising n treatments with fixed time interval I separating each treatment; that is, $T = (n-1)I$ (the time taken to deliver the radiation dose is considered to be negligible compared to I). The effect per treatment, E_n, is calculated as follows:

$$E_n = \frac{E}{n} = \alpha D_n + \beta D_n^2. \qquad (11.61)$$

Following the initial treatment, the surviving fraction is $S(0) = e^{-E_n}$. During intertreatment intervals, the clonogenic tumor cell population expands exponentially at rate γ. Therefore, using the notation S^- to represent the surviving fraction before treatment and S^+ to represent the surviving fraction immediately after treatment, we have

$$S^-(I) = e^{-E_n + \gamma I},$$
$$S^+(I) = e^{-2E_n + \gamma I},$$
$$S^-(2I) = e^{-2E_n + 2\gamma I},$$
$$S^+(2I) = e^{-3E_n + 2\gamma I},$$

$$\vdots$$

$$S^-((n-1)I) = e^{-(n-1)E_n + (n-1)\gamma I},$$
$$S^+((n-1)I) = e^{-nE_n + (n-1)\gamma I} = e^{-nE_n + \gamma T}.$$

The modified effect, E, is

$$E = nE_n - \gamma T = n(\alpha D_n + \beta D_n^2) - \gamma T, \qquad (11.62)$$

and the modified BED becomes

$$BED = \frac{E}{\alpha} = D\left(1 + \frac{D_n}{\alpha/\beta}\right) - \frac{\gamma T}{\alpha}. \qquad (11.63)$$

The term γ/α is frequently represented by the single parameter K. Jones et al. [21] suggest that reasonable values of K are between 0.5 and 0.9 Gy/day for squamous cell carcinomas, 0.25 and 0.5 Gy/day for rapidly proliferating glioblastoma and ovarian cancers, and perhaps 0.1 Gy/day for slowly growing tumors. These values are probably the best available if estimates for α and γ are unavailable.

For fast growing tumors the inclusion of a time-factor in the BED is essential. Accelerated fractionation (AF) schedules have been proposed, where the total treatment time is reduced to (hopefully) limit tumor regrowth during treatment. The modified BED successfully predicts that AF regimes are better for fast growing tumors, while the unmodified BED fails. The modified BED can also be directly employed in the clinic to determine how missed doses should be properly compensated (see Jones et al. [21] for examples).

During radiation therapy, a phenomenon known as accelerated tumor repopulation occurs, where the tumor growth rate increases as the tumor is shrunk by radiation. Such a dynamic naturally follows from sigmoidal growth models for tumor growth, such as the Gompertz or von Bertalanffy models, discussed extensively in Chapter 2. While the *effective tumor doubling times* for human cancers before treatment are often quite long (studies using serial radiographs and an exponential model of tumor growth suggest doubling times between several weeks and *several years* [39][4]), this is often due to a

---

[4]There was significant interest by radiologists in estimating tumor doubling times beginning in the late 1950s. Some of these early efforts are reviewed, and the validity of the exponential model is criticized, by Steel and Lamerton, 1966 [39].

high rate of cell loss, not low proliferation. The *potential tumor doubling time* gives the rate at which a tumor could grow if all cells survived, and this may be on order of days. Poor perfusion and hypoxia are a major reason that cells die in tumors, but as the tumor shrinks during radiotherapy, perfusion and oxygenation typically improve, thus increasing the rate at which tumor regrowth can occur. Thus, radiotherapy can "unmask" the potential doubling time, and increasing doses of radiation may be needed to control tumor with a small potential doubling time.

Jones and Dale [17] suggested a modification of the LQ model to account for accelerated regrowth. We define the cell loss factor, ϕ, as the fraction of cells produced in a tumor that fail to survive. Defining T_D as the effective tumor doubling time and T_{pot} as the potential doubling time, we have

$$\phi = 1 - \frac{T_{pot}}{T_D}. \tag{11.64}$$

If $\phi = 0$ then $T_D = T_{pot}$. By the hypothesis being modeled, ϕ decreases with increased treatment as tumor perfusion improves, which Jones and Dale approximated by assuming ϕ decreases exponentially with time, giving:

$$\phi(t) = \phi_0 e^{-zt}, \tag{11.65}$$

where z is the rate of decay in ϕ. Assuming exponential tumor regrowth,

$$\frac{\ln 2}{T_{pot}}(1 - \phi(t)) = \frac{\ln 2}{T_{pot}}\left(1 - \phi_0 e^{-zt}\right). \tag{11.66}$$

We integrate over the total time T to determine total repopulation, R:

$$R = \int_0^T \frac{\ln 2}{T_{pot}}\phi_0 e^{-zt}\, dt = \frac{\ln 2}{T_{pot}}\left(T + \frac{\phi}{z}\left(e^{-zT} - 1\right)\right). \tag{11.67}$$

Factoring this into the survival fraction gives a modified effect,

$$E = n(\alpha D_n + \beta D_n^2) - \frac{\ln 2}{T_{pot}}\left(T + \frac{\phi}{z}\left(e^{-zT} - 1\right)\right), \tag{11.68}$$

and the BED follows as E/α. It is also straightforward to include a delay of time τ between the start of therapy and the onset of accelerated regrowth. In this case we have $\phi(t) = \phi_0$ when $\tau \leq T$ and $\phi(t) = \phi_0 \exp[-z(T - \tau)]$ when $\tau > T$. Calculations similar to those above show the modified effect in this case to be

$$E = n\left(\alpha D_n + \beta D_n^2\right) - \frac{\ln 2}{T_{pot}}\left(T - \phi_0\tau + \frac{\phi}{z}\left(e^{-z(T-\tau)} - 1\right)\right). \tag{11.69}$$

11.8.3 Hypoxia

Hypoxia is an important cause of chemoresistance, as production of reactive oxygen species by radiation is limited in a low oxygen environment. Recall that the effect of hypoxia is quantified by the *oxygen enhancement ratio* (OER). The OER can be incorporated into the LQ model most simply by assuming that OER modifies the dose, giving the effect under severe hypoxia as [19, 47]:

$$E = nD_n \left(\frac{\alpha}{OER_\alpha} + \frac{\beta D_n}{OER_\beta^2} \right). \tag{11.70}$$

The use of OER_α and OER_β for the linear and quadratic terms, respectively, accounts for slight variations between the OER at low and high radiation doses. Under the LQ model, the linear term dominates at low doses, so we take OER_α to be the OER at low doses, which is about 2.5 [47]. The quadratic term is dominant at high doses, and we take OER_β to be OER at high dose, about 3.0 [47]. We can also take heterogenous oxygenation and intermediate hypoxia into account by making the OER a function of the local P_{O2} [47].

Several authors have studied the effect of hypoxia with the LQ model. For example, Wouters and Brown [47] applied a modified LQ model to a simple tumor cord geometry that considered diffusion of oxygen from a central vessel into the surrounding tissue. They predicted that, in a clinical rather than experimental setting, moderately hypoxic cells may be more important in determining the response to therapy than a severely hypoxic fraction that is highly radioresistant. Daşu and Denenkamp [8] also modeled response to radiation under hypoxia and normoxia using several versions of the LQ model and predicted that the OER depends upon the fractionation scheme and decreases for low, clinically relevant doses.

11.8.4 Radiation with chemotherapy

We end this chapter with a brief discussion of models combining chemo- and radiotherapy using the LQ framework. In 2005, Jones and Dale [20] proposed a method to quantify chemotherapy's contribution to cell kill in a combined chemo-radiotherapy regimen. They start by assuming that drugs either:

1. Potentiate the effect of radiation. In this case a dose D is modified to the equivalent dose sD, where s is the "dose enhancement factor."

2. Independently kill cells. In this case, the radiation dose D is not modified, but an additional term is added to the total effect, E.

In the first case (sensitization), the modified effect for a fractionated regimen is calculated as

$$E = n \left(\alpha D_n^2 s + \beta D_n^2 s^2 \right), \tag{11.71}$$

and as usual the BED is E/α. We can also calculate the TCP in the presence (TCP_2) and absence (TCP_1) of a sensitizing agent:

$$TCP_1 = \exp\left(-C_0 e^{-\alpha D\left(1+\frac{\beta D_n}{\alpha}\right)}\right), \qquad (11.72)$$

$$TCP_2 = \exp\left(-C_0 e^{-\alpha Ds\left(1+\frac{\beta D_n s}{\alpha}\right)}\right), \qquad (11.73)$$

where $D = nD_n$ is the total dose delivered. Dividing TCP_1/TCP_2 and doing some algebra gives an expression for s:

$$s = \frac{\sqrt{\alpha^2 D^2(\alpha/\beta)^2 - 4\alpha D_n D\left(\alpha D D_n - \alpha D(\alpha/\beta) - (\alpha/\beta)\ln\left(\frac{\ln TCP_1}{\ln TCP_2}\right)\right)} - \alpha D(\alpha/\beta)}{2\alpha D_n D}.$$
$$(11.74)$$

This expression allows s to be estimated from clinical survival data.

The case where chemotherapy independently kills cells is a simpler matter. One just adds an effect E_C to the total effect equation:

$$E_T = n\left(\alpha D_n^2 s + \beta D_n^2 s^2\right) + E_C, \qquad (11.75)$$

assuming a simple log-kill model for the effect of chemotherapy (see Chapters 2 and 9). TCPs in the presence and absence of chemotherapy can be similarly defined and used to calculate the expression for E_C:

$$E_C = \ln\left(\frac{\ln TCP_1}{\ln TCP_2}\right). \qquad (11.76)$$

This value can be converted to a BED for a given radiation regimen by dividing by α as usual. The value of this approach comes from the fact that clinical trials report survival or tumor control data, which give the tumor control probability for different treatment arms. Since the radiobiological parameters α and α/β and the fractionation schedule are known, the effect of chemotherapy on cell survival can be calculated and cast in terms of the BED.

Conversely, if the efficacy of a course of chemotherapy can be estimated, this model provides a simple framework to estimate the impact of including it with various courses of radiotherapy. In the case of cytotoxic therapy, E_C, is the $\log_e$ cell kill. As an example, suppose a tumor with parameter values $C_0 = 5 \times 10^9$ cells, $\alpha = 0.3$ Gy^{-1}, $\alpha/\beta = 10$ Gy, and a constant exponential growth rate $\gamma = 0.0693$ day^{-1} (i.e., a doubling time of 10 day). Delivering a standard radiation course of 35F $\times$ 2 Gy = 70 Gy in 7 weeks with regrowth gives total treatment effect $E = 21.8$. We calculate the expected number of cells following treatment to be $C_T = 1.704$ cells, corresponding to a TCP of 18.2%. If $E_C = 3$, corresponding to a cell kill of 3 $\log_e$ and a BED of 10 Gy, TCP increases to 91.9%.

An earlier model by Jones and Dale [18] considered radiation combined with either cytostatic therapy that inhibits tumor regrowth but is neither directly toxic nor a radiosensitizing agent. This model also considered accelerated regrowth of the tumor using the model in [17] and discussed above. Moreover, as increased tumor perfusion leads to accelerated regrowth, it may also improve the delivery of drugs to the tumor mass, and this was also taken into account.

Several later works have quantified the effect of chemotherapy in clinical cancers. For example, Jones and Sanghera [22] estimated the BED of temozolomide chemotherapy to be 11.03 Gy for $\alpha/\beta = 9.3$ Gy in high-grade malignant gliomas treated by chemoradiation. In an unrelated effort, Rockne et al. also used the LQ formalism in models of malignant glioma growth [35, 36].

References

[1] Barendsen GW: Dose fractionation, dose rate and iso-effect relationships for normal tissue responses. *Int J Radiat Oncol Biol Phys* 1982, 8:1981–1997.

[2] Brenner DJ: The linear-quadratic model is an appropriate methodology for determining isoeffective doses at large doses per fraction. *Semin Radiat Oncol* 2008, 18:234–239.

[3] Brenner DJ, Hlatky LR, Hahnfeldt PJ, Huang Y, Sachs RK: The linear-quadratic model and most other common radiobiological models result in similar predictions of time-dose relationships. *Radiat Res* 1998, 150:83–91.

[4] Cucinotta FA, Nikjoo H, O'Neill P, Goodhead DT: Kinetics of DSB rejoining and formation of simple chromosome exchange aberrations. *Int J Radiat Biol* 2000, 76:1463–1474.

[5] Cucinotta FA, Pluth JM, Anderson JA, Harper JV, O'Neill P: Biochemical kinetics model of DSB repair and induction of gamma-H2AX foci by non-homologous end joining. *Radiat Res* 2008, 169:214–222.

[6] Curtis SB: Lethal and potentially lethal lesions induced by radiation—a unified repair model. *Radiat Res* 1986, 106:252–270.

[7] Dale RG, Fowler JF, Jones B: A new incomplete-repair model based on a 'reciprocal-time' pattern of sublethal damage repair. *Acta Oncol* 1999, 38:919–929.

[8] Daşu A, Denekamp J: New insights into factors influencing the clinically relevant oxygen enhancement ratio. *Radiother Oncol* 1998, 46:269–277.

[9] Dienes GJ: A kinetic model of biological radiation response. *Radiat Res* 1966, 28:183–202.

[10] Fowler JF: Differences in survival curve shapes for formal multi-target and multi-hit models. *Phys Med Biol* 1964, 9:177–188.

[11] Fowler JF: The linear-quadratic formula and progress in fractionated radiotherapy. *Br J Radiol* 1989, 62:679–694.

[12] Fowler JF: Is repair of DNA strand break damage from ionizing radiation second-order rather than first-order? A simpler explanation of apparently multiexponential repair. *Radiat Res* 1999, 152:124–136.

[13] Goodhead DT: Saturable repair models of radiation action in mammalian cells. *Radiat Res Suppl* 1985, 8:S58–S67.

[14] Gulston M, de Lara C, Jenner T, Davis E, O'Neill P: Processing of clustered DNA damage generates additional double-strand breaks in mammalian cells post-irradiation. *Nucleic Acids Res* 2004, 32:1602–1609.

[15] Haynes RH: The interpretation of microbial inactivation and recovery phenomena. *Radiat Res Suppl* 1966, 6:1–29.

[16] Horiot JC, Le Fur R, N'Guyen T, Chenal C, Schraub S, Alfonsi S, Gardani G, Van Den Bogaert W, Danczak S, Bolla M, et al.: Hyperfractionation versus conventional fractionation in oropharyngeal carcinoma: Final analysis of a randomized trial of the EORTC cooperative group of radiotherapy. *Radiother Oncol* 1992, 25:231–241.

[17] Jones B, Dale RG: Cell loss factors and the linear-quadratic model. *Radiother Oncol* 1995, 37:136–139.

[18] Jones B, Dale RG: Inclusion of molecular biotherapies with radical radiotherapy: Modeling of combined modality treatment schedules. *Int J Radiat Oncol Biol Phys* 1999, 45:1025–1034.

[19] Jones B, Dale RG. Mathematical models of tumour and normal tissue response. *Acta Oncol* 1999, 38:883–893.

[20] Jones B, Dale RG: The potential for mathematical modelling in the assessment of the radiation dose equivalent of cytotoxic chemotherapy given concomitantly with radiotherapy. *Br J Radiol* 2005, 78:939–944.

[21] Jones B, Dale RG, Deehan C, Hopkins KI, Morgan DA: The role of biologically effective dose (BED) in clinical oncology. *Clin Oncol (R Coll Radiol)* 2001, 13:71–81.

[22] Jones B, Sanghera P: Estimation of radiobiologic parameters and equivalent radiation dose of cytotoxic chemotherapy in malignant glioma. *Int J Radiat Oncol Biol Phys* 2007, 68:441–448.

[23] Kogelnik HD: Inauguration of radiotherapy as a new scientific speciality by Leopold Freund 100 years ago. *Radiother Oncol* 1997, 42:203–211.

[24] Krebs JS: The response of erythropoietic stem cells of mice to irradiation with fission neutrons. *Radiat Res* 1967, 31:796–807.

[25] Maciejewski B, Withers HR, Taylor JM, Hliniak A: Dose fractionation and regeneration in radiotherapy for cancer of the oral cavity and oropharynx: Tumor dose-response and repopulation. *Int J Radiat Oncol Biol Phys* 1989, 16:831–843.

[26] Metzger L, Iliakis G: Kinetics of DNA double-strand break repair throughout the cell cycle as assayed by pulsed field gel electrophoresis in CHO cells. *Int J Radiat Biol* 1991, 59:1325–1339.

[27] Nias AHW: *An introduction to radiobiology, 2nd ed.* New York: Wiley, 1998.

[28] Pouget JP, Mather SJ: General aspects of the cellular response to low- and high-LET radiation. *Eur J Nucl Med* 2001, 28:541–561.

[29] Pusey WA: Case of sarcoma and of Hodgkin's disease treated by exposures to X-rays—A preliminary report. *JAMA* 1902, 38:166–169.

[30] Radivoyevitch T, Hoel DG, Hahnfeldt PJ, Rydberg B, Sachs RK: Recent data obtained by pulsed-field gel electrophoresis suggest two types of double-strand breaks. *Radiat Res* 1998, 149:52–58.

[31] Reddy NM, Mayer PJ, Lange CS: The saturated repair kinetics of Chinese hamster V79 cells suggests a damage accumulation–interaction model of cell killing. *Radiat Res* 1990, 121:304–311.

[32] Röntgen WC: On a new kind of rays. *Nature* 1896, 53:274–276.

[33] Röntgen WK: On a new kind of rays. *CA Cancer J Clin* 1972, 22:153–157.

[34] Wilhelm Konrad Röntgen (1845-1923). *CA Cancer J Clin* 1972, 22:151–152.

[35] Rockne R, Alvord EC Jr, Rockhill JK, Swanson KR: A mathematical model for brain tumor response to radiation therapy. *J Math Biol* 2009, 58:561–578.

[36] Rockne R, Rockhill JK, Mrugala M, Spence AM, Kalet I, Hendrickson K, Lai A, Cloughesy T, Alvord EC Jr, Swanson KR: Predicting the efficacy of radiotherapy in individual glioblastoma patients in vivo: A mathematical modeling approach. *Phys Med Biol* 2010, 55:3271–285.

[37] Sachs RK, Hahnfeld P, Brenner DJ: The link between low-LET dose-response relations and the underlying kinetics of damage production/repair/misrepair. *Int J Radiat Biol* 1997, 72:351–374.

[38] Sánchez-Reyes A: A simple model of radiation action in cells based on a repair saturation mechanism. *Radiat Res* 1992, 130:139–147.

[39] Steel GG, Lamerton LF: The growth rate of human tumours. *Br J Cancer* 1966, 20:74-86.

[40] Stewart RD: Two-lesion kinetic model of double-strand break rejoining and cell killing. *Radiat Res* 2001, 156:365–378.

[41] Sutherland BM, Bennett PV, Cintron-Torres N, Hada M, Trunk J, Monteleone D, Sutherland JC, Laval J, Stanislaus M, Gewirtz A: Clustered DNA damages induced in human hematopoietic cells by low doses of ionizing radiation. *J Radiat Res* 2002, 43 Suppl:S149–S152.

[42] Tobias CA: The repair-misrepair model in radiobiology: Comparison to other models. *Radiat Res Suppl* 1985, 8:S77–S95.

[43] Valerie K, Povirk LF: Regulation and mechanisms of mammalian double-strand break repair. *Oncogene* 2003, 22:5792–5812.

[44] Weterings E, Chen DJ: The endless tale of non-homologous end-joining. *Cell Res* 2008, 18:114–124.

[45] Wheeler KT, Nelson GB: Saturation of a DNA repair process in dividing and nondividing mammalian cells. *Radiat Res* 1987, 109:109–117.

[46] Williams MV, Denekamp J, Fowler JF: A review of alpha/beta ratios for experimental tumors: Implications for clinical studies of altered fractionation. *Int J Radiat Oncol Biol Phys* 1985, 11:87–96.

[47] Wouters BG, Brown JM: Cells at intermediate oxygen levels can be more important than the "hypoxic fraction" in determining tumor response to fractionated radiotherapy. *Radiat Res* 1997, 147:541–550.

Chapter 12

Chemical Kinetics

12.1 Introduction and the law of mass action

By chemical kinetics, we mean the study of the rates of chemical reactions. Chemical kinetics are fundamental to understanding biochemical and enzymatic reactions in living systems, and ideas from chemical kinetics have been widely used to describe (with varying validity) interactions between other biological agents, such as infection of healthy cells by virions. Similar discussions of this topic can be found in other textbooks, such as Keener and Sneyd [1] or Nelson and Cox [2]. Of central importance is the *law of mass action*, which essentially states that two chemical species react at a rate proportional to their concentration. In the simple case of species A and B reacting to form a product C,

$$A + B \rightarrow C, \tag{12.1}$$

we have the rate, or velocity, of the reaction, v, as

$$v = \frac{dC}{dt} = k[A][B], \tag{12.2}$$

where k is the *rate* or *mass-action constant*, and the brackets in $[A]$ and $[B]$ denote concentration. The law may be justified in terms of collision theory: rate of reaction is proportional to the rate of collisions between the molecules and the probability that such collisions are sufficiently energetic to overcome the energy of activation for the reaction. In a well-mixed system, the rate of molecular collisions increase in direct proportion to an increase in either reactant.

We believe it is instructive to sketch the historical development of this law. The basic question of chemical kinetics is, simply, why do chemicals react? Much early thought, beginning around the time of Newton, attempted to frame chemical reactions in terms of forces between chemicals similar to the force of gravity [3]. Such forces were believed particular to the chemicals involved, and in 1780, Bergman advanced the theory that all substances have a particular intrinsic affinity for every other substance that is independent of mass. Consider the case of two species, A, and B, that react to form products, C and D:

$$A + B \to C + D. \tag{12.3}$$

According to Bergman's theory, this reaction (and all others) should progress to completion and be wholly irreversible. At the turn of the 18th century, Berthollet proposed that the affinities between two substances are modified by their physical characteristics and, most importantly, their respective masses. Thus, a reaction like the one above will not necessarily go to completion, as long as there is some "affinity," even very small, between C and D which forces the reaction backwards. Thus, the reaction will reach an equilibrium point based on opposing forces. This was a dramatic conceptual improvement, as it explained the fact that many reactions do not, in fact, proceed to completion, and the reverse reaction can occur. Thus, our model system becomes reversible:

$$A + B \rightleftarrows C + D. \tag{12.4}$$

In the second half of the 19th century, quantitative chemical kinetics came about, drawing on both the earlier idea of a chemical affinity force driving reactions and the newly emerging field of thermodynamics. In 1850, Wilhemy quantitatively studied the rate of inversion of sucrose. In 1864, building on 1862 work on esterification reactions by Bertholet and Giles, Waage and Guldberg [5] presented the first statement of the law of mass action:

> The substitution [reaction] force, other conditions being equal, is directly proportional to the product of the masses provided each is raised to a particular exponent.

This was supplemented with a law of volume action, implying that concentration rather than absolute mass is the determining factor. That is, the force causing A and B to react is:

$$k[A]^{\alpha}[B]^{\beta}. \tag{12.5}$$

Note that the law as originally imagines that some "affinity force" drives reaction, and criticisms of this ignore the conceptual environment in which they worked. Waage and Guldberg initially considered k, α, and β to be empirically derived constants, but for elementary reactions, α and β are equal to the stoichiometric coefficients of A and B. In subsequent papers (see Lund [4] for a review), they considered reaction rates to be proportional to the chemical force, and cast their ideas in terms of the "active mass" (concentration). Consider the reaction scheme:

$$A + B \underset{k_-}{\overset{k_+}{\rightleftarrows}} C + D \tag{12.6}$$

where k_+ and k_- are the constants for the forward and reverse reactions, and let us assume that α and β equal 1, i.e. this is an elementary reaction. Let x be the concentration of reacted substrate for both A and B (the concentrations are necessarily equal), and let p, q, s, and u represent the original concentrations of A, B, C, and D, respectively. Then we have that the velocity of the forward reaction is:

$$v = \frac{dx}{dt} = k_+(p-x)^\alpha (q-x)^\beta \tag{12.7}$$

and the velocity, $\acute{v}$, of the reverse reaction is:

$$\acute{v} = \frac{dx}{dt} = k_-(s+x)^\gamma (u+x)^\delta. \tag{12.8}$$

At equilibrium, the rates of the forward and reverse reactions are equal, implying $dx/dt = 0$, and we have

$$\frac{k_+}{k_-} = \frac{(s+x)^\gamma (u+x)^\delta}{(p-x)^a (q-x)^b} = \frac{[C]_{eq}[D]_{eq}}{[A]_{eq}[B]_{eq}} = K_c \tag{12.9}$$

where $[A]_{eq}$, $[B]_{eq}$, $[C]_{eq}$, and $[D]_{eq}$ are the equilibrium concentrations of the respective species. The ratio k_+/k_- is referred to as the equilibrium constant, K, K_c or K_{eq}. In their final publication, in 1879, Waage and Guldberg [7] justified the law of mass action in terms of collision theory. Consider an elementary reaction with arbitrary stoichiometric coefficients:

$$\alpha A + \beta B \underset{k_-}{\overset{k_+}{\rightleftharpoons}} \gamma C + \delta D. \tag{12.10}$$

From collision theory, the forward reaction rate is $k_+[A]^\alpha[B]^\beta$ and the reverse is $k_-[C]^\gamma[D]^\delta$, and we have

$$K_c = \frac{k_+}{k_-} = \frac{[C]_{eq}^\gamma [D]_{eq}^\delta}{[A]_{eq}^\alpha [B]_{eq}^\beta}. \tag{12.11}$$

In sum, Waage and Guldberg were the first to propose the law of mass action, and were first to determine the general form of the equilibrium equation. Their derivation was based on general arguments using the notion of a chemical force [4]. This concept of a chemical force has since been abandoned, and the modern derivation of the equilibrium constant is based on the Gibbs Free Energy of the chemical system.

Elementary reactions. As mentioned above, elementary reactions are those that occur in a single mechanistic step, i.e. they proceed directly from a collision between the reactants. Elementary reactions follow the law of mass action. Many reactions proceed by multiple steps, each consisting of an elementary reaction, and the overall reaction does not necessarily (and likely does not) follow mass action kinetics.

12.1.1 Dissociation constant

Consider a single chemical with the general formula A_xB_y, that dissociates into its subunits completely and irreversibly:

$$A_xB_y \underset{k_a}{\overset{k_d}{\rightleftharpoons}} xA + yB. \tag{12.12}$$

Here the dissociation constant, K_d, is defined as:

$$K_d = \frac{[A]^x[B]^y}{[A_xB_y]} = \frac{k_a}{k_d}. \tag{12.13}$$

Note that this is simply the inverse of the equilibrium constant (also called the association constant, K_a, in this context). Consider the biochemical example of a protein, P, binding a single ligand, S:

$$P + S \underset{k_a}{\overset{k_d}{\rightleftharpoons}} PS. \tag{12.14}$$

Then we have the dissociation constant:

$$K_d = \frac{[P][S]}{[PS]}. \tag{12.15}$$

We can express the fraction of protein that is ligand-bound. θ, as:

$$\theta = \frac{[PS]}{[PS] + [P]}. \tag{12.16}$$

Substituting $K_d/([P][S])$ for $[PS]$ and rearranging gives [2]:

$$\theta = \frac{[S]}{[S] + K_d}. \tag{12.17}$$

This is also known as the Michaelis-Henri equation and is related to the Hill equation, which we discuss in Section 12.5. Note that when $[S] = K_d$, then $\theta = 1/2$, and half the ligand binding sites are occupied.

12.2 Enzyme kinetics

Enzymes catalyze chemical reactions and are essential to life. Most biologically useful reactions do not occur at appreciable rates in the absence of enzymes, which increase the speed of such reactions by many orders of magnitude by lowering the activation energy. Enzymes are generally unchanged by the reaction, although so-called *suicide enzymes* are consumed in the reactions they catalyze. Essentially, enzymes allow reactions to be regulated (a reaction that always occurs everywhere is not much use in most biological systems), and enzyme expression and activity is generally regulated by multiple systems of positive and negative feedback. Nearly all enzymes are proteins, the only known exceptions being catalytic RNAs.

Enzymatic reactions do not follow the law of mass action, as they are the summation of several elementary reactions. In 1913, Michaelis and Menten [9] first described such kinetics quantitatively. Consider an enzyme E that catalyzes the conversion of a single substrate S to a product, P. An essential idea in enzyme kinetics is that the substrate and enzyme combine to form an enzyme-substrate (ES) complex; this complex then dissociates to yield the product and free enzyme. Schematically:

$$S + E \underset{k_{-1}}{\overset{k_1}{\rightleftarrows}} C \overset{k_2}{\longrightarrow} P + E. \tag{12.18}$$

Note that the second step is, in general, reversible, but we disregard the reverse reaction for simplicity. The Michaelis-Menten equation describes the rate of this reaction:

$$V = \frac{d[P]}{dt} = \frac{V_{max}[S]}{K + [S]}. \tag{12.19}$$

The parameter K can have two different meanings, depending on the derivation of the equation. From the law of mass action, we define a system of differential equations:

$$\frac{d[S]}{dt} = k_{-1}[C] - k_1[S][E] \tag{12.20}$$

$$\frac{d[E]}{dt} = k_{-1}[C] + k_2[C] - k_1[S][E] \tag{12.21}$$

$$\frac{d[C]}{dt} = k_1[S][E] - k_{-1}[C] - k_2[C] \tag{12.22}$$

$$\frac{d[P]}{dt} = k_2[C]. \tag{12.23}$$

12.2.1 Equilibrium approximation

In Michaelis and Menten's original analysis [9], they made use of the *equilibrium approximation*, which assumes that the enzyme and substrate are always in equilibrium. This equilibrium is (necessarily) assumed to be achieved instantaneously, implying that $d[S]/dt = 0$, and we have

$$k_1[S][E] = k_{-1}[C] \Rightarrow \frac{[S][E]}{[C]} = \frac{k_{-1}}{k_1} = K_d. \tag{12.24}$$

Note that such an assumption demands that the dissociation of the enzyme-substrate complex to yield product be the rate-limiting step. Now, observe that the total amount of enzyme, E_t, either complexed or free, is constant, implying that $[E] + [C] = [E_t]$. Substituting $[E_t] - [C]$ for $[E]$ and doing some algebra gives

$$[C] = \frac{[E_t][S]}{K_d + [S]} \tag{12.25}$$

with $K_d = k_{-1}/k_1$. The velocity of the overall reaction, V, is the rate at which product is formed, and we have

$$V = \frac{d[C]}{dt} = k_2[C] = \frac{k_2[E_t][S]}{K_d + [S]} = \frac{V_{max}[S]}{K_d + [S]} \tag{12.26}$$

where $V_{max} = k_2[E_t]$.

12.3 Quasi-steady-state approximation

In 1925, Briggs and Haldane [10] proposed an alternative theoretical derivation of the Michaelis-Menten equation based on the *quasi-steady-state assumption (QSSA)*, which is generally considered the modern basis for the equation. Briggs and Haldane assumed that the total amount of enzyme is negligible compared to the total amount of substrate, except at the very beginning of the reaction when the substrate is "loading" onto the enzyme and $[C]$ increases rapidly from 0. Now, following this initial loading phase, $d[C]/dt$ is strictly negative, and since $[C]$ is negligible compared to $[S]$ and $[P]$, it must also hold that $d[C]/dt$ is negligible compared to $d[S]/dt$ and $d[P]/dt$. If not, the reaction would rapidly cease as all enzyme would be set free [10]. This line of argument justifies the QSSA that $d[C]/dt = 0$, and we have that the rates of complex formation and breakdown are equal:

$$k_1[S][E] = k_{-1}[C] + k_2[C]. \tag{12.27}$$

Again substituting $[E_t] - [C]$ for $[E]$ and doing some algebra gives:

$$[C] = \frac{[E_t][S]}{\frac{k_{-1} + k_2}{k_1} + [S]} = \frac{[E_t][S]}{K_m + [S]} \tag{12.28}$$

where $K_m = (k_{-1} + k_2)/k_1$ is the so-called *Michaelis constant*. Similar to the equilibrium approximation,

$$V = \frac{d[C]}{dt} = k_2[C] = \frac{k_2[E_t][S]}{K_m + [S]} = \frac{V_{max}[S]}{K_m + [S]}. \tag{12.29}$$

Thus, we get an equation of the same form under either derivation, but with differing meanings for the constant K. If we assume that step 2 is rate-limiting in the sense that $k_2 \ll k_{-1}$, then $K_m \approx K_d$, as $K_m/K_d = 1 + k_2/k_{-1}$. In this case, the half-maximal substrate concentration depends only on the enzyme-substrate binding kinetics. However, regardless of the relative magnitudes of the rate constants, for a sufficiently high substrate concentration the reaction rate is always limited by $[E_t]$ and k_2, i.e. $V_{\max}$.

Figure 12.1 illustrates how the reaction velocity changes with substrate concentration, and also shows graphically how the Michaelis-Menten equation approximates the rate of product formation very well, except for an early transient when substrate is loading onto the enzyme.

12.3.1 Turnover number

The turnover number, k_{cat}, can be defined as the maximum number of substrate molecules that a single enzyme molecule can convert to product per unit time. For the Michaelis-Menten equation, $k_{\mathrm{cat}} = V_{\max}/[E_t]$, and if the two-step reaction discussed above holds, then it is also true that $k_{\mathrm{cat}} = k_2$.

12.3.2 Specificity constant

The specificity constant, k_{cat}/K_m, is a measure of how efficiently an enzyme converts substrate to product. When the enzyme is far from saturated, i.e. $[S] \ll K_m$, then we can approximate V as:

$$V = \frac{V_{\max}}{K_m}[S] = \frac{k_{\mathrm{cat}}}{K_m}[E_t][S] \tag{12.30}$$

and k_{cat}/K_m is the approximate second-order rate constant for the reaction [2].

12.3.3 Lineweaver-Burk equation

To determine the kinetic parameters K_m and $V_{\max}$ from experimental data, it is convenient to take the reciprocal of both sides of the Michaelis-Menten equation, which gives the *Lineweaver-Burk equation*:

$$\frac{1}{V} = \frac{1}{V_{\max}} + \frac{K_m}{V_{\max}}\frac{1}{[S]}. \tag{12.31}$$

Plotting $1/V$ versus $1/[S]$ gives a straight line with slope $K_m/V_{\max}$, y-intercept $1/V_{\max}$, and x-intercept $-1/K_m$. In general, the *initial* substrate concentrations and reaction velocities from *multiple* experiments are used to generate the plot, rather than using multiple data points from a single experimental run, as it is difficult to measure changes in substrate concentration with time.

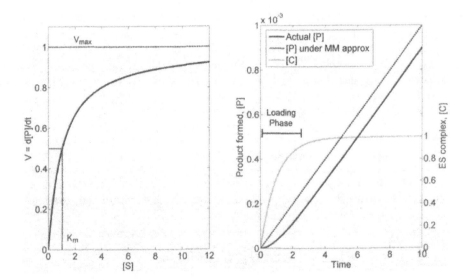

FIGURE 12.1: The left panel shows the reaction velocity, $V = d[P]/dt$, under the Michaelis-Menten equation as a function of substrate concentration, $[S]$. The right panel gives the total product formed as a function of time under direct numerical solution of the two-step Michaelis-Menten system and the product formed under the Michaelis-Menten approximation. Early in time there is a loading phase, where the enzyme-substrate complex is being formed and the QSSA assumption is violated. The enzyme-substrate concentration, $[C]$, is also shown.

12.4 Enzyme inhibition

Enzyme activity may be inhibited in several ways. The first is *competitive inhibition*, where some inhibitor competes with the normal substrate for the active site on the enzyme. Unlike competitive inhibitors, *allosteric inhibitors* bind at a site different from the active site, an allosteric site, which affects the activity of the enzyme.

12.4.1 Competitive inhibition

Consider the simpler case of competitive inhibition where the inhibitor, I, binds to the free enzyme and no further reaction takes place. Letting C_1 represent the ES complex and C_2 represent the EI complex, we have:

$$S + E \underset{k_{-1}}{\overset{k_1}{\rightleftharpoons}} C_1 \overset{k_2}{\longrightarrow} P + E \tag{12.32}$$

$$I + E \underset{k_{-3}}{\overset{k_3}{\rightleftharpoons}} C_2. \tag{12.33}$$

As before, we can easily derive the complete system of differential equations from the law of mass action, but we need only consider those governing C_1 and C_2:

$$\frac{d[C_1]}{dt} = k_1[S][E] - k_{-1}[C_1] - k_2[C_1] \tag{12.34}$$

$$\frac{d[C_2]}{dt} = k_3[I][E] - k_{-3}[C_2]. \tag{12.35}$$

In this case, $[E] + [C_1] + [C_2] = [E_t]$, with $[E_t]$ constant. Applying the QSSA that $d[C_1]/dt = 0$ and $d[C_2]/dt = 0$ and substituting $[E_t] - [C_1] - [C_2]$ for $[E]$, we solve for $[C_1]$ and $[C_2]$ (at quasi-steady state):

$$[C_1] = \frac{[E_t][S]}{K_m\left(1 + \frac{[I]}{K_I}\right) + [S]}, \tag{12.36}$$

$$[C_2] = \frac{[E_t][I]}{K_I\left(1 + \frac{[S]}{K_m}\right) + [I]} \tag{12.37}$$

where $K_I = k_{-3}/k_3$. The reaction velocity is

$$V = \frac{d[P]}{dt} = \frac{k_2[E_t][S]}{K_m\left(1 + \frac{[I]}{K_I}\right) + [S]} = \frac{V_{\max}[S]}{\alpha K_m + [S]} \tag{12.38}$$

where

$$\alpha = \left(1 + \frac{[I]}{K_I}\right).$$

Thus, the presence of a competitive inhibitor modifies K_m by the factor α, but the maximum reaction rate, $V_{\max}$ is unchanged. This makes intuitive sense, as if enough substrate is introduced into the system it will nearly completely out-compete the inhibitor for active sites, causing the probability of the inhibitor binding to approach zero.[1]

12.4.2 Allosteric inhibition

Allosteric inhibitors bind to an allosteric site different from the active site. *Uncompetitive inhibitors* are a type of allosteric inhibitor that only binds the enzyme when it is complexed with substrate. Letting C_3 represent the EIS complex, we have

$$S + E \underset{k_{-1}}{\overset{k_1}{\rightleftarrows}} C_1 \overset{k_2}{\longrightarrow} P + E \tag{12.39}$$

$$I + C_1 \underset{k_{-4}}{\overset{k_4}{\rightleftarrows}} C_3. \tag{12.40}$$

Also, let $\acute{K}_I = k_{-4}/k_4$. Once again we use the law of mass action to derive a system of equations and apply the QSSA, $d[C_1]/dt = d[C_3]/dt = 0$, and after some work arrive at the reaction velocity:

$$V = \frac{d[P]}{dt} = k_2 C_1 = \frac{k_2[E_t][S]}{K_m + [S]\left(1 + \frac{[I]}{K_I}\right)} = \frac{V_{\max}[S]}{K_m + \acute{\alpha}[S]} \tag{12.41}$$

where

$$\acute{\alpha} = 1 + \frac{[I]}{\acute{K}_I}.$$

In this case, as $[S] \to \infty$, $V \to V_{\max}/\acute{\alpha}$, and therefore uncompetitive inhibition decreases the maximum reaction rate. Note that half-maximal substrate concentration is also decreased by the factor $\acute{\alpha}$ to $K_m/\acute{\alpha}$. Since uncompetitive inhibitors affect $V_{\max}$, while competitive inhibitors do not, these two types of inhibition can be distinguished experimentally using a Lineweaver-Burk plot.

Finally, we consider the case of a *mixed inhibitor*, which can bind the enzyme whether it is complexed with its substrate or not, giving the scheme shown in Figure 12.2. While this system can be analyzed using the QSSA, the solution

---

[1]A common textbook example of competitive inhibition is treatment of methanol poisoning. Methanol ("wood alcohol") is metabolized to formaldehyde by the same enzyme (alcohol dehydrogenase) that metabolizes ethanol ("drinking alcohol"). Therefore, ethanol is used as a competitive inhibitor to slow the rate of formaldehyde formation while methanol is excreted by the kidneys.

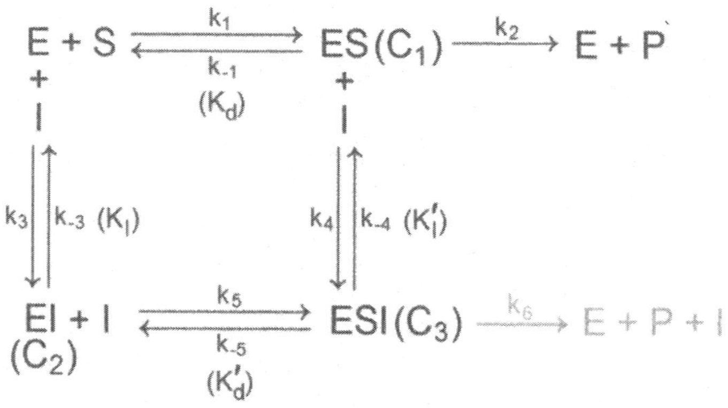

$$E + S \underset{k_{-1}}{\overset{k_1}{\rightleftharpoons}} ES(C_1) \xrightarrow{k_2} E + P$$

(diagram of reaction scheme)

FIGURE 12.2: Schematic for a reaction involving an enzyme, substrate, and mixed inhibitor. The inhibitor binds both the free enzyme and the enzyme-substrate complex. The grayed reaction is for the case when the enzyme-substrate-inhibitor complex still has catalytic activity.

for V is horribly complex, and one should only attempt to solve it with the aid of a computer algebra system.[2] A more tractable approach is to use the equilibrium approximation. We assume that the 4 protein-ligand interactions in the reaction scheme are at equilibrium, and we define the dissociation constants as $K_d = k_{-1}/k_1$, $K_I = k_{-3}/k_3$, $\acute{K}_d = k_{-5}/k_5$, and $\acute{K}_I = k_{-4}/k_4$. Assuming equilibrium, we have from mass action:

$$K_d = \frac{[S][E]}{[C_1]} \tag{12.42}$$

$$K_I = \frac{[I][E]}{[C_2]} \tag{12.43}$$

$$\acute{K}_d = \frac{[S][C_2]}{[C_3]} \tag{12.44}$$

$$\acute{K}_I = \frac{[I][C_1]}{[C_3]}. \tag{12.45}$$

Substituting $[E_t] - [C_1] - [C_2] - [C_3]$ for $[E]$, as usual, gives a linear system of equations that is overdetermined unless we set $\acute{K}_I = K_I$ and $\acute{K}_d = K_d$, in which case the system has rank 3 and we can find a unique solution for $[C_1]$, $[C_2]$, and $[C_3]$. After more tedium, we get

---

[2]As Gottfried Wilhelm von Leibniz once remarked [11], "It is unworthy of excellent men to lose hours like slaves in the labour of calculation which could safely be relegated to anyone else if machines were used."

$$[C_1] = [E_t]\frac{[S]}{K_d + [S]}\frac{K_I}{K_I + [I]} = [E_t]\frac{[S]}{\alpha K_d + \alpha[S]} \tag{12.46}$$

with

$$\alpha = 1 + \frac{[I]}{K_I}.$$

And, of course

$$V = k_2[C_1] = k_2[E_t]\frac{[S]}{K_d + [S]}\frac{K_I}{K_I + [I]} = V_{\max}\frac{[S]}{\alpha K_d + \alpha[S]}. \tag{12.47}$$

From this, we have that mixed inhibition reduces the maximum reaction rate to $V_{\max}/\alpha$. Perhaps surprisingly, the half-maximal rate occurs when $[S] = K_d$.

We can generalize a little from this model to an inhibitor that modifies the reaction rate, rather than block the reaction entirely. Enzyme activity is frequently modified allosterically, most frequently by phosphorylation (addition of a phosphate group), and such modifications can either increase or decrease enzyme activity. Suppose that the EIS complex retains enzyme activity and dissociates with rate constant k_6 to yield free product, inhibitor, and enzyme. The rate equations for $[C_3]$, $[P]$, and $[I]$ will be modified accordingly. The reaction rate becomes $V = k_2[C_1] + k_6[C_3]$, and the solution for $[C_3]$ is:

$$[C_3] = [E_t]\frac{[S]}{K_d + [S]}\frac{[I]}{K_I + [I]} \tag{12.48}$$

and the reaction velocity is modified to:

$$V = k_2[C_1] + k_6[C_3] = [E_t]\frac{[S]}{K_d + [S]}\left(\frac{k_2 K_I + k_6[I]}{K_I + [I]}\right). \tag{12.49}$$

12.5   Hemoglobin and the Hill equation

Many proteins bind multiple ligands, and it is often the case that the binding of one ligand modifies the kinetics of further binding. Hemoglobin, which carries oxygen in the blood, is the classical example of such behavior. Each hemoglobin molecule is made up of four subunits, each of which can bind one molecule of oxygen. Hemoglobin faces a challenge in that it must avidly bind oxygen diffusing from the air while circulating through the lungs, where oxygen concentration is high, but it must easily release to peripheral tissues, where oxygen concentration is low. Recall from Section 12.1.1 that, for a single ligand the fraction of ligand bound to a protein, θ is:

$$\theta = \frac{[S]}{[S] + K_d}. \tag{12.50}$$

If this were the case, the oxygen saturation curve would be hyperbolic, and hemoglobin would be unable to both tightly bind oxygen in the lungs and deliver it to tissue [2] (see Figure 12.3). Instead, hemoglobin oxygen saturation changes sigmoidally. This is a consequence of cooperativity: when an oxygen molecule binds to a single subunit, the overall conformation of the molecule is changed so that it becomes much easier for oxygen to bind to the other subunits.

The simplest and earliest model for cooperative binding was proposed by Hill in 1910 [8]. Hill was concerned with the binding of oxygen to aggregates of hemoglobin subunits in solution, and considered the case of an n subunit aggregate, Hb_n, binding n oxygen molecules:

$$Hb_n + nO_2 \xrightleftharpoons{K_d} Hb_nO_{2n} \tag{12.51}$$

where K_d is the dissociation constant for Hb-O_2 binding. Hill determined that an equation of the type

$$\theta = \frac{x^n}{x^n + K}, \tag{12.52}$$

where x is O_2 concentration, could explain observed oxygen-dissociation curves. He did not ascribe any mechanistic meaning to the equation, i.e. it is empirical. Following Nelson and Cox [2], the Hill equation can be taken to represent the scenario of perfect cooperation in ligand binding: either all ligands bind simultaneously and saturate the protein, or none do. To see this, consider n ligands, S, binding a protein:

$$P + nS \xrightleftharpoons{K_d} PS. \tag{12.53}$$

Following reasoning similar to that in Section 12.1.1, we arrive at the *Hill equation*:

$$\theta = \frac{[S]^n}{[S]^n + K_d}. \tag{12.54}$$

We can rewrite this as:

$$\log\left(\frac{\theta}{1 - \theta}\right) = n \log[S] - \log K_d. \tag{12.55}$$

A plot of $\log(\theta/(1 - \theta))$ versus $\log[S]$ is called a *Hill plot*, and if cooperation is perfect, then the slope, n_H, should equal n. In practice, cooperation is never perfect ($n_H < n$), and n_H, called the *Hill coefficient*, approximates the degree of cooperativity. If $n_H > 1$, then there is positive cooperativity. If $n_H < 1$, a rare occurrence, then cooperativity is negative, i.e. binding of one ligand inhibits bind by others.

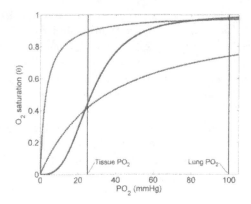

FIGURE 12.3: Hypothetical oxygen-hemoglobin dissociation curves. The two dotted curves are examples of no cooperativity: the top curve binds oxygen adequately in the lung, but does not release oxygen in the tissues, where it is needed. The lower curve releases oxygen to tissue, but binds it poorly in the lungs. The sigmoidal curve, which approximates actual dissociation curves, performs both tasks adequately, and is the solution to a Hill equation with $K_m = 27$ and $n = 3$. Thus, cooperativity is essential to proper hemoglobin function.

12.6 Monod-Wyman-Changeux model

In 1965, Monod, Wyman, and Changeux [12] proposed a model (MWC model) for cooperative binding of ligands to protein, with hemoglobin principally in mind. We define an *oligomer* as a polymeric protein with several identical subunits, and define a single subunit as a *protomer*. Monod *et al.* based their model on the following assumptions:

1. An allosteric protein is an oligomer, and all its promoters occupy equivalent positions, implying that there exists at least one axis of symmetry.

2. There is exactly one ligand binding site on each protomer, and each site is identical.

3. The oligomer has at least two conformational states, which it may reversibly transition between.

4. When the protein goes from one state to the other, its molecular symmetry is conserved. That is, all protomers must always be identical.

Following Monod *et al.*'s original derivation and notation, we consider the case of a single ligand, F, binding an oligomer. We assume that oligomer has exactly two states, a "tight" state that has low ligand affinity, and a "relaxed" state that has high affinity for the ligand. As a consequence of the symmetry assumption, a change in protein conformation changes the ligand affinity of all binding sites. When free of all ligands, these two states are in equilibrium with equilibrium constant L. Let n represent the number of protomers and hence the number of homologous binding sites. A oligomer with i ligands bound is represented by R_i or T_i, whether it is in the tight or relaxed state. Let $K_R = kr_d/kr_a$ and $K_T = kt_d/kt_a$ be the dissociation constants for ligand binding to a *single* stereospecific site. Importantly, the model does not allow transition between the R_i and T_i for $i > 0$. Therefore, cooperativity in this model refers to the ability of ligand binding to (temporarily) "lock" the protein into the relaxed (or tight) state. We have the reaction scheme:

$$R_0 \underset{}{\overset{L}{\rightleftharpoons}} T_0$$

$$R_0 + F \underset{kr_d}{\overset{nkr_a}{\rightleftharpoons}} R_1 \qquad\qquad T_0 + F \underset{kt_d}{\overset{nkt_a}{\rightleftharpoons}} T_1$$

$$R_1 + F \underset{2kr_d}{\overset{(n-1)kr_a}{\rightleftharpoons}} R_2 \qquad\qquad T_1 + F \underset{2kt_d}{\overset{(n-1)kt_a}{\rightleftharpoons}} T_2$$

$$R_2 + F \underset{3kr_d}{\overset{(n-2)kr_a}{\rightleftharpoons}} R_3 \qquad\qquad T_2 + F \underset{3kt_d}{\overset{(n-2)kt_a}{\rightleftharpoons}} T_3$$

$$\cdots \qquad\qquad\qquad \cdots$$

$$R_{n-1} + F \underset{nkr_d}{\overset{kr_a}{\rightleftharpoons}} R_n \qquad\qquad T_{n-1} + F \underset{nkt_d}{\overset{kt_a}{\rightleftharpoons}} T_n.$$

It is important to note that the rate constants vary, reflecting the number of binding sites available and the number of ligands currently bound. We assume that the system is in equilibrium, yielding the equilibrium equations from mass action:

$$T_0 = LR_0 \tag{12.56}$$

$$R_1 = R_0 n \frac{F}{K_R} \qquad T_1 = T_0 n \frac{F}{K_T} \tag{12.57}$$

$$R_2 = R_1 \frac{n-1}{2} \frac{F}{K_R} \qquad T_2 = T_1 \frac{n-1}{2} \frac{F}{K_T} \tag{12.58}$$

$$\cdots \qquad\qquad\qquad \cdots \tag{12.59}$$

$$R_n = R_{n-1} \frac{1}{n} \frac{F}{K_R} \qquad T_n = T_{n-1} \frac{1}{n} \frac{F}{K_T}. \tag{12.60}$$

Finally, the fraction of protein in the R state is given by

$$\bar{R} = \frac{(R_1 + R_2 + \ldots + R_n)}{(R_1 + R_2 + \ldots + R_n) + (T_1 + T_2 + \ldots + T_n)}, \tag{12.61}$$

and the fraction of actual ligand-binding sites bound is:

$$\bar{Y}_F = \frac{(R_1 + 2R_2 + \ldots + nR_n) + (T_1 + 2T_2 + \ldots + nT_n)}{n(R_1 + R_2 + \ldots + R_n) + n(T_1 + T_2 + \ldots + T_n)}. \tag{12.62}$$

Using the equilibrium equations, we arrive at

$$\bar{R} = \frac{(1+\alpha)^n}{L(1+c\alpha)^n + (1+\alpha)^n} \tag{12.63}$$

$$\bar{Y}_F = \frac{Lc\alpha(1+c\alpha)^{n-1} + \alpha(1+\alpha)^{n-1}}{L(1+c\alpha)^n + (1+\alpha)^n} \tag{12.64}$$

where

$$\alpha = \frac{F}{K_R}$$

$$c = \frac{K_R}{K_T}.$$

Monod *et al.* determined graphically that cooperativity, as measured by the curvature of the lower part of a $\bar{Y}_F$ versus α curve, is most pronounced when c is small and L is large. If c is small, then $K_R \ll K_T$ and ligand binding in the relaxed state is much stronger. If L is large, then the free oligomer has a strong tendency to be in the tight state. Ligand binding "locks" the previously free protein into the relaxed state, subsequent ligand binding will occur far more readily, and cooperativity is therefore strong. Thus, these two observations make physical sense.

Note that if $c = 1$, i.e. ligand binding occurs at the same rate in the tight and relaxed states ($K_R = K_T$), then regardless of the value of L, $\bar{Y}_F$ reduces to the case where there is no cooperativity:

$$\bar{Y}_F = \frac{F}{K_R + F}. \tag{12.65}$$

This also makes perfect physical sense.

References

[1] Keener J, Sneyd J: *Mathematical physiology*. New York: Springer Science+Business Media, 2004.

[2] Nelson DL, Cox MM: *Lehninger Principles of Biochemistry* (5th ed.). New York: W.H. Freeman, 2008.

[3] Quílez J: A historical approach to the development of chemical equilibrium through the evolution of the affinity concept. *Chem Educ Res Pract* 2004, 5:69–87.

[4] Lund EW: Guldberg and Waage and the law of mass action. *J Chem Educ* 1965, 42:548–550.

[5] Guldberg CM, Waage P: Avhandl. Norske Videnskaps-Akademi Oslo. *Mat Natorv Kl* 1864, 35.

[6] Guldberg CM, Waage P: *Etudes sur les Affinites chimique.* Ghent, Belgium: Brogger & Christie, 1867.

[7] Guldberg CM, Waage P: *J Prakt Chem* 1879, 19, 69.

[8] AV Hill: The possible effects of the aggregation of the molecules of haemoglobin on its dissociation curves. *J Physiol.* 1910, 40:iv–vii.

[9] Michaelis L, Menten ML: Kinetik der Invertinwirkung. *Biochem Z* 1913, 49:333-369.

[10] Briggs GE, Haldane JBS: A note on the kinetics of enzyme action. *Biochem J* 1925, 19:338-339.

[11] Sawday J: *Engines of the Imagination: Renaissance Culture and the Rise of the Machine* (p. 240). New York: Routledge, 2007.

[12] Monod J, Wyman J, Changeux JP: On the nature of allosteric transitions: A plausible model. *J Mol Biol* 1965, 12:88–118.

Chapter 13

Epilogue: Toward a Quantitative Theory of Oncology

Medicine is both art and science. As promised in the introduction, this book has focused primarily on the scientific side of the medical subfield of oncology. Specifically, we explored applications of mathematics, primarily dynamical systems, to cancer biology and treatment. We have attempted to give readers a glimpse of a few major research threads in the developing field of mathematical oncology, with an emphasis on seminal work, applications to treatment, and biological background and motivations.

The growth of the field in the last 10 years alone precludes any attempt on our part at a comprehensive survey, and for this we owe many of our colleagues, whose work we deeply appreciate, an apology. We hope that anyone who feels slighted in these pages will recognize and take comfort in our attempt to give students the tools necessary to study and place into context any work we have omitted.

But in addition to these conscious (and unconscious) omissions, this book is missing something vital, because the field itself is missing a critical piece connecting the art and science of medicine. We realize that this sounds like hyperbole. It is not. As we pointed out in Chapter 1, medical practice is considered evidence-based when the art is informed by science. Science's main product is theory. Here we use "theory" not in the journalistic sense of a hunch or shot-in-the-dark, but in the sense expressed by the U.S. National Academy of Sciences as "[a] well-substantiated explanation of some aspect of the natural world that can incorporate facts, [natural] laws, inferences and tested hypotheses" [19, p. 2], or in the substantially similar but more complete expression by Gerald Holton and Stephen Brush [11, p. 27]:

> [A theory] is a conceptual scheme we initially invent or postulate in order to explain to ourselves, and to others, observed phenomena and the relationships among them, thereby bringing together into one structure the concepts, laws, principles, hypothesis and observations from often very widely different fields.

In other words, theory expresses our current understanding of the natural world and is generally "well-substantiated." When we say that science informs the medical art, we mean that medical practice refers primarily to the best theory currently available.

And here we find the missing piece. Medicine largely lacks coherent, quantitative—or better, mathematical—theories. One may wonder how we can say such a thing at the end of a book full of medical theory couched in mathematics, but the key word in the claim is *coherent*. Scientific medicine simply has not produced mathematical theory applicable to broad areas of medicine. To be clear, we presume no sort of "grand unified medical theory"; indeed, such a pretension seems, to us, preposterous. But what has become glaringly obvious to us while putting this text together is the ad hoc manner in which mathematical models are applied to biological and physical phenomena relevant to medical oncology. And we thoroughly expect that deepening the connections among traditionally disjointed theoretical constructs will increase the power theory will have to inform medical art.

In this day and age, one may counter-argue that medicine, and oncology in particular, needs no coherent mathematical theory. Medical oncology already boasts one of the best theories in all of medicine, expressed by Hanahan and Weinberg as the "hallmarks of cancer," a list of necessary and sufficient phenotypes a cell must acquire to become malignant [9, 10]. The theory is completely coherent, highly successful, becoming increasingly comprehensive and yet devoid of mathematics. From a more general perspective, medicine looks toward genomics for its theory, and to argue that genomics is not a sufficient source of coherent theory is to argue against the dominant intellectual tradition of early 21st century biology. But this is precisely what we argue here. In fact, we go so far as to claim that genomics is the opposite of theory. Theory identifies patterns within chaos. Genomics generates chaos.

The general genomics program, as applied to a given species, attempts to generate a consensus sequence for the entire genome of that species, characterize all or most variation in homologous DNA sequences among individuals, and identify the function(s) of all functional DNA sequences. Clearly this is an ideal. However, as of this writing the genomic program has produced draft genomes, with accounts of variation, for humans—both modern and Neanderthal—many important human pathogens and an array of species of varying importance to human health. And, as a generator of information and knowledge, the genomics program has enjoyed many outstanding successes. Genomics has identified hundreds (or more) of genetic variations— genotypes—associated with cancer progression, pathogenesis and treatment resistance, some of which we exemplify in this book. Genomics has shown us broad patterns of gene regulation, like upregulation of ribosome structural genes, that led to new hypotheses about the biological nature of the disease [5, 6, 13]. Most importantly, it has unveiled the fundamental complexity of life at the level of macromolecules. Genomic organization, for one thing, is not as originally envisioned. The early notion of "one gene, one molecule" (i.e., protein) is deeply flawed—there appear to be around 20,000 traditional coding genes in the human genome, but our cells make more than 100,000 *proteins* by traditional transcription and translation, let alone other molecular products of transcription, like rRNA, iRNA, eRNA and others. Furthermore, the idea that a gene—however one wishes to define such a thing—occupies a specific locus in the genome is also contradicted. Even the most traditional of genes is regulated by enhancers scattered in remote locations of the genome. To give a single, medically relevant example, one genotype causing lactose persistence in humans—i.e., expression of the enzyme lactase-phlorizin hydrolase, which declines significantly as individuals with "adult lactose intolerance" age—is not determined by the sequence of the *LCT* (lactase) gene. Rather, the persistence mutation[1] is found more than 10,000 base pairs upstream of the *LCT* promoter in a gene called *MCM6*. The "normal" gene product of *MCM6* has nothing to do with lactose metabolism. However, a region within it, where the mutation is found, acts as an enhancer for *LCT*. This sequence in *MCM6* is therefore a very real part of the *LCT* gene, although tradition says it is not. This tradition is likely to be abandoned as the dominant role of enhancer and other regulatory regions becomes more clear (see, e.g., [1, 20]).

As these few examples illustrate, genomics has been successful in that it has forced us to reorganize our thinking—that is, it has forced us to build theory to explain the cascade of new observations. But this theory has been by and large ad hoc and essentially qualitative. Nevertheless, the successes have given medical science a case of "-omics fever," a proliferation of research programs attempting, like their genomic parent, to produce enormous lists of all intracellular proteins (proteome), RNAs (transcriptome), smaller molecules (metabolome) and molecular interactions (interactome), among other "-omes."

As useful and interesting as they are, such gobs of data are not theory and therefore are not considered an end product by the research community. The main theoretical attacks employed in the classical -omics program involve cataloguing and cross-referencing all the data, with some attempt to identify patterns (bioinformatics), and an attempt to understand the cell from an engineering perspective, in which one identifies the general behavior of sets of interacting molecules without slogging through all the details of every interaction (systems biology). The product of the systems biology program would be what is now dubbed the "systeome," another list, this time of all systems in the cell, their properties and their connections to other systems. Beyond this, the ultimate end point—the final theory—generated by this program is only ever vaguely expressed at best. The literature expresses little discomfort with the end goal being yet another list—there seems to be an assumption that eventually we will hit on a list so simple that the patterns will be obvious from inspection. But at some point we must move beyond lists. One such path might be thought to lead to a brute-force integration of the "systeome"

---

[1]Note that we call lactase persistence a mutation even though it is considered the normal, healthy state. The state we think of as the "disease"—adult lactose intolerance—is conserved among non-human mammals and therefore probably the ancestral state in humans. Indeed, the majority of humans alive today are "lactose intolerant."

into an immense computational program simulating all molecular activities within a cell.

Neither of these end points—another list or a brute-force simulation—are satisfactory. The former is obviously not an end-goal. At some point we have to integrate the elements of the list—molecules or systems or whatever—into a comprehensible formulation, one also capable of describing variation among the 200 different cell types in a healthy mammalian body, not to mention all pathological cells. And while a complete simulation of the cell would without question be such an integration, it is not practical. Certainly, such a simulation at the molecular level is absurd. It would require estimates of the number of each chemical species as they vary over time. Due to intracellular concentration and compartmentalization, these concentrations would be best represented as random variables, requiring stochastic characterization in the models. In addition, all parameters governing all possible dynamic interactions among these molecules would have to be accurately assigned. If molecular interactions require on average p unique parameters to be correctly modeled in a single cell, and each molecule interacts on average with n others, then one must accurately estimate on the order of $p(n-1)\,10^5$ parameters to simulate *only protein interactions in a single cell*. Even if p and n are modest (10 or less), an accurate model of the proteomes of the 200 types of cells in the *healthy* human body could easily require hundreds of millions to billions of parameter estimates before one even begins to address the transcriptome, the metabolome or any other -ome. More would be needed to characterize diseased cells and pathogens. And we have not even mentioned intracellular interactions, which are clearly critical in many if not most disease states.

This is chaos (in the literary, but probably also mathematical, sense). Detailed modeling of chaos like this has never, in the history of science, been the way forward. Recognition of this fact was one factor motivating the systems biology program, which in part attempts to represent cascades of molecular interactions as individual "systems" with relatively simple inputs and outputs that can be easily modeled. And, although this program has generated insight and testable explanations or descriptions (the primary goals of theory), how these systems are to be integrated into simple patterns, common themes, understandable descriptions and explanations that fit together into a more-or-less comprehensive narrative (the ultimate goal of theory) remains an open question.

One should not mistake our message. We claim that -omics and systems biology programs have immense value. One must have an accurate, clear view of the chaos before one can begin to discern patterns. Hence, our claim above that genomics generates chaos means that it provides the fodder needed to sustain theory-builders. Systems biology is a step in the right direction. However, at some point we must begin to emphasize a mathematical and computational program that rises above the *ad hoc* models of the past and connects them in a single, descriptive and explanatory framework.

Luckily, classical biology has such a theoretical framework ready-made—

evolution. Another general lesson we learned while putting this book together is that the mechanisms of evolution, particularly natural selection and genetic drift, arise either explicitly or implicitly, but always naturally, in nearly every explanation of cancer etiology, pathogenesis and treatment. This includes both empirical and theoretical studies, and this book provides many examples of the latter. Revealingly, in an astonishing number of cases, when researchers recognized the evolutionary theme in their own results, they gave every appearance of reinventing Darwin's wheel. For example, Darwin's evolutionary insight is more frequently referred to as "mutation" than "natural selection," especially in the empirical literature. Also, many researchers avoided the well-defined, well-understood term "genetic drift" in favor of a new term they themselves coin, usually built around either "random" or "stochastic." Imagine reading a research article in which the authors never use the word "cell," but talk instead about "living boxes" (equivalent to calling drift, "random change") or using the word "phenotype" to mean a DNA sequence (equivalent to conflating mutation with selection). If they otherwise describe and explain everything accurately, we might conclude that the researchers correctly identified a classic, well-known pattern but for some reason did not know what to call it. This is the situation we find in so many studies of cancer biology and treatment. Evolution is so obvious to these researchers that, for whatever reason, they fail to couch their explanations in classical terms, even though they get the explanation right.

Observations like these, and the ubiquity of evolutionary themes in this book, suggest that evolution can serve as a unifying, quantitative theory of oncology, and we are by no means the first to hit on this suggestion. That evolution, and natural selection in particular, underlies malignant transformation and treatment resistance are old, well-established ideas [14, 18, 21]. But the last decade has seen an explosion of interest in evolutionary oncology [2, 3, 4, 7, 8, 12, 15, 16, 17]. These references (and more), and indeed this text, make clear that evolution touches essentially every aspect of oncology— etiology, pathogenesis, tumor progression, morphology, prognosis and, perhaps most importantly, treatment. And the concept of evolution even subsumes the beautiful theory of Hanahan and Weinberg—cancer hallmarks are phenotypes that are caused by evolutionary forces, primarily natural selection and perhaps genetic drift. These traits then alter tumor ecology by changing the environment, both locally (in tissues within and immediately around the tumor—it's microenvironment) and globally (the host physiology). We suggest, therefore, that cancer theory should explicitly strive to build an evolutionary narrative that connects genetic and genomic alterations with phenotypic traits and their interactions with tumor ecology. We say *explicitly* because this appears to be the path the field is taking naturally. However, many cancer researchers appear not to realize it, treating evolutionary ecology as a fringe idea instead of the core notion of oncological theory.

Finally, we observe that this theory "narrative" will be written in the language of mathematics. Evolution, by definition, is dynamic. Mathematics, including numerical schemes implemented on computers, provides the most powerful tools ever devised by humans to handle the theories of dynamic processes. Without doubt, evolutionary theory applied to oncology and medicine in general promises to produce not only a coherency to scientific medicine, but also beautiful new mathematics.

References

[1] Andersson R, Gebhard C, Miguel I, *et al.*: An atlas of active enhancers across human cell types and tissues. *Nature* 2014, 507:455–461.

[2] Crespi B, Foster K, Úbeda F: First principles of Hamiltonian medicine. *Phil Trans R Soc B* 2014, 369:20130366.

[3] Crespi B, Summers K: Evolutionary biology of cancer. *Trends Ecol Evol* 2005, 10:545–552.

[4] Davies PCW, Lineweaver CH: Cancer tumors as Metazoa 1.0: Tapping genes of ancient ancestors. *Phys Biol* 2011, 8:015001.

[5] Elser JJ, Kyle MM, Smith MS, Nagy JD: Biological stoichiometry in human cancer. *PLoS ONE* 2007, 2:e1028.

[6] Elser JJ, Nagy JD, Kuang Y: Biological stoichiometry: An ecological perspective on tumor dynamics. *BioScience* 2003, 53:1112–1120.

[7] Gatenby, RA, Silva AS, Gillies RJ, Frieden BR: Adaptive therapy. *Cancer Res* 2009, 69:4894–4903.

[8] Greaves M, Maley CC: Clonal evolution in cancer. *Nature* 2012, 481:306–313.

[9] Hanahan D, Weinberg RA: The hallmarks of cancer. *Cell* 2000, 100:57–70.

[10] Hanahan D, Weinberg RA: The hallmarks of cancer: The next generation. *Cell* 2011, 144:646–674.

[11] Holton GJ, Brush SG: *Physics, the Human Adventure: From Copernicus to Einstein and Beyond*, 3rd ed. New Brunswick, NJ: Rutgers University Press, 2001.

[12] Korolev KS, Xavier JB, Gore J: Turning ecology and evolution against cancer. *Nat Rev Cancer* 2014, 14:371–380.

[13] Kuang Y, Nagy JD, Elser JJ: Biological stoichiometry of tumor dynamics: Mathematical models and analysis. *Disc Cont Dyn Sys B* 2004, 4:221–240.

[14] Law LW: Origin of the resistance of leukaemic cells to folic acid antagonists. *Nature* 1952, 169:628–629.

[15] Leroi AM, Koufopanou, Burt A: Cancer selection. *Nat Rev Cancer* 2003, 3:226–231.

[16] Merlo LM, Pepper JW, Reid BJ, Maley CC: Cancer as an evolutionary and ecological process. *Nat Rev Cancer* 2006, 6:924–935.

[17] Nagy JD: The ecology and evolutionary biology of cancer: A review of mathematical models of necrosis and tumor cell diversity. *Math Biosci Eng* 2005, 2:381–418.

[18] Nowell PC: The clonal evolution of tumor cell populations. *Science* 1976, 194:23–28.

[19] Steering Committee on Science and Creationism, National Academy of Sciences: *Science and creationism: A view from the National Academy of Sciences*, 2nd ed. Washington, DC: National Academies Press, 1999.

[20] Visel A, Rubin EM, Pennacchio LA: Genomic views of distant-acting enhancers. *Nature* 2009, 461:199–205.

[21] Weinberg, RA: *The Biology of Cancer*. New York: Garland Press, 2007.

Index